AF598024

Methods in Molecular Biology™

Series Editor
John M. Walker
School of Life Sciences
University of Hertfordshire
Hatfield, Hertfordshire, AL10 9AB, UK

For further volumes:
http://www.springer.com/series/7651

Lymphoma

Methods and Protocols

Edited by

Ralf Küppers

Institute of Cell Biology (Cancer Research), University of Duisburg-Essen, Medical School, Essen, Germany

Editor
Ralf Küppers
Institute of Cell Biology (Cancer Research)
University of Duisburg-Essen, Medical School
Essen, Germany

ISSN 1064-3745 ISSN 1940-6029 (electronic)
ISBN 978-1-62703-268-1 ISBN 978-1-62703-269-8 (eBook)
DOI 10.1007/978-1-62703-269-8
Springer New York Heidelberg Dordrecht London

Library of Congress Control Number: 2012951649

Printed on acid-free paper

Humana Press is a brand of Springer
Springer is part of Springer Science+Business Media (www.springer.com)

Preface

Lymphomas are lymphoid malignancies derived from B or T lymphocytes. Although they are only listed around position 10–12 among the most frequent cancers in the Western world, their study has been and still is paradigmatic for many aspects of cancer research. The aim of this book is to present and discuss key methods that are used in lymphoma research. These methods are partly specific for lymphoma research, but many can be adopted to the study of other cancers. Several chapters describe assays based on the usage of the highly diverse B cell and T cell receptor gene rearrangements in B and T cell lymphomas, respectively. These somatic gene rearrangements are ideal clonal markers to study minimal residual disease and intraclonal tumor diversification. Moreover, chromosomal translocations involving the immunoglobulin or T cell receptor gene loci are frequent oncogenic events in many lymphomas, and the cloning of translocation breakpoint regions has led to the identification of numerous novel oncogenes. The study of deregulated mRNA and miRNA expression has revealed many novel insights into lymphoma pathogenesis and has led to the identification of disease subsets. Further topics include the analysis of epigenetic alterations, the search for viruses in lymphoma cells, the role of the B cell receptor in driving B cell lymphoma proliferation, and the usage of cell lines and mouse models in lymphoma research. By covering this broad variety of molecular studies of lymphomas, this book should be of interest not only for hematologists, hematopathologists, and immunologists but also for scientists interested in other fields of cancer research as well as human genetics.

Essen, Germany *Ralf Küppers*

Contents

Contributors

CONSTANCE BAER • *Department of Epigenomics and Cancer Risk Factors, German Cancer Research Center (DKFZ), Heidelberg, Germany*

KATIA BASSO • *Institute for Cancer Genetics, Columbia University, New York, NY, USA*

Department of Pathology and Cell Biology, Columbia University, New York, NY, USA

The Herbert Irving Comprehensive Cancer Center, Columbia University, New York, NY, USA

ANKE VAN DEN BERG • *Department of Pathology and Medical Biology, University of Groningen, University Medical Center Groningen, Groningen, The Netherlands*

MAGDALENA A. BERKOWSKA • *Department of Immunology, Erasmus MC, University Medical Center, Rotterdam, The Netherlands*

ELKE BOONE • *Heilig Hart Ziekenhuis, Roeselare, Belgium*

SEBASTIAN BÖTTCHER • *Second Department of Medicine, University Hospital of Schleswig-Holstein, Kiel, Germany*

MONIKA BRÜGGEMANN • *Second Department of Medicine, University Hospital Schleswig-Holstein, Kiel, Germany*

NIKOS DARZENTAS • *Medical Genomics Research Group, Molecular Medicine Program, CEITEC/Central European Institute of Technology, Masaryk University, Brno, Czech Republic*

Institute of Agrobiotechnology, Center for Research and Technology Hellas, Thessaloniki, Greece

JACQUES J.M. VAN DONGEN • *Department of Immunology, Erasmus MC, University Medical Center, Rotterdam, The Netherlands*

MARTIN J.S. DYER • *Department of Cancer Studies and Molecular Medicine, University of Leicester, Leicester, UK*

JONATHAN R. FROMM • *Department of Laboratory Medicine, University of Washington, Seattle, WA, USA*

ALICE GALLAGHER • *Centre for Virus Research, MRC - University of Glasgow, Glasgow, UK*

DEREK GATHERER • *Centre for Virus Research, MRC - University of Glasgow, Glasgow, UK*

MACIEJ GIEFING • *Institute of Human Genetics, University Hospital Schleswig-Holstein, Christian-Albrechts University, Kiel, Germany*

Institute of Human Genetics, Polish Academy of Sciences, Poznan, Poland

MARTIN-LEO HANSMANN • *Senckenberg Institute of Pathology, University of Frankfurt, Frankfurt/Main, Germany*

RUTH F. JARRETT • *Centre for Virus Research, MRC - University of Glasgow, Glasgow, UK*

ULF KLEIN • *Departments of Pathology and Cell Biology and Microbiology & Immunology, Columbia University, New York, NY, USA*

Herbert Irving Comprehensive Cancer Center, Columbia University, New York, NY, USA

JOOST KLUIVER • *Department of Pathology and Medical Biology, University of Groningen, University Medical Center Groningen, Groningen, The Netherlands*

MICHAEL KNEBA • *Second Department of Medicine, University Hospital of Schleswig-Holstein, Kiel, Germany*

JULIANE KOFER • *Research Group Molecular Immunology, Max Planck Institute for Infection Biology, Berlin, Germany*

RALF KÜPPERS • *Institute of Cell Biology (Cancer Research), Medical School, University of Duisburg-Essen, Essen, Germany*

ANTON W. LANGERAK • *Department of Immunology, Erasmus MC, University Medical Center, Rotterdam, The Netherlands*

VU N. NGO • *Division of Hematopoietic Stem Cell and Leukemia Research, Beckman Research Institute of City of Hope, Duarte, CA, USA*

CHRISTOPH PLASS • *Department of Epigenomics and Cancer Risk Factors, German Cancer Research Center (DKFZ), Heidelberg, Germany*

CHRISTIANE POTT • *Second Department of Medicine, University Hospital Schleswig-Holstein, Kiel, Germany*

MICHAEL REHLI • *Department of Hematology and Oncology, University Hospital, Regensburg, Germany*

MATTHIAS RITGEN • *Second Department of Medicine, University Hospital of Schleswig-Holstein, Kiel, Germany*

ROLAND SCHMITZ • *Metabolism Branch, Center for Cancer Research, National Cancer Institute, NIH, Bethesda, MD, USA*

MARKUS SCHNEIDER • *Institute of Cell Biology (Cancer Research), Medical School, University of Duisburg-Essen, Essen, Germany*

RENÉ SCHOLTYSIK • *Institute of Cell Biology (Cancer Research), Medical School, University of Duisburg-Essen, Essen, Germany*

MARC SEIFERT • *Institute of Cell Biology (Cancer Research), Medical School, University of Duisburg-Essen, Essen, Germany*

REINER SIEBERT • *Institute of Human Genetics, University Hospital Schleswig-Holstein, Christian-Albrechts University, Kiel, Germany*

IZABELLA SLEZAK-PROCHAZKA • *Department of Pathology and Medical Biology, University of Groningen, University Medical Center Groningen, Groningen, The Netherlands*

MIRIAM SONNET • *Department of Epigenomics and Cancer Risk Factors, German Cancer Research Center (DKFZ), Heidelberg, Germany*

KOSTAS STAMATOPOULOS • *Hematology Department and HCT Unit, G. Papanicolaou Hospital, Thessaloniki, Greece; Institute of Agrobiotechnology, Center for Research and Technology Hellas, Institute of Applied Biosciences, Thessaloniki, Greece*

LOUIS M. STAUDT • *Metabolism Branch, Center for Cancer Research, National Cancer Institute, NIH, Bethesda, MD, USA*

BAO TRAN • *Center for Cancer Research Sequencing Facility, SAIC-F Advanced Technology Program, National Cancer Institute, NIH, Frederick, MD, USA*
VINCENT H.J. VAN DER VELDEN • *Department of Immunology, Erasmus MC, University Medical Center, Rotterdam, The Netherlands*
BRENDA VERHAAF • *Deptartment of Immunology, Erasmus MC, University Medical Center, Rotterdam, The Netherlands*
HEDDA WARDEMANN • *Research Group Molecular Immunology, Max Planck Institute for Infection Biology, Berlin, Germany*
DIETER WEICHENHAN • *Department of Epigenomics and Cancer Risk Factors, German Cancer Research Center (DKFZ), Heidelberg, Germany*
BRENT L. WOOD • *Department of Laboratory Medicine, University of Washington, Seattle, WA, USA*
DAVID WU • *Department of Laboratory Medicine, University of Washington, Seattle, WA, USA*
WENMING XIAO • *Bioinformatics and Molecular Analysis Section, Division of Computational Bioscience, Center for Information Technology, National Institutes of Health, Bethesda, MD, USA*
MENNO C. VAN ZELM • *Department of Immunology, Erasmus MC, University Medical Center, Rotterdam, The Netherlands*

Chapter 1

Origin and Pathogenesis of B Cell Lymphomas

Marc Seifert, René Scholtysik, and Ralf Küppers

Abstract

Immunoglobulin (Ig) gene remodeling by V(D)J recombination plays a central role in the generation of normal B cells, and somatic hypermutation and class switching of Ig genes are key processes during antigen-driven B cell differentiation. However, errors of these processes are involved in the development of B cell lymphomas. Ig locus-associated translocations of proto-oncogenes are a hallmark of many B cell malignancies. Additional transforming events include inactivating mutations in various tumor suppressor genes, and also latent infection of B cells with viruses, such as Epstein–Barr virus. Many B cell lymphomas require B cell antigen receptor expression, and in several instances chronic antigenic stimulation plays a role in sustaining tumor growth. Often, survival and proliferation signals provided by other cells in the microenvironment are a further critical factor in lymphoma development and pathophysiology. Many B cell malignancies derive from germinal center B cells, most likely because of the high proliferation rate of these cells and the high activity of mutagenic processes.

Key words: B cells, B cell lymphoma, Clonality, Chromosomal translocation, Germinal center, Hodgkin's lymphoma, Immunoglobulin genes, V gene recombination, Somatic hypermutation

1. B Cell Development and Differentiation

1.1. Introduction

B cells are lymphocytes that confer efficient and long-lasting adaptive immunity by the generation of high-affinity antibodies against antigens. These cells form an essential part of the humoral immune response and play a central role in immunologic memory. Beyond this, B lymphocytes participate in a broad range of immunological functions, including antigen presentation, immune regulation, and provision of a cellular and humoral pre-immune repertoire. Their contribution to the immune system is complex and multilayered.

Ralf Küppers (ed.), *Lymphoma: Methods and Protocols*, Methods in Molecular Biology, vol. 971,
DOI 10.1007/978-1-62703-269-8_1,

1.2. B Cell Diversity and Antibody Structure

All mature B cells express a membrane-bound antibody with individual specificity. This immunoglobulin (Ig) is associated with cofactors, and together these molecules form the B cell receptor (BCR). The cofactors immunoglobulin alpha and beta (Igα/Igβ) participate in signal transduction of this surface receptor. The diversity of immunologically competent B cells results from the variability of their BCR. This is a consequence of recombination processes during B lymphocyte development in which gene segments located in the Ig loci are joined to give rise to new and individually generated Ig genes. Antibodies are composed of four polypeptides, two identical heavy chains (IgH) and two identical light chains (IgL), that are linked by disulfide bonds. The IgL chains are of either κ or λ isotype. All these polypeptides consist of a carboxyterminal constant (C) and an aminoterminal variable (V) fragment. The V region includes four framework regions, each separated by hypervariable regions, the complementarity determining regions 1, 2, and 3 (CDRI to CDRIII). Whereas the V_H region gene is generated by the recombination of three independent gene segments, the variable (V_H), diversity (D_H), and joining (J_H) segments, the light-chain V region genes are composed of only two segments, namely, the V_L and J_L segment (1). The somatic recombination of these segments is catalyzed by the enzymes RAG1 and RAG2. These enzymes recognize recombination signal sequences flanking the gene segments, cut the DNA at these sites, and build hairpin structures at the coding ends (2). The hairpin structures can be resolved in different ways to generate (palindromic) P elements. Moreover, exonucleases can act arbitrarily to remove nucleotides from the ends of the rearranging gene segments. The enzyme terminal deoxynucleotidyltransferase (TdT) randomly adds (non-germline-encoded) N nucleotides to the ends of the rearranging gene segments before they are joined, and DNA repair factors finally complete the recombination process (1).

1.3. B Cell Development and Differentiation

The development of B cells is initiated in the fetal liver and relocated to the bone marrow during maturation of mammalian embryos. Throughout the differentiation processes, the microenvironment of the respective tissues (the microenvironmental niche) plays an essential role in providing nutrition, survival, and developmental stimuli.

Multipotent hematopoietic stem cells give rise to lymphoid precursors that initiate an irreversible differentiation program. The development of B cells from lymphoid precursors is orchestrated by several key transcription factors that determine B cell fate. Early B cell factor 1 (EBF1), E2A, and PAX5 are the three main transcription factors for early B cell development (3). The production of a functional and unique BCR through V(D)J recombination is the central process for the generation of a mature B cell (4). Hence, selection processes for appropriate receptor molecules play a key role during B cell development, as nonfunctional or autoreactive B cells have to be eliminated. B cell development is regulated by an ordered rearrangement of antigen receptor gene segments, and can be divided into distinct steps according to the rearrangement status of the Ig loci and phenotypical features. The initial step in B cell development is a D_H-to-J_H gene rearrangement at the IgH locus on human chromosome 14. In humans, 27 D_H segments and six J_H gene segments are available for this rearrangement (Fig. 1) (5, 6) that can occur on both alleles. B lymphocyte precursors carrying D_HJ_H joints are called pro B cells. Subsequently, one of about 120 V_H segments is rearranged to the D_HJ_H joint (Fig. 1) (7). The newly generated V_H chain is expressed and paired to a surrogate light chain. The so formed pre-BCR is tested for functional competence. If functional, recombination processes of the second allele are suppressed (allelic exclusion), and the B lymphocyte precursor reaches the stage of the pre-B cell (4, 8). However, there are several possibilities to generate a nonfunctional pre-BCR: e.g., one of approximately 80 nonfunctional V_H segments encoded in the human genome can be recombined to the D_HJ_H joint (7). As well, nucleotide insertions or deletions occurring during the rearrangement process can cause frameshifts of the IgH gene, or the expressed V_H chain cannot bind properly to the surrogate light chain and fails to form a stable pre-BCR. In case of a nonfunctional pre-BCR, a rearrangement of the second IgH allele or the potential use of V-gene replacement (recombination of further upstream located V_H gene segments to the existing $V_HD_HJ_H$ joint) is an alternative for the B lymphocyte precursor to generate a functionally competent pre-BCR (9). In case these escape-mechanisms are unsuccessful, the respective B lymphocyte precursor will undergo apoptosis (10). Only those B cell precursors that survive the selection for a functional pre-BCR start rearranging V_L-to-J_L light-chain genes in order to generate an immunoglobulin light chain. Light chain recombination starts at the κ loci on chromosome 2. In the human, depending on the haplotype, 30–35 functional V_κ gene segments and five J_κ segments are available for recombination (11, 12). In case of nonfunctional V_κ rearrangements on both alleles, the λ loci on chromosome 22 can be rearranged subsequently with 30–37 functional V_λ and four J_λ gene segments available (13, 14). B cells express either κ or λ light chains, a phenomenon called isotype exclusion. Only in very rare instances (<2% of total B cells), two different light chains (κ and/or λ) are expressed by a single B cell at the same time (4, 15).

When a B cell precursor expresses a functional heavy and light chain that can appropriately pair to form a stable BCR, the stage of the immature B cell is reached. At this stage of B cell development, the BCR is expressed exclusively with the IgM isotype of the heavy chain and the cell is now counterselected for autoreactivity (4). If the BCR of an immature B cell shows reactivity to autoantigens,

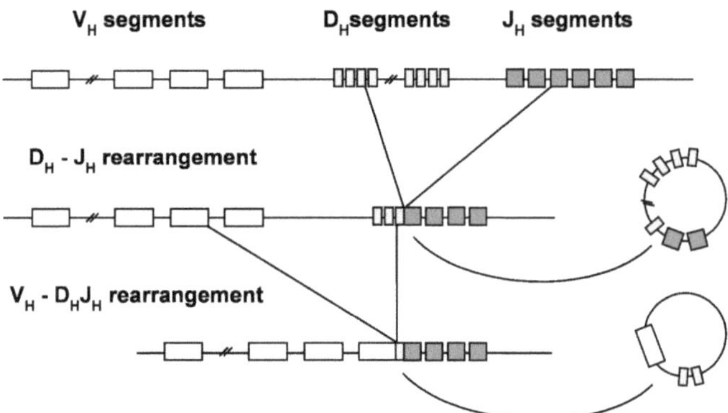

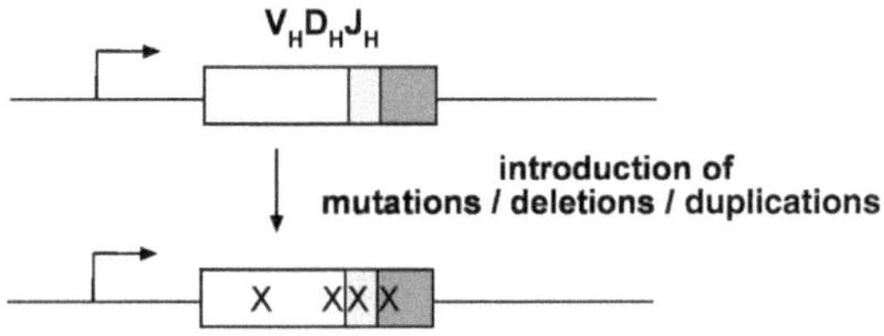

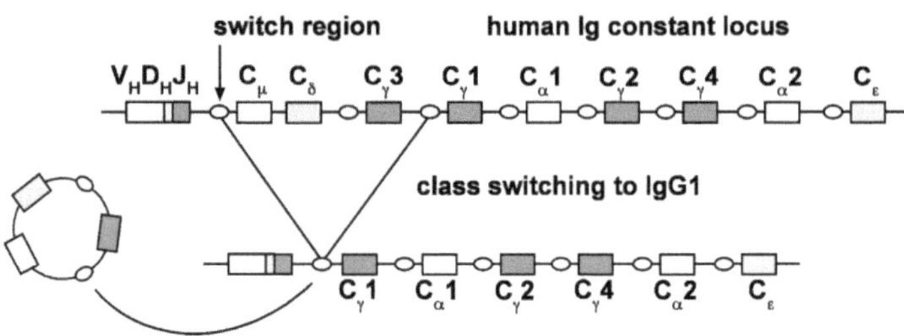

Fig. 1. Ig gene remodeling processes in B cells. (**a**) Shown is a schematic presentation of the stepwise rearrangement of V_H-, D_H-, and J_H-segments during B cell ontogeny. Excision circle by-products are depicted to the *right*. (**b**) The introduction of somatic mutations into a transcribed $V_HD_HJ_H$ gene by the somatic hypermutation machinery active in GC B lymphocytes. Each "X" denotes an independent mutational event. (**c**) Schematic presentation of class switch recombination from Cμ to Cγ1 on the human IgH chain constant region locus. An excised switch circle is shown on the *left*.

the corresponding immature B cell has the opportunity to escape counterselection by receptor editing. This includes either the usage of so far unrearranged Ig loci or potentially the additional rearrangement of upstream V with downstream J gene segments in the light-chain loci (16, 17). Negatively selected B cells are excluded from the B cell pool by conversion into an anergic state (i.e., immunological unresponsiveness) or by apoptosis. Those B cells with positively selected functional BCR are called mature, naive B cells. These cells coexpress IgM and IgD through differential splicing of the V_H region exon to the Cμ and Cδ constant region exons coding for the IgM and IgD isotypes. Mature B cells leave the bone marrow microenvironment and circulate as small, resting lymphocytes in peripheral blood and secondary lymphoid tissues.

1.4. The Germinal Center Reaction

Upon contact with cognate antigen, mature B cells are activated and migrate into the T cell zones of secondary lymphatic organs. The interaction with activated T helper cells provides further stimulation to the B lymphocytes and induces proliferation, leading to the formation of primary foci. A fraction of the proliferating B cells differentiates into short-lived plasma cells that secrete antibodies of mostly IgM isotype and hence provide an initial wave of low-affinity antibodies. Some B cells, however, migrate together with activated T helper cells into B cell follicles and initiate a germinal center (GC) reaction (4, 18).

In the GC, the highly proliferative GC B cells interact in immunological synapses with follicular T helper cells and follicular dendritic cells (FDC). These tightly regulated mechanisms of proliferation and cellular interaction lead to a histological structure that is characteristic for all GCs. A loosely associated network of FDCs, follicular T helper cells, and GC B cells (centrocytes) form the so-called light zone, whereas a dense area of quickly dividing B cells (centroblasts) can be recognized as dark zone. The expanding GC displaces the locally residing, resting B lymphocytes and compacts them into a mantle zone surrounding the GC. GC B cells circulate mainly within but also between the two zones (4, 18). The underlying mechanisms that drive these processes are currently under thorough investigation (19–21).

In the dark zone, the process of somatic hypermutation (SHM) introduces point mutations as well as small deletions, insertions, or duplications into the rearranged IgV genes (Fig. 1) (4, 22, 23). The SHM process is strictly dependent on transcriptional activity of the affected template and mainly takes place in a regionally defined area of 1–2 kb downstream of the V segment promoters (24). Moreover, special sequence motifs (SHM hot spots) are preferentially targeted and nucleotide exchanges have a transition over transversion bias when compared to a random mutation process (24, 25). The SHM mechanism is essentially dependent on the enzyme activation-induced cytidine deaminase (AID) that converts cytidine into uracil in the affected DNA strand (24). As uracil is not a normal component of DNA, these sites are subsequently targeted by error-prone DNA repair mechanisms that will produce somatic mutations (26). SHM also affects some non-Ig genes, although at a much lower rate (27, 28).

The somatic IgV gene mutations may lead to a change in BCR affinity. The centroblasts that acquired mutations migrate into the light zone and compete with other GC B cells for survival signals from FDC and T helper cells. The amount of survival signal correlates with improved or at least retained affinity to the cognate antigen: comparably low affinity is counterselected by induction of apoptosis in the corresponding centrocytes (4, 22, 29). GC B cells undergo multiple cycles of proliferation, mutation, and selection. This iterative process leads to a stepwise improvement of the affinity of the BCR to its cognate antigen.

Another important DNA recombination process occurs in the centrocytes: the constant region of the antibody heavy chain (C_H) may be exchanged by class switch recombination. The C_H region of naive B cells is initially expressed with IgM and IgD isotype due to alternative splicing. During class switching, the Cμ and Cδ gene segments can be replaced by one of the two Cα, one Cε, or four Cγ gene segments in humans (Fig. 1). This recombination process is dependent on AID and mediated by DNA double-strand breaks in specialized switch regions upstream of the C_H gene segments (24). Upon deletion of Cμ and Cδ the $V_HD_HJ_H$ exon will now be expressed as part of a heavy chain with the C_H gene segment that replaced the Cμ gene. Class switching leads to changes in BCR signaling competence and modified effector functions of the antibody (30). Notably, class switching is not an obligatory feature of GC B cells, as part of the GC B cell progeny—mostly early descendants—leave the GC as non-class switched lymphocytes (31, 32).

After several cycles of proliferation, mutation, and positive selection, GC B cells differentiate into either antibody secreting plasma cells or resting memory B cells and leave the GC microenvironment (33).

1.5. Immunologic Memory

Memory B cells and post-GC plasma cells provide the two most important functions of B cell adaptive immunity: first of all, post-GC plasma cells antagonize the invaded pathogens with a potent wave of high-affinity antibodies. Second, both cell types provide the organism with the potential of enhanced and improved immune responses upon reencounter of antigen. These quick and effective secondary immune responses constitute humoral immunological memory (34). High-affinity antibodies are secreted over long periods by long-lived post-GC plasma cells that reside in specialized niches in the bone marrow (35). Quiescent long-lived memory B cells can easily be reactivated to take part in improved secondary responses (34).

1.6. T-Independent Immune Responses

Whereas for the induction of a GC reaction, a B lymphocyte is dependent on the interaction and communication with T cells, the activation of B lymphocytes and their differentiation into plasma cells may also occur without T cell help in T cell-independent (TI) immune responses. There are two types of TI activations of B cells: either antigens (mitogens) that trigger conserved pattern recognition receptors (e.g., Toll-like receptors) provoke a polyclonal B cell response (TI-1) (36) or antigens with highly repetitive structure (e.g., bacterial capsules) activate specific B lymphocytes by intensive BCR crosslinking (TI-2) (37). Typically, TI immune responses do not give rise to memory B cells. Plasma cells generated in TI immune responses are short-lived and unmutated (38). Class switching can also take place during TI immune responses, mainly to the IgG2 isotype in humans.

2. Cellular Origin of Human B Cell Lymphomas

When B cells undergo malignant transformation, they usually retain key features of their cell of origin, including specific characteristics of the particular differentiation stage of the lymphoma precursor (33, 39). Histological and immunohistochemical studies of lymphomas have hence been very important to classify B cell malignancies and determine the cellular derivation of these tumors. For example, in follicular lymphoma the tumor cells morphologically resemble GC B cells, they express typical markers of GC B cells, and they grow in follicular structures that resemble GC and that harbor GC T helper cells and FDC networks (40). Thus, all these features point to a GC B cell derivation of follicular lymphomas.

When it became feasible to comprehensively study Ig gene rearrangements by polymerase chain reaction (PCR) and sequencing, the histopathological evaluations of lymphomas were complemented by V gene analyses. Such studies, for example, validated the GC B cell origin of follicular lymphomas by showing that the lymphoma cells carry somatically mutated Ig V genes and show intraclonal diversity as a sign of ongoing SHM throughout clonal expansion, further characteristic features of GC B cells (41).

The cellular origin of Burkitt lymphomas could not be clarified based on the histological picture, because this highly aggressive lymphoma shows a disruption of the normal lymph node structure and growth in a diffuse pattern. However, the lymphoma cells morphologically resemble centroblasts, they express key GC B cell markers, and they have somatically mutated Ig V genes with ongoing hypermutation in a fraction of cases (42). Thus, Burkitt lymphomas are derived from GC B cells, too.

The development of genome-wide gene expression tools by microarrays enabled a much more comprehensive comparison of the gene expression of human B cell lymphomas to normal B cell subsets than was previously possible by immunohistochemical staining for single or few markers. Such comparisons were consequently widely used in recent years to identify the cellular origin of lymphomas. One landmark finding was that the heterogenous group of diffuse large B cell lymphomas (DLBCL) can be subdivided into several subgroups (43, 44). The two main subgroups were defined by a high similarity of the lymphoma cells to GC B cells (GCB-DLBCL) or to in vitro-activated B cells (ABC-DLBCL) (43, 44). As the GCB-DLBCL also often showed ongoing somatic hypermutation, this lymphoma is now considered as a further GC B cell lymphoma (45). ABC-DLBCL have a highly activated phenotype, carry somatically mutated V genes, but lack most specific features of GC B cells. This lymphoma is most similar to post-GC immunoblasts.

A special case is classical Hodgkin lymphoma, because the tumor cells of this malignancy, the Hodgkin and Reed/Sternberg (HRS) cells, express only few B cell markers and express multiple markers of various other hematopoietic cell types (46). However, the B cell origin of HRS cells from mature B cells was unequivocally shown by the demonstration that these cells carry clonally rearranged and somatically mutated V region genes (47, 48). Surprisingly, in a quarter of cases, destructive somatic mutations were found in originally productive V gene rearrangements (47, 48). Based on this finding, it was proposed that HRS cells derive from the pool of pre-apoptotic GC B cells that acquired unfavorable mutations and that normally would have undergone apoptosis (47).

As a main finding of the numerous studies to reveal the cellular origin of human B cell lymphomas, it can be concluded that the majority of these lymphomas is derived from GC or post-GC B cells (Fig. 2). This is remarkable, because B cells reside only for a few days to weeks in a GC, and more than half of the B cell pool is represented by naive B cells. The vigorous proliferation of the GC B cells may represent one critical factor why these cells become transformed, permanently proliferating cells. Moreover, the genetic processes of SHM and class switching are mutagenic processes that strongly increase the risk for a B cell to undergo malignant transformation, as discussed in the following paragraph. It should, however, also be stressed that lymphoma development is a multistep process, and that genetic lesions may be acquired in a lymphoma precursor over multiple differentiation steps. It is a matter of discussion which of these intermediate steps to define as the cell of origin. For example, in follicular lymphomas, the prototypic GC B cell lymphoma, the *t*(14;18) *BCL2*/IgH chromosomal translocation, found in nearly all cases, occurs at the pro B cell stage of B cell development during misguided V gene recombination. However, this translocation becomes pathogenetically relevant only much later in B cell development in GC B cells when BCL2 is normally downregulated.

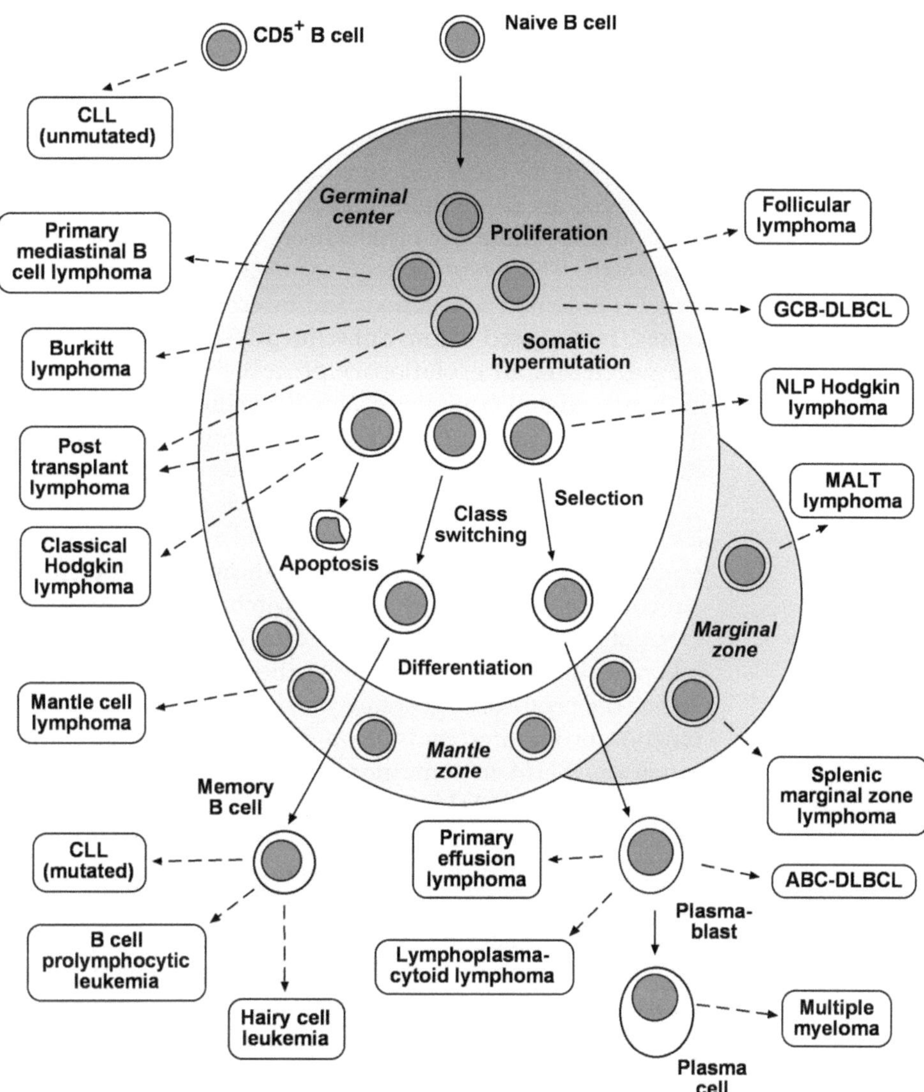

Fig. 2. Germinal center reaction and cellular origin of human B cell lymphomas. Shown are the main steps in mature B cell differentiation in the GC and the presumed cellular origin of human B cell lymphomas. Antigen-activated mature B cells are driven into a GC reaction when T cell help is available. The GC B cells undergo massive clonal expansion in the dark zone of the GC and activate the process of SHM. Mainly in the light zone, mutated B cells are selected for affinity-increasing IgV gene mutations. Positively selected cells will undergo multiple rounds of proliferation, mutation, and selection before they differentiate into memory B cells or plasmablasts and exit the GC. The majority of GC B cells will acquire disadvantageous mutations and undergo apoptosis. Many GC B cells perform class switch recombination in the light zone. Most lymphomas are derived from GC B cells or from post-GC B cells. Also in the latter types of lymphomas, decisive transforming events have presumably occurred in premalignant GC B cell precursors of these lymphomas. CLL with unmutated IgV genes is derived from CD5+ B cells. Mutated CLL is presumably derived from a small subset of CD5+ memory B cells. Most mantle cell lymphomas carry unmutated IgV genes and are presumably derived from (CD5+) mantle zone B cells. However, 20–30% of these lymphomas carry mutated V genes, suggesting a GC experience. Primary mediastinal large B cell lymphoma is likely derived from GC-experienced thymic B cells. A clear marginal zone is present around B cell follicles in the spleen. Marginal zone B cells are likely the origin of splenic marginal zone B cells, although it is puzzling that normal marginal B cells harbor mutated IgV genes, whereas a considerable fraction of splenic marginal zone B cell lymphomas has unmutated V genes (148). The tumor cells of classical Hodgkin lymphoma and some posttransplant lymphomas carry destructive IgV gene mutations, indicating a derivation from pre-apoptotic GC B cells.

3. Genetic Lesions in the Pathogenesis of B Cell Lymphomas

To reach full malignancy, several security checks in a B cell have to fail. The current theory of cancer development promulgates that a normal cell needs multiple "hits" that change its normal functions regarding the control of proliferation, apoptosis, and regulation by other cells. These hits are acquired in B cell malignancies both by largely random genetic lesions and by erronous B cell-specific processes, as discussed below. Subsequently, cells with genetic lesions are selected in an evolutionary process for environmental fitness and survival and have the chance to acquire additional changes. The nature of these hits is diverse, including viral infection, genetic mutations, and epigenetic restructurations (Table 1).

A hallmark of many B cell lymphomas are chromosomal translocations involving one of the Ig loci and a proto-oncogene. These translocations happen as by-products during the processes of V(D)J recombination, SHM, and class switching (49). Apparently, the DNA strand breaks occurring in each of these processes bear an inherent risk of generating translocations. The strand breaks in the loci of the proto-oncogenes may be random, involve some recombination-prone sites, and in some instances are due to off-target activity of SHM in some non-Ig genes (e.g., *BCL6*) (50). Off-target activity of SHM can also cause point mutations in proto-oncogenes and is most frequently seen in DLBCL (50).

The translocation of a proto-oncogene into an Ig locus in B cell lymphomas deregulates the oncogene expression through the associated Ig enhancers, which are highly active in the B cells. For example, in over 90% of follicular lymphomas, *BCL2* is found translocated into the heavy-chain locus, sustaining the continuous expression of the *BCL2* oncogene (51, 52). The situation is similar in Burkitt and mantle cell lymphoma, where *MYC* and *CCND1*, respectively, are translocated into an Ig locus in virtually every case (53–55). MYC and CCND1 both cause uncontrolled proliferation of the cells harboring the respective translocation.

Other lymphoma entities do not show such monotypic translocation patterns as Burkitt lymphoma or mantle cell lymphoma. Although diverse translocations were described in DLBCL, mucosa-associated lymphatic tissue (MALT) lymphoma, and multiple myeloma, none of these is found in the majority of cases (Table 1). This may also reflect the more diverse phenotypes of these lymphomas.

The same processes that mediate a translocation into an Ig locus sometimes cause non-Ig translocations. For some of them, the postulated effect is deregulation by a similar mechanism as in Ig translocation. Here, a regulatory element on the translocation partner overrides the normal expression pattern of a gene. This is for example the case in a diverse range of *BCL6* translocations in

Table1
Genetic lesions in human B cell lymphomas

Entity	Chromosomal translocations	Tumor-suppressor gene mutations/deletions/silencing	Additional alterations
Mantle cell lymphoma	*CCND1*-IgH (95%) (55)	*ATM* (40%) (98, 99) *DLEU2*/miR-15a/16-1 (Del13q14, 50–70%) (64) *CDKN2A* (16–31%) (101) *TP53* (13–45%) (101) *TNFAIP3* (40%) (102)	*NOTCH1* mutations (12%) (100)
Chronic lymphocytic leukemia		*ATM* (30%) (103, 104) *TP53* (15%) (107) *DLEU2*/miR-15a/16-1 (Del13q14, 60%) (66)	*NOTCH1* mutations (10–20%) (105, 106) *SF3B1* mutations (10–20%) (108, 109)
Follicular lymphoma	*BCL2*-IgH (90%) (51, 52)	*MLL2* (89%) (60) *CREBBP* (33%) (61) *EP300* (9%) (61)	
Diffuse large B-cell lymphoma	*BCL6*-(many loci) (35%) (56, 58) *BCL2*-IgH (15–30%) (111) *MYC*-(IgH or IgL) (15%) (113)	*CD95* (10–20%) (110) *ATM* (15%) (112) *TP53* (25%) (114, 115) *BLIMP1* (20% of ABC-DLBCL) (116) *CREBBP* (20%) (61) *EP300* (10%) (61) *MLL2* (32%) (60) *CDKN2A* (35%) (117) *TNFAIP3* (38%) (118)	Mutation of proto-oncogenes (by aberrant somatic hypermutation) (50%) (50)
Primary mediastinal B-cell lymphoma		*SOCS1* (40%) (119) *STAT6* (36%) (121) *TNFAIP3* (36%) (122)	Mutation of proto-oncogenes (by aberrant somatic hypermutation) (70%) (120)
Burkitt lymphoma	*MYC*-(IgH or IgL) (100%) (53, 54)	*TP53* (40%) (107) *RB2* (20–80%) (123)	

(continued)

Table 1 (continued)

Entity	Chromosomal translocations	Tumor-suppressor gene mutations/deletions/silencing	Additional alterations
Posttransplant lymphoma	–	–	–
Classical Hodgkin lymphoma	–	*NFKBIA* (10–20%) (124–126) *NFKBIE* (10%) (128) *CD95* (<10%) (130) *TNFAIP3* (44%) (122) *SOCS1* (42%) (133)	*REL* amplifications (50%) (127) *JAK2* gains (25%) (129) *MAP3K14* gains (131, 132)
Lymphocyte-predominant Hodgkin lymphoma	*BCL6*-(many loci) (48%) (57)	–	–
Splenic marginal-zone lymphoma	–	–	Del 7q22–36 (40%) (134)
MALT lymphoma	*API2-MALT1* (30%) (59) *BCL10*-IgH (5%) (135, 136) *MALT1*-IgH (15–20%) (138) *FOXP1*-IgH (10%) (139)	*CD95* (5–80%) (110) *CDKN2A* (60%) (137)	–
Lymphoplasmacytoid lymphoma	*PAX5*-IgH (50%) (140)	–	–
Primary effusion lymphoma	–	–	–
Multiple myeloma	*CCND1*-IgH (15–20%) (141) *FGFR3*-IgH (10%) (144) *MAF*-IgH (5–10) (146)	*CD95* (10%) (142) *DLEU2*/miR-15a/16-1 (Del13q14, 50%) (65)	*MYC* alterations (40%) (143) *RAS* mutations (40%) (145) Gains of 5q (50%) (147

DLBCL and lymphocyte predominant Hodgkin lymphoma, leading to a constitutive overexpression of this gene (56–58).

A second class of translocations leads to the formation of fusion genes. These are transcribed as fusion transcripts that span the translocation breakpoint, allowing the translation of proteins that combine parts of two separate genes. In MALT lymphoma, about one-third of cases harbors a translocation that leads to the expression of the fusion protein API2-MALT1 (59). This protein, a combination of the amino terminus of API2 with the carboxy terminus of MALT1, has acquired the function to noncanonically activate the NFκB pathway by cleavage of NFκB-inducing kinase MAP3K14 (NIK) into a constitutively active form.

Probably because of the tight surveillance of normal B cell populations, most B cell lymphomas acquire changes during their clonal evolution that allow them to bypass intrinsic or extrinsic signals that normally trigger apoptosis. A central player in the apoptosis network is *TP53*. This gene encodes the tumor suppressor protein p53, a transcription factor influencing the expression of genes that partake directly or indirectly in apoptosis, senescence, and cell cycle progression. Alterations in this single gene have been found in a variety of lymphomas (Table 1) and can have a profound impact on the cell's reaction to a multitude of stimuli, for example DNA damage or cell surface receptor signaling.

Another important apoptosis-related pathway that is often blocked is the sensing of cellular stress and DNA damage during cell cycle checkpoints. For example, the checkpoint guardians ATM and CDKN2A are often inactivated by mutation, deletion, or epigenetic silencing in lymphomas (Table 1). This allows cell growth and proliferation despite the accumulation of DNA damage during tumor progression.

The NFκB pathway is important for many cell types during inflammation, including B cells. It is exploited by some tumors to gain resistance to external and internal danger signals. Hodgkin lymphoma, primary mediastinal B cell lymphoma, ABC-DLBCL, MALT lymphoma, and mantle cell lymphoma seem to sustain a gene expression program resembling continuous inflammation by acquiring inactivating mutations in inhibitors of the NFκB complex (e.g., *TNFAIP3*, *NFKBIA*, and *NFKBIE*) and/or activating mutations, e.g., in *REL*, a component of the NFκB complex itself (Table 1).

A cellular function that is not completely understood but nevertheless found to be recurrently deregulated in several types of lymphomas is chromatin remodeling. Follicular lymphoma and DLBCL show recurrent inactivation of MLL2, a histone methyltransferase, and CREBBP, a histone acetyltransferase (60, 61). Artificial inhibition of these genes in cell lines has shown that their activity controls the chromatin state of a variety of genetic loci.

This probably allows the malignant cell to silence tumor-suppressor genes and to reactivate epigenetically silenced oncogenes by rewriting histone marks. Indeed, many lymphomas show an abnormal pattern of histone marks and/or DNA methylation when compared to their nonmalignant counterparts (62, 63).

In recent years, with advances in the understanding of microRNAs, it became clear that this class of small noncoding RNAs is another frequent target of deregulation in lymphomas. For example the microRNA cluster discovered in the minimal deleted region of chronic lymphocytic leukemia (comprising *miR-15a* and *miR-16-1*) is also recurrently deleted in mantle cell lymphoma and multiple myeloma (64–66). Both this cluster and the single protein coding gene *DLEU2* in this region on chromosome 13 have been shown to act as tumor suppressors (67), providing a selective advantage for tumor clones that have lost one or both copies of this region.

4. Viruses in B Cell Lymphomas

In several types of B cell lymphomas viruses are implicated in their pathogenesis. In B cell lymphomagenesis mainly two members of the γ herpes virus family are involved, i.e., Epstein–Barr virus (EBV) and human herpes virus 8 (HHV8) (68, 69). EBV is found in the tumor cells in nearly all cases of endemic Burkitt lymphoma in Africa and about 30% of sporadic Burkitt lymphoma in other parts of the world. EBV is also found in most lymphomas in post-transplant patients (posttransplant lymphoproliferative disease, PTLD) and in a fraction of classical Hodgkin lymphoma. About 30% of classical Hodgkin lymphoma in the Western world show an EBV association, and pediatric cases of this disease in Central America are EBV positive in more than 80% of cases. Moreover, EBV is found in some diffuse large B cell lymphomas (in particular plasmablastic lymphomas), and primary central nervous system lymphomas. EBV establishes a latent infection in B cells, and different forms of latency are seen in the EBV-associated lymphomas (69). In Burkitt lymphoma, EBV typically shows latency form I, which means that only one EBV-encoded protein is expressed, i.e., EBV nuclear antigen 1 (EBNA1). EBNA1 is essential for the replication of the episomal EBV genome in replicating cells. Whether EBNA1 also has oncogenic functions is still debated. In the HRS cells of Hodgkin lymphoma, besides EBNA1, the two latent membrane proteins 1 and 2a (LMP1 and LMP2a, respectively) of EBV are expressed, a pattern which is called latency II. LMP1 mimics an active CD40 receptor and is a classical oncogene (70). LMP2a mimics an active BCR. Hence, two main survival signals for B cells in the GC are provided by the virus.

Indeed, it has been shown that EBV can rescue BCR-deficient GC B cells from apoptosis, supporting an important role of the virus in the transformation of such B cells (71, 72). In line with this view, all cases of classical Hodgkin lymphoma in which the HRS cells carried crippling mutations preventing expression of a BCR were found to be EBV positive (73).

In PTLD, nine viral proteins are usually expressed (latency pattern III). Besides EBNA1, LMP1, and LMP2a also LMP2b and the EBNAs 2, 3a, 3b, and 3c and LP are expressed (69). EBNA2 is important to drive the proliferation of the transformed B cells. EBV encodes also multiple noncoding RNAs, including the EBV-encoded RNAs (EBERs) and numerous miRNAs. EBERs are expressed by all EBV-infected B cells, which is often used to detect such cells by in situ hybridization for these transcripts.

HHV8 is found in virtually all cases of primary effusion lymphoma, which is a very rare lymphoma mainly occurring in acquired immune deficiency syndrome (AIDS) patients (68). Similar to EBV, HHV8 establishes a latent infection in B cells and persists as an episome in the cells. Several viral proteins are implicated in the pathogenetic role of HHV8 in primary effusion lymphomas, including LNA1, which inhibits p53 and Rb, a viral cyclin, and a viral FLICE inhibitory protein (68). Notably, a fraction of primary effusion lymphomas is co-infected with HHV8 and EBV (68).

Hepatitis C virus (HCV) does not establish a latent infection in B cells, but is also implicated in B cell lymphomagenesis. In chronic HCV carriers, HCV appears to drive B cell lymphomagenesis by two means. On the one hand, HCV may act as a chronic antigenic stimulus for HCV-specific B cells, which is indicated by the specificity of the BCR of some HCV-associated lymphomas for viral antigens (74). On the other hand, HCV can upregulate activation-induced cytidine deaminase, the key factor of SHM, in B cells and also causes the production of reactive oxygen species, two factors that can be mutagenic for B cells (75, 76).

Several additional viruses have been discussed as potential factors in B cell lymphoma pathogenesis (77, 78), but further studies are needed to validate these findings.

5. Microenvironmental Interactions in B Cell Lymphomas and the Role of the BCR

In many B cell lymphomas, interaction of tumor cells with other cells in the microenvironment plays an essential role for the survival and proliferation of the tumor cells. A prototypical example for this is follicular lymphoma; here the transformed GC B cells grow in follicular structures resembling normal GC in close association with FDC and follicular T helper cells. The dependency of follicular lymphoma cells on these normal constituents of a GC is suggested from in vitro studies. The malignant cells survive longer when cocultured with CD4+ T cells or stromal cells and with stimulation of the CD40 receptor, a main survival receptor for GC B cells (79, 80). As normal GC B cells require BCR triggering for their survival, it was an important finding that the BCRs of follicular lymphomas often acquire replacement mutations in their rearranged IgV genes that allow linking of sugar moieties to the V regions (81). There is evidence that sugar-binding receptors (lectins) on stromal cells can constantly trigger the BCR of the lymphoma cells, and hence replace the original foreign antigen that would normally be only transiently available to stimulate the GC B cells (82).

A similar situation as seen in follicular lymphoma is given in CLL, where proliferation mostly takes place in specialized structures in lymph nodes (83). In these proliferation centers, CD40-expressing CLL cells are in contact with stromal and CD4+ T cells, partly expressing CD40L (84). The relevance of these interactions for the survival of CLL cells is indicated from the observation that CLL cells survive only for a short time when cultured alone, but their survival is significantly improved upon coculture with stromal cells or CD40 stimulation (85). Moreover, there is now evidence that the BCR of CLL cells are often polyreactive against autoantigens, including modified components from apoptotic cells (86, 87). Whether and how such autoantigens are presented in the proliferation centers is currently unclear.

Gastric MALT lymphomas are very frequently associated with chronic *Helicobacter pylori* infections. The lymphoma cells themselves usually do not recognize these bacteria via their BCR, but often express autoreactive surface Ig (88). However, CD4+ T cells in the MALT lymphoma microenvironment are activated upon interaction with *H. pylori*, and it thus appears that the *H. pylori*-activated T cells stimulate the lymphoma B cells (89). The importance of this interaction is evident from the fact that elimination of *H. pylori* by antibiotic treatment often leads to regression of the lymphoma (90). Interestingly, also in splenic B cell lymphomas in patients chronically infected with HCV, elimination of the virus by antiviral treatment resulted in regression of the lymphoma (91), representing a further example for chronic antigenic stimulation as critical survival signal for lymphoma cells.

Regarding the extend (and possibly the role) of the microenvironment in lymphomas, classical Hodgkin lymphoma is a rather unique lymphoma, because here the tumor cells usually account for only about 1% of cells in the tissue. The vast majority of cells in the lymphoma is composed of various types of nonmalignant cells, including T cells, B cells, eosinophils, neutrophils, plasma cells, mast cells, and others (46). The normal histological structure of the affected lymph nodes is disturbed. CD4+ T helper and regulatory T cells are usually the most frequent cells in the Hodgkin lymphoma microenvironment, and HRS cells are typically directly surrounded by CD4+ T cells (92). Many of the infiltrating cells are most likely beneficial for the tumor, and it seems that HRS cells actively attract such cells by secretion of a multitude of chemokines and cytokines (46). $CD4^+$ T cells may stimulate HRS cells through CD40 ligand–CD40 interaction and CD28–CD80 interaction, as HRS cells express CD40 and CD80 (46). The regulatory T cells may play a role in the rescue of HRS cells from an attack by cytotoxic T cells and natural killer cells (92). Other immunosuppressive factors secreted by the HRS cells include TGFβ and galectin-1, and HRS cells also express the T cell-suppressive ligand PD-1 ligand (93–96). Eosinophils on the one hand may also contribute to an immunosuppressive microenvironment by secretion of TGFβ, and on the other hand may stimulate the CD30-positive HRS cells through interaction with CD30 ligand, which is expressed by the eosinophils (46).

6. Conclusions

B cell lymphoma pathogenesis is a multistep process involving various genetic and epigenetic lesions, viruses, and distorted microenvironmental and antigenic interactions. Often, the specific hit is less important than the affected pathway, because most gene products exert their function not directly but involve a network of relays and switches. In this way, different combinations of hits can lead to a phenotypically similar lymphomas, as the resulting deregulation of cellular processes is identical. Numerous hallmark chromosomal translocations in B cell lymphomas occur as errors of the Ig gene remodeling processes of V gene recombination, SHM, and class switching. Most B cell lymphomas derive from GC B cells, which is explainable by the highly proliferative activity of these cells and the mutagenic potential of SHM and class switch recombination. A number of mutations in GC B cells appear to function by freezing GC B cells in their proliferative state and preventing their differentiation into resting post GC memory or plasma cells. *BCL6* translocations and inactivating mutations in BLIMP1, the master regulator of the plasma cell program, are prototypical examples for this (97).

The study of rearranged Ig V genes and chromosomal translocations in B cell lymphomas has not only been of major relevance to define the cellular origin of these tumors and identify main transforming events, but it is also a very important research tool. Ig V gene rearrangements and chromosomal translocations have been used as genetic markers to study lymphoma cell dissemination, and to identify and quantify minimal residual disease. They are also valuable to search for lymphoma precursor cells, to define the replication history of B cells, and to determine BCR specificity and hence potential antigen-triggering of the lymphoma cells, to name several key applications.

Acknowledgements

Own work discussed in this review was supported by grants from the Deutsche Forschungsgemeinschaft (TRR60, KU1315/7-1, KU1315/8-1, GK1431), the Deutsche Krebshilfe, the Wilhelm Sander Foundation, the José Carreras Leukemia Foundation, and the BMBF (Haematosys consortium). We thank the other members of our group and Martin-Leo Hansmann for many stimulating discussions.

References

1. Tonegawa S (1983) Somatic generation of antibody diversity. Nature 302:575–581
2. van Gent DC, Ramsden DA, Gellert M (1996) The RAG1 and RAG2 proteins establish the 12/23 rule in V(D)J recombination. Cell 85:107–113
3. Medina KL, Singh H (2005) Genetic networks that regulate B lymphopoiesis. Curr Opin Hematol 12:203–209
4. Rajewsky K (1996) Clonal selection and learning in the antibody system. Nature 381:751–758
5. Corbett SJ, Tomlinson IM, Sonnhammer EL, Buck D, Winter G (1997) Sequence of the human immunoglobulin diversity (D) segment locus: a systematic analysis provides no evidence for the use of DIR segments, inverted D segments, "minor" D segments or D–D recombination. J Mol Biol 270:587–597
6. Ravetch JV, Siebenlist U, Korsmeyer S, Waldmann T, Leder P (1981) Structure of the human immunoglobulin mu locus: characterization of embryonic and rearranged J and D genes. Cell 27:583–591
7. Cook GP, Tomlinson IM (1995) The human immunoglobulin VH repertoire. Immunol Today 16:237–242
8. Alt FW, Rathbun G, Oltz E, Taccioli G, Shinkai Y (1992) Function and control of recombination-activating gene activity. Ann N Y Acad Sci 651:277–294
9. Zhang Z (2007) VH replacement in mice and humans. Trends Immunol 28:132–137
10. Tiegs SL, Russell DM, Nemazee D (1993) Receptor editing in self-reactive bone marrow B cells. J Exp Med 177:1009–1020
11. Hieter PA, Maizel JV Jr, Leder P (1982) Evolution of human immunoglobulin kappa J region genes. J Biol Chem 257:1516–1522
12. Schäble KF, Zachau HG (1993) The variable genes of the human immunoglobulin kappa locus. Biol Chem Hoppe Seyler 374: 1001–1022
13. Kawasaki K, Minoshima S, Nakato E, Shibuya K, Shintani A, Schmeits JL, Wang J, Shimizu N (1997) One-megabase sequence analysis of the human immunoglobulin lambda gene locus. Genome Res 7:250–261
14. Vasicek TJ, Leder P (1990) Structure and expression of the human immunoglobulin lambda genes. J Exp Med 172:609–620
15. Bräuninger A, Goossens T, Rajewsky K, Küppers R (2001) Regulation of immunoglobulin light chain gene rearrangements during early B cell development in the human. Eur J Immunol 31:3631–3637
16. Nadel B, Tang A, Feeney AJ (1998) V(H) replacement is unlikely to contribute significantly to receptor editing due to an ineffectual embedded recombination signal sequence. Mol Immunol 35:227–232
17. Zhang Z, Zemlin M, Wang YH, Munfus D, Huye LE, Findley HW, Bridges SL, Roth DB, Burrows PD, Cooper MD (2003) Contribution of Vh gene replacement to the primary B cell repertoire. Immunity 19:21–31
18. MacLennan IC (1994) Germinal centers. Annu Rev Immunol 12:117–139
19. Allen CD, Ansel KM, Low C, Lesley R, Tamamura H, Fujii N, Cyster JG (2004) Germinal center dark and light zone organization is mediated by CXCR4 and CXCR5. Nat Immunol 5:943–952
20. Hauser AE, Junt T, Mempel TR, Sneddon MW, Kleinstein SH, Henrickson SE, von Andrian UH, Shlomchik MJ, Haberman AM (2007) Definition of germinal-center B cell migration in vivo reveals predominant intrazonal circulation patterns. Immunity 26:655–667
21. Schwickert TA, Lindquist RL, Shakhar G, Livshits G, Skokos D, Kosco-Vilbois MH, Dustin ML, Nussenzweig MC (2007) In vivo

imaging of germinal centres reveals a dynamic open structure. Nature 446:83–87

22. Küppers R, Zhao M, Hansmann ML, Rajewsky K (1993) Tracing B cell development in human germinal centres by molecular analysis of single cells picked from histological sections. EMBO J 12:4955–4967
23. Goossens T, Klein U, Küppers R (1998) Frequent occurrence of deletions and duplications during somatic hypermutation: Implications for oncogene translocations and heavy chain disease. Proc Natl Acad Sci U S A 95:2463–2468
24. Pavri R, Nussenzweig MC (2011) AID targeting in antibody diversity. Adv Immunol 110:1–26
25. Neuberger MS (2008) Antibody diversification by somatic mutation: from Burnet onwards. Immunol Cell Biol 86:124–132
26. Di Noia JM, Neuberger MST (2007) Molecular mechanisms of antibody somatic hypermutation. Annu Rev Biochem 76:1–22
27. Pasqualucci L, Migliazza A, Fracchiolla N, William C, Neri A, Baldini L, Chaganti RSK, Klein U, Küppers R, Rajewsky K, Dalla-Favera R (1998) BCL-6 mutations in normal germinal center B cells: evidence of somatic hypermutation acting outside Ig loci. Proc Natl Acad Sci U S A 95:11816–11821
28. Liu M, Duke JL, Richter DJ, Vinuesa CG, Goodnow CC, Kleinstein SH, Schatz DG (2008) Two levels of protection for the B cell genome during somatic hypermutation. Nature 451:841–845
29. Liu YJ, Joshua DE, Williams GT, Smith CA, Gordon J, MacLennan IC (1989) Mechanism of antigen-driven selection in germinal centres. Nature 342:929–931
30. Manis JP, Tian M, Alt FW (2002) Mechanism and control of class-switch recombination. Trends Immunol 23:31–39
31. Klein U, Rajewsky K, Küppers R (1998) Human immunoglobulin (Ig)M+IgD+ peripheral blood B cells expressing the CD27 cell surface antigen carry somatically mutated variable region genes: CD27 as a general marker for somatically mutated (memory) B cells. J Exp Med 188:1679–1689
32. Seifert M, Küppers R (2009) Molecular footprints of a germinal center derivation of human IgM+(IgD+)CD27+ B cells and the dynamics of memory B cell generation. J Exp Med 206:2659–2669
33. Klein U, Dalla-Favera R (2008) Germinal centres: role in B-cell physiology and malignancy. Nat Rev Immunol 8:22–33
34. McHeyzer-Williams M, Okitsu S, Wang N, McHeyzer-Williams L (2012) Molecular programming of B cell memory. Nat Rev Immunol 12:24–34
35. Manz RA, Hauser AE, Hiepe F, Radbruch A (2005) Maintenance of serum antibody levels. Annu Rev Immunol 23:367–386
36. Han JH, Akira S, Calame K, Beutler B, Selsing E, Imanishi-Kari T (2007) Class switch recombination and somatic hypermutation in early mouse B cells are mediated by B cell and Toll-like receptors. Immunity 27:64–75
37. Mond JJ, Lees A, Snapper CM (1995) T cell-independent antigens type 2. Annu Rev Immunol 13:655–692
38. Toellner KM, Jenkinson WE, Taylor DR, Khan M, Sze DM, Sansom DM, Vinuesa CG, MacLennan IC (2002) Low-level hypermutation in T cell-independent germinal centers compared with high mutation rates associated with T cell-dependent germinal centers. J Exp Med 195:383–389
39. Küppers R (2005) Mechanisms of B-cell lymphoma pathogenesis. Nat Rev Cancer 5: 251–262
40. de Jong D (2005) Molecular pathogenesis of follicular lymphoma: a cross talk of genetic and immunologic factors. J Clin Oncol 23:6358–6363
41. Bende RJ, Smit LA, van Noesel CJ (2007) Molecular pathways in follicular lymphoma. Leukemia 21:18–29
42. Küppers R, Klein U, Hansmann M-L, Rajewsky K (1999) Cellular origin of human B-cell lymphomas. N Engl J Med 341: 1520–1529
43. Alizadeh AA, Eisen MB, Davis RE, Ma C, Lossos IS, Rosenwald A, Boldrick JC, Sabet H, Tran T, Yu X, Powell JI, Yang L, Marti GE, Moore T, Hudson J Jr, Lu L, Lewis DB, Tibshirani R, Sherlock G, Chan WC, Greiner TC, Weisenburger DD, Armitage JO, Warnke R, Levy R, Wilson W, Grever MR, Byrd JC, Botstein D, Brown PO, Staudt LM (2000) Distinct types of diffuse large B-cell lymphoma identified by gene expression profiling. Nature 403:503–511
44. Rosenwald A, Wright G, Chan WC, Connors JM, Campo E, Fisher RI, Gascoyne RD, Muller-Hermelink HK, Smeland EB, Giltnane JM, Hurt EM, Zhao H, Averett L, Yang L, Wilson WH, Jaffe ES, Simon R, Klausner RD, Powell J, Duffey PL, Longo DL, Greiner TC, Weisenburger DD, Sanger WG, Dave BJ, Lynch JC, Vose J, Armitage JO, Montserrat E, Lopez-Guillermo A, Grogan TM, Miller TP, LeBlanc M, Ott G, Kvaloy S, Delabie J, Holte H, Krajci P, Stokke T, Staudt LM (2002) The use of molecular profiling to predict survival after chemotherapy for diffuse large-B-cell lymphoma. N Engl J Med 346:1937–1947

45. Lossos IS, Alizadeh AA, Eisen MB, Chan WC, Brown PO, Botstein D, Staudt LM, Levy R (2000) Ongoing immunoglobulin somatic mutation in germinal center B cell-like but not in activated B cell-like diffuse large cell lymphomas. Proc Natl Acad Sci U S A 97:10209–10213
46. Küppers R (2009) The biology of Hodgkin's lymphoma. Nat Rev Cancer 9:15–27
47. Kanzler H, Küppers R, Hansmann ML, Rajewsky K (1996) Hodgkin and Reed-Sternberg cells in Hodgkin's disease represent the outgrowth of a dominant tumor clone derived from (crippled) germinal center B cells. J Exp Med 184:1495–1505
48. Küppers R, Rajewsky K, Zhao M, Simons G, Laumann R, Fischer R, Hansmann ML (1994) Hodgkin disease: Hodgkin and Reed-Sternberg cells picked from histological sections show clonal immunoglobulin gene rearrangements and appear to be derived from B cells at various stages of development. Proc Natl Acad Sci U S A 91:10962–10966
49. Küppers R, Dalla-Favera R (2001) Mechanisms of chromosomal translocations in B cell lymphomas. Oncogene 20:5580–5594
50. Pasqualucci L, Neumeister P, Goossens T, Nanjangud G, Chaganti RS, Küppers R, Dalla-Favera R (2001) Hypermutation of multiple proto-oncogenes in B-cell diffuse large-cell lymphomas. Nature 412:341–346
51. Jäger U, Bocskor S, Le T, Mitterbauer G, Bolz I, Chott A, Kneba M, Mannhalter C, Nadel B (2000) Follicular lymphomas' BCL-2/IgH junctions contain templated nucleotide insertions: novel insights into the mechanism of t(14;18) translocation. Blood 95:3520–3529
52. Tsujimoto Y, Gorham J, Cossman J, Jaffe E, Croce CM (1985) The t(14;18) chromosome translocations involved in B-cell neoplasms result from mistakes in VDJ joining. Science 229:1390–1393
53. Dalla-Favera R, Martinotti S, Gallo RC, Erikson J, Croce CM (1983) Translocation and rearrangements of the c-myc oncogene locus in human undifferentiated B-cell lymphomas. Science 219:963–967
54. Taub R, Kirsch I, Morton C, Lenoir G, Swan D, Tronick S, Aaronson S, Leder P (1982) Translocation of the c-myc gene into the immunoglobulin heavy chain locus in human Burkitt lymphoma and murine plasmacytoma cells. Proc Natl Acad Sci U S A 79:7837–7841
55. Vaandrager JW, Schuuring E, Zwikstra E, de Boer CJ, Kleiverda KK, van Krieken JH, Kluin-Nelemans HC, van Ommen GJ, Raap AK, Kluin PM (1996) Direct visualization of dispersed 11q13 chromosomal translocations in mantle cell lymphoma by multicolor DNA fiber fluorescence in situ hybridization. Blood 88:1177–1182
56. Baron BW, Nucifora G, McCabe N, Espinosa R 3rd, Le Beau MM, McKeithan TW (1993) Identification of the gene associated with the recurring chromosomal translocations t(3;14)(q27;q32) and t(3;22)(q27;q11) in B-cell lymphomas. Proc Natl Acad Sci U S A 90:5262–5266
57. Wlodarska I, Nooyen P, Maes B, Martin-Subero JI, Siebert R, Pauwels P, De Wolf-Peeters C, Hagemeijer A (2003) Frequent occurrence of BCL6 rearrangements in nodular lymphocyte predominance Hodgkin lymphoma but not in classical Hodgkin lymphoma. Blood 101:706–710
58. Ye BH, Rao PH, Chaganti RS, Dalla-Favera R (1993) Cloning of bcl-6, the locus involved in chromosome translocations affecting band 3q27 in B-cell lymphoma. Cancer Res 53:2732–2735
59. Dierlamm J, Baens M, Wlodarska I, Stefanova-Ouzounova M, Hernandez JM, Hossfeld DK, De Wolf-Peeters C, Hagemeijer A, Van den Berghe H, Marynen P (1999) The apoptosis inhibitor gene API2 and a novel 18q gene, MLT, are recurrently rearranged in the t(11;18)(q21;q21) associated with mucosa-associated lymphoid tissue lymphomas. Blood 93:3601–3609
60. Morin RD, Mendez-Lago M, Mungall AJ, Goya R, Mungall KL, Corbett RD, Johnson NA, Severson TM, Chiu R, Field M, Jackman S, Krzywinski M, Scott DW, Trinh DL, Tamura-Wells J, Li S, Firme MR, Rogic S, Griffith M, Chan S, Yakovenko O, Meyer IM, Zhao EY, Smailus D, Moksa M, Chittaranjan S, Rimsza L, Brooks-Wilson A, Spinelli JJ, Ben-Neriah S, Meissner B, Woolcock B, Boyle M, McDonald H, Tam A, Zhao Y, Delaney A, Zeng T, Tse K, Butterfield Y, Birol I, Holt R, Schein J, Horsman DE, Moore R, Jones SJ, Connors JM, Hirst M, Gascoyne RD, Marra MA (2011) Frequent mutation of histone-modifying genes in non-Hodgkin lymphoma. Nature 476:298–303
61. Pasqualucci L, Dominguez-Sola D, Chiarenza A, Fabbri G, Grunn A, Trifonov V, Kasper LH, Lerach S, Tang H, Ma J, Rossi D, Chadburn A, Murty VV, Mullighan CG, Gaidano G, Rabadan R, Brindle PK, Dalla-Favera R (2011) Inactivating mutations of acetyltransferase genes in B-cell lymphoma. Nature 471:189–195
62. Ammerpohl O, Haake A, Pellissery S, Giefing M, Richter J, Balint B, Kulis M, Le J, Bibikova M,

Drexler HG, Seifert M, Shaknovic R, Korn B, Küppers R, Martin-Subero JI, Siebert R (2011) Array-based DNA methylation analysis in classical Hodgkin lymphoma reveals new insights into the mechanisms underlying silencing of B cell-specific genes. Leukemia 26:185–188

63. Martin-Subero JI, Kreuz M, Bibikova M, Bentink S, Ammerpohl O, Wickham-Garcia E, Rosolowski M, Richter J, Lopez-Serra L, Ballestar E, Berger H, Agirre X, Bernd HW, Calvanese V, Cogliatti SB, Drexler HG, Fan JB, Fraga MF, Hansmann ML, Hummel M, Klapper W, Korn B, Küppers R, Macleod RA, Moller P, Ott G, Pott C, Prosper F, Rosenwald A, Schwaenen C, Schubeler D, Seifert M, Sturzenhofecker B, Weber M, Wessendorf S, Loeffler M, Trümper L, Stein H, Spang R, Esteller M, Barker D, Hasenclever D, Siebert R (2009) New insights into the biology and origin of mature aggressive B-cell lymphomas by combined epigenomic, genomic, and transcriptional profiling. Blood 113:2488–2497

64. Cuneo A, Bigoni R, Rigolin GM, Roberti MG, Bardi A, Campioni D, Minotto C, Agostini P, Milani R, Bullrich F, Negrini M, Croce C, Castoldi G (1999) 13q14 deletion in non-Hodgkin's lymphoma: correlation with clinicopathologic features. Haematologica 84:589–593

65. Kuehl WM, Bergsagel PL (2002) Multiple myeloma: evolving genetic events and host interactions. Nat Rev Cancer 2:175–187

66. Liu Y, Hermanson M, Grander D, Merup M, Wu X, Heyman M, Rasool O, Juliusson G, Gahrton G, Detlofsson R, Nikiforova N, Buys C, Soderhall S, Yankovsky N, Zabarovsky E, Einhorn S (1995) 13q deletions in lymphoid malignancies. Blood 86:1911–1915

67. Klein U, Lia M, Crespo M, Siegel R, Shen Q, Mo T, Ambesi-Impiombato A, Califano A, Migliazza A, Bhagat G, Dalla-Favera R (2010) The DLEU2/miR-15a/16-1 cluster controls B cell proliferation and its deletion leads to chronic lymphocytic leukemia. Cancer Cell 17:28–40

68. Carbone A, Gloghini A (2008) KSHV/HHV8-associated lymphomas. Br J Haematol 140:13–24

69. Küppers R (2003) B cells under influence: transformation of B cells by Epstein-Barr virus. Nat Rev Immunol 3:801–812

70. Kilger E, Kieser A, Baumann M, Hammerschmidt W (1998) Epstein-Barr virus-mediated B-cell proliferation is dependent upon latent membrane protein 1, which simulates an activated CD40 receptor. EMBO J 17:1700–1709

71. Bechtel D, Kurth J, Unkel C, Küppers R (2005) Transformation of BCR-deficient germinal-center B cells by EBV supports a major role of the virus in the pathogenesis of Hodgkin and posttransplantation lymphomas. Blood 106:4345–4350

72. Mancao C, Hammerschmidt W (2007) Epstein-Barr virus latent membrane protein 2A is a B-cell receptor mimic and essential for B-cell survival. Blood 110:3715–3721

73. Bräuninger A, Schmitz R, Bechtel D, Renné C, Hansmann M-L, Küppers R (2006) Molecular biology of Hodgkin and Reed/Sternberg cells in Hodgkin's lymphoma. Int J Cancer 118:1853–1861

74. Quinn ER, Chan CH, Hadlock KG, Foung SK, Flint M, Levy S (2001) The B-cell receptor of a hepatitis C virus (HCV)-associated non-Hodgkin lymphoma binds the viral E2 envelope protein, implicating HCV in lymphomagenesis. Blood 98:3745–3749

75. Machida K, Cheng KT, Sung VM, Lee KJ, Levine AM, Lai MM (2004) Hepatitis C virus infection activates the immunologic (type II) isoform of nitric oxide synthase and thereby enhances DNA damage and mutations of cellular genes. J Virol 78:8835–8843

76. Machida K, Cheng KT, Sung VM, Shimodaira S, Lindsay KL, Levine AM, Lai MY, Lai MM (2004) Hepatitis C virus induces a mutator phenotype: enhanced mutations of immunoglobulin and protooncogenes. Proc Natl Acad Sci U S A 101:4262–4267

77. Lacroix A, Collot-Teixeira S, Mardivirin L, Jaccard A, Petit B, Piguet C, Sturtz F, Preux PM, Bordessoule D, Ranger-Rogez S (2010) Involvement of human herpesvirus-6 variant B in classic Hodgkin's lymphoma via DR7 oncoprotein. Clin Cancer Res 16:4711–4721

78. Maggio E, Benharroch D, Gopas J, Dittmer U, Hansmann ML, Küppers R (2007) Absence of measles virus genome and transcripts in Hodgkin-Reed/Sternberg cells of a cohort of Hodgkin lymphoma patients. Int J Cancer 121:448–453

79. Johnson PW, Watt SM, Betts DR, Davies D, Jordan S, Norton AJ, Lister TA (1993) Isolated follicular lymphoma cells are resistant to apoptosis and can be grown in vitro in the CD40/stromal cell system. Blood 82:1848–1857

80. Umetsu DT, Esserman L, Donlon TA, DeKruyff RH, Levy R (1990) Induction of proliferation of human follicular (B type) lymphoma cells by cognate interaction with CD4+ T cell clones. J Immunol 144:2550–2557

81. Zhu D, McCarthy H, Ottensmeier CH, Johnson P, Hamblin TJ, Stevenson FK (2002) Acquisition of potential N-glycosylation sites in the immunoglobulin variable region by somatic mutation is a distinctive feature of follicular lymphoma. Blood 99:2562–2568
82. Coelho V, Krysov S, Ghaemmaghami AM, Emara M, Potter KN, Johnson P, Packham G, Martinez-Pomares L, Stevenson FK (2010) Glycosylation of surface Ig creates a functional bridge between human follicular lymphoma and microenvironmental lectins. Proc Natl Acad Sci U S A 107:18587–18592
83. Schmid C, Isaacson PG (1994) Proliferation centres in B-cell malignant lymphoma, lymphocytic (B-CLL): an immunophenotypic study. Histopathology 24:445–451
84. Ghia P, Strola G, Granziero L, Geuna M, Guida G, Sallusto F, Ruffing N, Montagna L, Piccoli P, Chilosi M, Caligaris-Cappio F (2002) Chronic lymphocytic leukemia B cells are endowed with the capacity to attract CD4+, CD40L+ T cells by producing CCL22. Eur J Immunol 32:1403–1413
85. Buske C, Gogowski G, Schreiber K, Rave-Frank M, Hiddemann W, Wormann B (1997) Stimulation of B-chronic lymphocytic leukemia cells by murine fibroblasts, IL-4, anti-CD40 antibodies, and the soluble CD40 ligand. Exp Hematol 25:329–337
86. Chu CC, Catera R, Zhang L, Didier S, Agagnina BM, Damle RN, Kaufman MS, Kolitz JE, Allen SL, Rai KR, Chiorazzi N (2010) Many chronic lymphocytic leukemia antibodies recognize apoptotic cells with exposed nonmuscle myosin heavy chain IIA: implications for patient outcome and cell of origin. Blood 115:3907–3915
87. Herve M, Xu K, Ng YS, Wardemann H, Albesiano E, Messmer BT, Chiorazzi N, Meffre E (2005) Unmutated and mutated chronic lymphocytic leukemias derive from self-reactive B cell precursors despite expressing different antibody reactivity. J Clin Invest 115:1636–1643
88. Bende RJ, Aarts WM, Riedl RG, de Jong D, Pals ST, van Noesel CJM (in press) Immunoglobulins of B-cell non Hodgkin's lymphomas: musosa-associated lymphoid tissue lymphomas express a distinctive repertoire with frequent rheumatoid factor reactivity. J Exp Med 201
89. Hussel T, Isaacson PG, Crabtree JE, Spencer J (1996) Helicobacter pylori-specific tumour-infiltrating T cells provide contact dependent help for the growth of malignant B cells in low-grade gastric lymphoma of mucosa-associated lymphoid tissue. J Pathol 178:122–127
90. Wotherspoon AC, Doglioni C, Diss TC, Pan L, Moschini A, de Boni M, Isaacson PG (1993) Regression of primary low-grade B-cell gastric lymphoma of mucosa-associated lymphoid tissue after eradication of Helicobacter pylori. Lancet 342:575–577
91. Hermine O, Lefrere F, Bronowicki JP, Mariette X, Jondeau K, Eclache-Saudreau V, Delmas B, Valensi F, Cacoub P, Brechot C, Varet B, Troussard X (2002) Regression of splenic lymphoma with villous lymphocytes after treatment of hepatitis C virus infection. N Engl J Med 347:89–94
92. Marshall NA, Christie LE, Munro LR, Culligan DJ, Johnston PW, Barker RN, Vickers MA (2004) Immunosuppressive regulatory T cells are abundant in the reactive lymphocytes of Hodgkin lymphoma. Blood 103:1755–1762
93. Chemnitz JM, Eggle D, Driesen J, Classen S, Riley JL, Debey-Pascher S, Beyer M, Popov A, Zander T, Schultze JL (2007) RNA fingerprints provide direct evidence for the inhibitory role of TGFbeta and PD-1 on CD4+ T cells in Hodgkin lymphoma. Blood 110:3226–3233
94. Gandhi MK, Moll G, Smith C, Dua U, Lambley E, Ramuz O, Gill D, Marlton P, Seymour JF, Khanna R (2007) Galectin-1 mediated suppression of Epstein-Barr virus specific T-cell immunity in classic Hodgkin lymphoma. Blood 110:1326–1329
95. Juszczynski P, Ouyang J, Monti S, Rodig SJ, Takeyama K, Abramson J, Chen W, Kutok JL, Rabinovich GA, Shipp MA (2007) The AP1-dependent secretion of galectin-1 by Reed Sternberg cells fosters immune privilege in classical Hodgkin lymphoma. Proc Natl Acad Sci U S A 104:13134–13139
96. Yamamoto R, Nishikori M, Kitawaki T, Sakai T, Hishizawa M, Tashima M, Kondo T, Ohmori K, Kurata M, Hayashi T, Uchiyama T (2008) PD-1–PD-1 ligand interaction contributes to immunosuppressive microenvironment of Hodgkin lymphoma. Blood 111:3220–3224
97. Shaffer AL, Young RM, Staudt LM (2012) Pathogenesis of human B cell lymphomas. Annu Rev Immunol 30:565–610
98. Camacho E, Hernandez L, Hernandez S, Tort F, Bellosillo B, Bea S, Bosch F, Montserrat E, Cardesa A, Fernandez PL, Campo E (2002) ATM gene inactivation in mantle cell lymphoma mainly occurs by truncating mutations and missense mutations involving the phosphatidylinositol-3 kinase domain and is associated with increasing numbers of chromosomal imbalances. Blood 99:238–244

99. Schaffner C, Idler I, Stilgenbauer S, Dohner H, Lichter P (2000) Mantle cell lymphoma is characterized by inactivation of the ATM gene. Proc Natl Acad Sci U S A 97:2773–2778
100. Kridel R, Meissner B, Rogic S, Boyle M, Telenius A, Woolcock B, Gunawardana J, Jenkins C, Cochrane C, Ben-Neriah S, Tan K, Morin RD, Opat S, Sehn LH, Connors JM, Marra MA, Weng AP, Steidl C, Gascoyne RD (2012) Whole transcriptome sequencing reveals recurrent NOTCH1 mutations in mantle cell lymphoma. Blood 119:1963–1971
101. Jares P, Colomer D, Campo E (2007) Genetic and molecular pathogenesis of mantle cell lymphoma: perspectives for new targeted therapeutics. Nat Rev Cancer 7:750–762
102. Chanudet E, Huang Y, Ichimura K, Dong G, Hamoudi RA, Radford J, Wotherspoon AC, Isaacson PG, Ferry J, Du MQ (2010) A20 is targeted by promoter methylation, deletion and inactivating mutation in MALT lymphoma. Leukemia 24:483–487
103. Schaffner C, Stilgenbauer S, Rappold GA, Dohner H, Lichter P (1999) Somatic ATM mutations indicate a pathogenic role of ATM in B-cell chronic lymphocytic leukemia. Blood 94:748–753
104. Stankovic T, Weber P, Stewart G, Bedenham T, Murray J, Byrd PJ, Moss PA, Taylor AM (1999) Inactivation of ataxia telangiectasia mutated gene in B-cell chronic lymphocytic leukaemia. Lancet 353:26–29
105. Fabbri G, Rasi S, Rossi D, Trifonov V, Khiabanian H, Ma J, Grunn A, Fangazio M, Capello D, Monti S, Cresta S, Gargiulo E, Forconi F, Guarini A, Arcaini L, Paulli M, Laurenti L, Larocca LM, Marasca R, Gattei V, Oscier D, Bertoni F, Mullighan CG, Foa R, Pasqualucci L, Rabadan R, Dalla-Favera R, Gaidano G (2011) Analysis of the chronic lymphocytic leukemia coding genome: role of NOTCH1 mutational activation. J Exp Med 208:1389–1401
106. Puente XS, Pinyol M, Quesada V, Conde L, Ordonez GR, Villamor N, Escaramis G, Jares P, Bea S, Gonzalez-Diaz M, Bassaganyas L, Baumann T, Juan M, Lopez-Guerra M, Colomer D, Tubio JM, Lopez C, Navarro A, Tornador C, Aymerich M, Rozman M, Hernandez JM, Puente DA, Freije JM, Velasco G, Gutierrez-Fernandez A, Costa D, Carrio A, Guijarro S, Enjuanes A, Hernandez L, Yague J, Nicolas P, Romeo-Casabona CM, Himmelbauer H, Castillo E, Dohm JC, de Sanjose S, Piris MA, de Alava E, San Miguel J, Royo R, Gelpi JL, Torrents D, Orozco M, Pisano DG, Valencia A, Guigo R, Bayes M, Heath S, Gut M, Klatt P, Marshall J, Raine K, Stebbings LA, Futreal PA, Stratton MR, Campbell PJ, Gut I, Lopez-Guillermo A, Estivill X, Montserrat E, Lopez-Otin C, Campo E (2011) Whole-genome sequencing identifies recurrent mutations in chronic lymphocytic leukaemia. Nature 475:101–105
107. Gaidano G, Ballerini P, Gong JZ, Inghirami G, Neri A, Newcomb EW, Magrath IT, Knowles DM, Dalla-Favera R (1991) p53 mutations in human lymphoid malignancies: association with Burkitt lymphoma and chronic lymphocytic leukemia. Proc Natl Acad Sci U S A 88:5413–5417
108. Quesada V, Conde L, Villamor N, Ordonez GR, Jares P, Bassaganyas L, Ramsay AJ, Bea S, Pinyol M, Martinez-Trillos A, Lopez-Guerra M, Colomer D, Navarro A, Baumann T, Aymerich M, Rozman M, Delgado J, Gine E, Hernandez JM, Gonzalez-Diaz M, Puente DA, Velasco G, Freije JM, Tubio JM, Royo R, Gelpi JL, Orozco M, Pisano DG, Zamora J, Vazquez M, Valencia A, Himmelbauer H, Bayes M, Heath S, Gut M, Gut I, Estivill X, Lopez-Guillermo A, Puente XS, Campo E, Lopez-Otin C (2011) Exome sequencing identifies recurrent mutations of the splicing factor SF3B1 gene in chronic lymphocytic leukemia. Nat Genet 44:47–52
109. Rossi D, Bruscaggin A, Spina V, Rasi S, Khiabanian H, Messina M, Fangazio M, Vaisitti T, Monti S, Chiaretti S, Guarini A, Del Giudice I, Cerri M, Cresta S, Deambrogi C, Gargiulo E, Gattei V, Forconi F, Bertoni F, Deaglio S, Rabadan R, Pasqualucci L, Foa R, Dalla-Favera R, Gaidano G (2011) Mutations of the SF3B1 splicing factor in chronic lymphocytic leukemia: association with progression and fludarabine-refractoriness. Blood 118:6904–6908
110. Gronbaek K, Straten PT, Ralfkiaer E, Ahrenkiel V, Andersen MK, Hansen NE, Zeuthen J, Hou-Jensen K, Guldberg P (1998) Somatic Fas mutations in non-Hodgkin's lymphoma: association with extranodal disease and autoimmunity. Blood 92:3018–3024
111. Weiss LM, Warnke RA, Sklar J, Cleary ML (1987) Molecular analysis of the t(14;18) chromosomal translocation in malignant lymphomas. N Engl J Med 317:1185–1189
112. Gronbaek K, Worm J, Ralfkiaer E, Ahrenkiel V, Hokland P, Guldberg P (2002) ATM mutations are associated with inactivation of the ARF-TP53 tumor suppressor pathway in diffuse large B-cell lymphoma. Blood 100:1430–1437

113. Ladanyi M, Offit K, Jhanwar SC, Filippa DA, Chaganti RS (1991) MYC rearrangement and translocations involving band 8q24 in diffuse large cell lymphomas. Blood 77:1057–1063
114. Koduru PR, Raju K, Vadmal V, Menezes G, Shah S, Susin M, Kolitz J, Broome JD (1997) Correlation between mutation in P53, p53 expression, cytogenetics, histologic type, and survival in patients with B-cell non-Hodgkin's lymphoma. Blood 90:4078–4091
115. Moller MB, Ino Y, Gerdes AM, Skjodt K, Louis DN, Pedersen NT (1999) Aberrations of the p53 pathway components p53, MDM2 and CDKN2A appear independent in diffuse large B cell lymphoma. Leukemia 13:453–459
116. Pasqualucci L, Compagno M, Houldsworth J, Monti S, Grunn A, Nandula SV, Aster JC, Murty VV, Shipp MA, Dalla-Favera R (2006) Inactivation of the PRDM1/BLIMP1 gene in diffuse large B cell lymphoma. J Exp Med 203:311–317
117. Jardin F, Jais JP, Molina TJ, Parmentier F, Picquenot JM, Ruminy P, Tilly H, Bastard C, Salles GA, Feugier P, Thieblemont C, Gisselbrecht C, de Reynies A, Coiffier B, Haioun C, Leroy K (2010) Diffuse large B-cell lymphomas with CDKN2A deletion have a distinct gene expression signature and a poor prognosis under R-CHOP treatment: a GELA study. Blood 116:1092–1104
118. Honma K, Tsuzuki S, Nakagawa M, Tagawa H, Nakamura S, Morishima Y, Seto M (2009) TNFAIP3/A20 functions as a novel tumor suppressor gene in several subtypes of non-Hodgkin lymphomas. Blood 114:2467–2475
119. Melzner I, Bucur AJ, Bruderlein S, Dorsch K, Hasel C, Barth TF, Leithäuser F, Möller P (2005) Biallelic mutation of SOCS-1 impairs JAK2 degradation and sustains phospho-JAK2 action in the MedB-1 mediastinal lymphoma line. Blood 105:2535–2542
120. Rossi D, Cerri M, Capello D, Deambrogi C, Berra E, Franceschetti S, Alabiso O, Gloghini A, Paulli M, Carbone A, Pileri SA, Pasqualucci L, Gaidano G (2005) Aberrant somatic hypermutation in primary mediastinal large B-cell lymphoma. Leukemia 19:2363–2366
121. Ritz O, Guiter C, Castellano F, Dorsch K, Melzner J, Jais JP, Dubois G, Gaulard P, Möller P, Leroy K (2009) Recurrent mutations of the STAT6 DNA binding domain in primary mediastinal B-cell lymphoma. Blood 114:1236–1242
122. Schmitz R, Hansmann ML, Bohle V, Martin-Subero JI, Hartmann S, Mechtersheimer G, Klapper W, Vater I, Giefing M, Gesk S, Stanelle J, Siebert R, Küppers R (2009) TNFAIP3 (A20) is a tumor suppressor gene in Hodgkin lymphoma and primary mediastinal B cell lymphoma. J Exp Med 206:981–989
123. Cinti C, Leoncini L, Nyongo A, Ferrari F, Lazzi S, Bellan C, Vatti R, Zamparelli A, Cevenini G, Tosi GM, Claudio PP, Maraldi NM, Tosi P, Giordano A (2000) Genetic alterations of the retinoblastoma-related gene RB2/p130 identify different pathogenetic mechanisms in and among Burkitt's lymphoma subtypes. Am J Pathol 156:751–760
124. Cabannes E, Khan G, Aillet F, Jarrett RF, Hay RT (1999) Mutations in the IkBa gene in Hodgkin's disease suggest a tumour suppressor role for IkappaBalpha. Oncogene 18:3063–3070
125. Krappmann D, Emmerich F, Kordes U, Scharschmidt E, Dörken B, Scheidereit C (1999) Molecular mechanisms of constitutive NF-kappaB/Rel activation in Hodgkin/Reed-Sternberg cells. Oncogene 18:943–953
126. Jungnickel B, Staratschek-Jox A, Bräuninger A, Spieker T, Wolf J, Diehl V, Hansmann ML, Rajewsky K, Küppers R (2000) Clonal deleterious mutations in the IkappaBalpha gene in the malignant cells in Hodgkin's lymphoma. J Exp Med 191:395–402
127. Martin-Subero JI, Gesk S, Harder L, Sonoki T, Tucker PW, Schlegelberger B, Grote W, Novo FJ, Calasanz MJ, Hansmann ML, Dyer MJ, Siebert R (2002) Recurrent involvement of the REL and BCL11A loci in classical Hodgkin lymphoma. Blood 99:1474–1477
128. Emmerich F, Theurich S, Hummel M, Haeffker A, Vry MS, Dohner K, Bommert K, Stein H, Dörken B (2003) Inactivating I kappa B epsilon mutations in Hodgkin/Reed-Sternberg cells. J Pathol 201:413–420
129. Schmitz R, Stanelle J, Hansmann ML, Küppers R (2009) Pathogenesis of classical and lymphocyte-predominant Hodgkin lymphoma. Annu Rev Pathol 4:151–174
130. Müschen M, Re D, Brauninger A, Wolf J, Hansmann ML, Diehl V, Küppers R, Rajewsky K (2000) Somatic mutations of the CD95 gene in Hodgkin and Reed-Sternberg cells. Cancer Res 60:5640–5643
131. Otto C, Giefing M, Massow A, Vater I, Gesk S, Schlesner M, Richter J, Klapper W, Hansmann M-L, Siebert R, Küppers R (2012) Genetic lesions of the TRAF3 and MAP3K14 genes in classical Hodgkin lymphoma. Br J Haematol 157:702–708
132. Steidl C, Telenius A, Shah SP, Farinha P, Barclay L, Boyle M, Connors JM, Horsman DE, Gascoyne RD (2010) Genome-wide copy number analysis of Hodgkin Reed-Sternberg

cells identifies recurrent imbalances with correlations to treatment outcome. Blood 116:418–427

133. Weniger MA, Melzner I, Menz CK, Wegener S, Bucur AJ, Dorsch K, Mattfeldt T, Barth TF, Möller P (2006) Mutations of the tumor suppressor gene SOCS-1 in classical Hodgkin lymphoma are frequent and associated with nuclear phospho-STAT5 accumulation. Oncogene 25:2679–2684
134. Mateo M, Mollejo M, Villuendas R, Algara P, Sanchez-Beato M, Martinez P, Piris MA (1999) 7q31-32 allelic loss is a frequent finding in splenic marginal zone lymphoma. Am J Pathol 154:1583–1589
135. Willis TG, Jadayel DM, Du MQ, Peng H, Perry AR, Abdul-Rauf M, Price H, Karran L, Majekodunmi O, Wlodarska I, Pan L, Crook T, Hamoudi R, Isaacson PG, Dyer MJ (1999) Bcl10 is involved in t(1;14)(p22;q32) of MALT B cell lymphoma and mutated in multiple tumor types. Cell 96:35–45
136. Zhang Q, Siebert R, Yan M, Hinzmann B, Cui X, Xue L, Rakestraw KM, Naeve CW, Beckmann G, Weisenburger DD, Sanger WG, Nowotny H, Vesely M, Callet-Bauchu E, Salles G, Dixit VM, Rosenthal A, Schlegelberger B, Morris SW (1999) Inactivating mutations and overexpression of BCL10, a caspase recruitment domain-containing gene, in MALT lymphoma with t(1;14)(p22;q32). Nat Genet 22:63–68
137. Takino H, Okabe M, Li C, Ohshima K, Yoshino T, Nakamura S, Ueda R, Eimoto T, Inagaki H (2005) p16/INK4a gene methylation is a frequent finding in pulmonary MALT lymphomas at diagnosis. Mod Pathol 18:1187–1192
138. Streubel B, Lamprecht A, Dierlamm J, Cerroni L, Stolte M, Ott G, Raderer M, Chott A (2003) T(14;18)(q32;q21) involving IGH and MALT1 is a frequent chromosomal aberration in MALT lymphoma. Blood 101:2335–2339
139. Streubel B, Vinatzer U, Lamprecht A, Raderer M, Chott A (2005) T(3;14)(p14.1;q32) involving IGH and FOXP1 is a novel recurrent chromosomal aberration in MALT lymphoma. Leukemia 19:652–658
140. Iida S, Rao PH, Nallasivam P, Hibshoosh H, Butler M, Louie DC, Dyomin V, Ohno H, Chaganti RS, Dalla-Favera R (1996) The t(9;14)(p13;q32) chromosomal translocation associated with lymphoplasmacytoid lymphoma involves the PAX-5 gene. Blood 88:4110–4117
141. Avet-Loiseau H, Li JY, Facon T, Brigaudeau C, Morineau N, Maloisel F, Rapp MJ, Talmant P, Trimoreau F, Jaccard A, Harousseau JL, Bataille R (1998) High incidence of translocations t(11;14)(q13;q32) and t(4;14)(p16;q32) in patients with plasma cell malignancies. Cancer Res 58:5640–5645
142. Landowski TH, Qu N, Buyuksal I, Painter JS, Dalton WS (1997) Mutations in the Fas antigen in patients with multiple myeloma. Blood 90:4266–4270
143. Shou Y, Martelli ML, Gabrea A, Qi Y, Brents LA, Roschke A, Dewald G, Kirsch IR, Bergsagel PL, Kuehl WM (2000) Diverse karyotypic abnormalities of the c-myc locus associated with c-myc dysregulation and tumor progression in multiple myeloma. Proc Natl Acad Sci U S A 97:228–233
144. Chesi M, Nardini E, Brents LA, Schrock E, Ried T, Kuehl WM, Bergsagel PL (1997) Frequent translocation t(4;14)(p16.3;q32.3) in multiple myeloma is associated with increased expression and activating mutations of fibroblast growth factor receptor 3. Nat Genet 16:260–264
145. Liu P, Leong T, Quam L, Billadeau D, Kay NE, Greipp P, Kyle RA, Oken MM, Van Ness B (1996) Activating mutations of N- and K-ras in multiple myeloma show different clinical associations: analysis of the Eastern Cooperative Oncology Group Phase III Trial. Blood 88:2699–2706
146. Chesi M, Bergsagel PL, Shonukan OO, Martelli ML, Brents LA, Chen T, Schrock E, Ried T, Kuehl WM (1998) Frequent dysregulation of the c-maf proto-oncogene at 16q23 by translocation to an Ig locus in multiple myeloma. Blood 91:4457–4463
147. Munshi NC, Avet-Loiseau H (2011) Genomics in multiple myeloma. Clin Cancer Res 17:1234–1242
148. Dunn-Walters DK, Isaacson PG, Spencer J (1995) Analysis of mutations in immunoglobulin heavy chain variable region genes of microdissected marginal zone (MGZ) B cells suggests that the MGZ of human spleen is a reservoir of memory B cells. J Exp Med 182:559–566

Chapter 2

Flow Cytometry for Non-Hodgkin and Classical Hodgkin Lymphoma

David Wu, Brent L. Wood, and Jonathan R. Fromm

Abstract

Multiparametric flow cytometry is a powerful diagnostic tool that permits rapid assessment of cellular antigen expression to quickly provide immunophenotypic information suitable for disease classification. This chapter describes a general approach for the identification of abnormal lymphoid populations by flow cytometry, including B, T, and Hodgkin lymphoma cells suitable for the clinical and research environment. Knowledge of the common patterns of antigen expression of normal lymphoid cells is critical to permit identification of abnormal populations at disease presentation and for minimal residual disease assessment. We highlight an overview of procedures for processing and immunophenotyping non-Hodgkin B- and T-cell lymphomas and also describe our strategy for the sensitive and specific diagnosis of classical Hodgkin lymphoma.

Key words: B cells, B-cell lymphoma, Clonality, Flow cytometry, Hodgkin lymphoma, Light chain restriction, T cells, T-cell lymphoma, T-cell receptor V-beta repertoire analysis

1. Introduction

Immunophenotypic analysis of both non-Hodgkin lymphomas (NHL) and Hodgkin lymphomas is commonly required for proper diagnostic classification and is currently accomplished by immunohistochemistry and flow cytometry (1). Multiparameter flow cytometry is an advanced diagnostic technology that is widely used for the clinical evaluation of complex cellular mixtures such as peripheral blood, bone marrow aspirates, body fluids, lymph node, and other tissue specimens (2). The ability of the technique to perform rapid characterization of numerous cellular populations for the simultaneous expression of multiple cell surface or cytoplasmic antigens on individual cells over several orders of magnitude of

Ralf Küppers (ed.), *Lymphoma: Methods and Protocols*, Methods in Molecular Biology, vol. 971,
DOI 10.1007/978-1-62703-269-8_2, © Springer Science+Business Media, LLC 2013

expression has led flow cytometry to be regarded as a "standard of care" for the evaluation of hematopoietic processes in the clinical laboratory.

In flow cytometry, a suspension of cells is injected into a fluid stream under laminar flow conditions (3) resulting in individual cells being directed through a quartz capillary tube (flow cell). Following illumination by one or more light sources, typically lasers, multiple cellular properties may be simultaneously assessed, including light scatter and the expression of surface and/or cytoplasmic antigens (3, 4), when cells are labeled with fluorochrome-conjugated antibodies directed against specific antigens (5). Both liquid (peripheral blood, bone marrow, cerebrospinal fluid, etc.) and tissue specimens are routinely analyzed to provide unique diagnostic information about hematopoietic populations (2). In general, flow cytometry permits the evaluation of multiple antigens on cells in a given experiment, typically 4–6 and increasingly in up to 10 or more antigens (4). Here, we discuss practical aspects of lymphoma diagnosis and classification by flow cytometry. As part of our approach, we assume that an abnormal cell population will have an aberrant immunophenotype as compared to background normal or reactive cells. As such, emphasis is placed on identifying aberrant immunophenotypes in a multiparametric analysis. As there have been numerous articles describing various strategies for flow cytometry, the reader is advised to access other articles in the field to fill deficiencies or intentional omissions herein.

2. Materials

Reagents are generally used as provided by the manufacturer with in-house validation prior to use. However, it is important to titer antibodies for optimal signal-to-noise response under the conditions to be used, and this may result in the use of antibodies at concentrations below that recommended by the manufacturer. The antigens that we target for routine B- and T-cell NHL and classical Hodgkin lymphoma (CHL) analysis are included in Table 1, with specific antibodies and fluorochromes employed using a modified LSRII flow cytometer (Becton-Dickinson).

2.1. Buffers and Cell Staining Reagents

1. PBS–BSA buffer: Dulbecco's Phosphate-Buffered Saline (GIBCO®) with 3% bovine serum albumin added. PBS contains 2.67 mM KCl, 1.47 mM KH_2PO_4, 137.9 mM NaCl, and 8.1 mM Na_2HPO_4.
2. RPMI 1640.

Table 1
Fluorochrome combinations used for lymphoma immunophenotyping at the University of Washington

Antigen	B cells	B-cell add-on for CLL/mantle cell	Add on tube for CLL prognosis	B-cell add-on for hairy cell leukemia	B-cell add-on for marginal zone	T cell	TCR V-beta	cHL (9-color)	cHL (6-color)
CD2						FITC			
CD3						PE-Cy7	PE-Cy7		APC-Cy7
CD4						A594	A594		
CD5	PE-Cy5.5	PE-Cy5.5	PE-Cy5			PE	ECD	APC-Cy7 (or ECD)	
CD7						APC	APC		
CD8						V450	APC-Cy7		
CD10	APC								
CD11c				APC	APC				
CD15								APC	
CD19	PE-Cy7	ECD	ECD	PE-Cy7	PE-Cy7				
CD20	V450							PE-Cy7	PE-Cy7
CD23		PE							
CD25				PE					
CD30						APC-A700		PE	PE
CD34						ECD			
CD38	A594		PE						
CD40								PE-Cy5.5	PE-Cy5.5

(continued)

Table 1 (continued)

Antigen	B cells	B-cell add-on for CLL/mantle cell	Add on tube for CLL prognosis	B-cell add-on for hairy cell leukemia	B-cell add-on for marginal zone	T cell	TCR V-beta	cHL (9-color)	cHL (6-color)
CD45	APC-H7			V450	APC-H7	APC-H7		ECD (or APC-H7)	
CD56						PE-Cy5	PE-Cy5		
CD64								FITC	FITC
CD71								APC-A700	
CD95								PB	APC
CD103				FITC					
HLA-DR							V450		
FMC-7		FITC							
ZAP-70			FITC						
Kappa	FITC				FITC				
Lambda	PE				PE				
V-beta isoform reagents							FITC, PE, and FITC and PE		

PB pacific blue, *FITC* fluorescein isothiocyanate, *PE* phycoerythrin, *ECD/PE-TR* PE-Texas Red, *PECy5* PE-Cyanine-5, *PE-Cy5.5* PE-Cyanine-5.5, *PE-Cy7* PE-Cyanine-7, *A594* AlexaFluor 594, *APC* allophycocyanin, *APC-A700* APC-AlexaFluor 700

3. Lysing/fixation solution: 0.15 mol/L NH_4Cl, pH 7.2 containing 0.25% ultrapure formaldehyde (Polysciences). The solution fixes cells and lyses red blood cells.
4. Medium A and B for cytoplasmic antibody staining are from Invitrogen (Fix & Perm®).

3. Methods

3.1. Sample Preparation

Tissue specimens are first disaggregated to create a single-cell suspension:

3.1.1. Disaggregation of Tissue Specimens

1. Mince the tissue with a scalpel in approximately 5 mL of RPMI 1640.
2. Filter the disaggregated cell suspension through a 40 μm filter.
3. Pellet the cells by centrifugation (550 × *g* for 5 min) and decant the remaining RPMI 1640.
4. Resuspend the cells in PBS–BSA or RPMI 1640, pellet the cells again by centrifugation (550 × *g* for 5 min), and resuspend in RPMI 1640 to a cell count of 10,000 cells/μL or less.
5. Add sufficient cell suspension to deliver up to one million cells in a volume of <200 μL.

3.1.2. Cell Surface Labeling of Cell Suspensions

1. Add appropriate fluorescently labeled, titered antibodies to cell suspension (from disaggregated tissue, bone marrow, blood, etc.), typically 5–20 μL of each antibody, and mix gently. The antibodies may be cocktailed prior to use for efficient delivery.
2. Incubate the labeled cells for 15 min at room temperature (RT) in the dark.
3. Add 1.5 mL of lysing/fixation solution.
4. Incubate for 15 min at RT in the dark.
5. Centrifuge the cells (550 × *g* for 5 min) and decant the supernatant.
6. Add 3 mL of PBS–BSA, centrifuge (550 × *g* for 5 min), and decant the supernatant.
7. Resuspend the cells in 100 μL of PBS–BSA.
8. Collect 150,000 events (if possible).

3.1.3. Cytoplasmic Labeling of Cell Suspensions (for ZAP-70 and Bcl-2)

1. Add appropriate fluorescently labeled, titered cell surface antibodies to cell suspension and mix gently.
2. Incubate the labeled cells for 15 min at room temperature in the dark.
3. After washing the cells twice with PBS–BSA, add 100 μL of Medium A and mix well.

4. Incubate for 15 min at room temperature in the dark.
5. Wash the cells twice with PBS–BSA, add 100 μL of Medium B, and mix well.
6. Add appropriate amount of cytoplasmic antibody, mix, and incubate the labeled cells for 30 min in the dark at room temperature.
7. Wash the cells twice with PBS–BSA and resuspend the cells in 100 μL of PBS–BSA.

3.1.4. Sample Preparation for Hodgkin Lymphoma

The sample processing and immunostaining protocols are the same as those used for NHL with the exception that more events should be collected, preferably ~500,000. This CHL assay has only been validated on lymph nodes and thus is only recommended for tissue specimens.

3.2. Gating Strategies, Data Analysis, and Interpretation

3.2.1. B-Cell Analysis

Compensated data files for B-cell NHL are first gated to exclude the so-called doublet events using a plot of forward scatter area versus forward scatter height. The doublet events represent coincident cells in the flow cell and need to be excluded as the antigenic profile derived from these events may result in an apparent composite immunophenotype due to the combined antigenic expression of two or more cells. Subsequently, nonviable events are excluded using forward and side light scatter gating. As cells degenerate, initially forward scatter decreases while side scatter increases and later both decrease in intensity. These nonviable events can be readily excluded by selective gating considering these findings.

The evaluation for B-cell NHL next involves identification of B cells with subsequent evaluation for the presence of surface immunoglobulin light chain-restricted populations with aberrant immunophenotypes (see Note 1). We currently gate lymphocytes with forward and side light scatter (see Note 2), computationally subtract the CD5-positive, CD19-negative T cells, and then isolate the B cells on a plot of CD19 versus side scatter. Data is subsequently evaluated by examining various antigens (Table 1) plotted against each other for both the B cells and lymphocytes. Evaluation of lymphocytes based on forward versus side scatter (in addition to CD19-positive B cells) prevents inadvertent exclusion of B-cell populations that have aberrant loss or decreased expression of CD19 (see Note 3). In general, the normal kappa-to-lambda ratio of the B cells is approximately 1.4; however, sole emphasis on a skewed light chain ratio for identifying an abnormal B-cell population is discouraged due to its poor sensitivity and specificity. Rather, an emphasis is placed on identifying abnormalities of other antigens assessed in these studies (CD5, CD10, CD19, CD20, CD38, and CD45) that are often under- or over-expressed relative to any normal population, helping to separate the abnormal and normal populations. Typical examples of CLL/SLL (Fig. 1), marginal

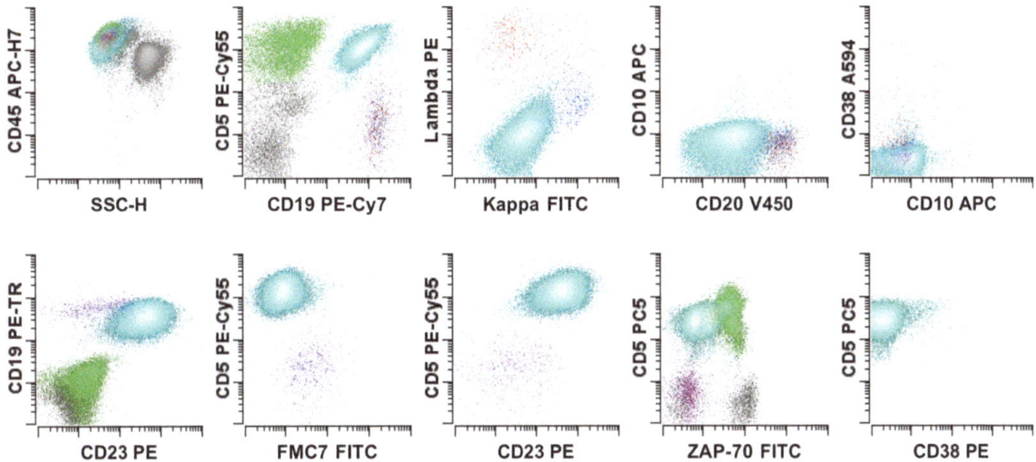

Fig. 1. Flow cytometric characterization of chronic lymphocytic leukemia/small lymphocytic lymphoma (CLL/SLL). The neoplastic population (*cyan*) demonstrates expression of CD45, CD19, CD5, low-level kappa surface light chains, decreased CD20, CD23, and no expression of FMC7 or CD10. The neoplastic population does not express ZAP-70 (positive cells determined by a discriminator which is negative on the normal B cells and positive on the normal T cells) or CD38, features associated with a more favorable prognosis in CLL/SLL (52, 53). The first dot plot of the *top panel* shows all cells, the second shows lymphocytes, and the last three shows only B cells. In the *bottom panel* of dot plots, all lymphocytes are shown in the first and fourth plots, while the second, third, and fifth plots show only B cells. A small population of reactive B cells identified by *blue* (kappa expressing) and *red* (lambda expressing) events. Reactive T cells are colored in *green*.

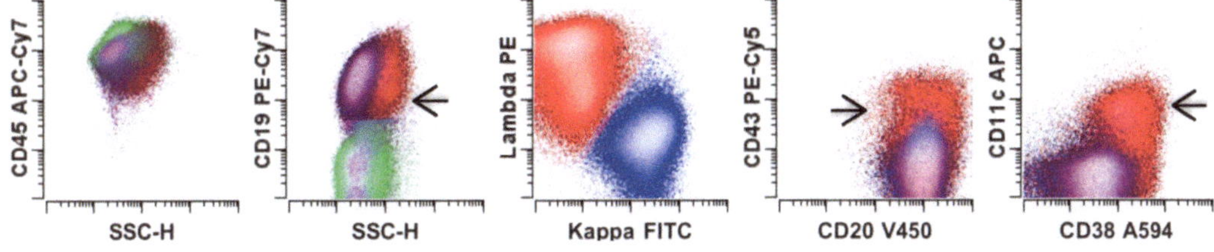

Fig. 2. Characterization of an abnormal B-cell population using CD11c and CD43. The right parotid gland biopsy demonstrated a decreased kappa-to-lambda ratio (0.83) but no other antigenic abnormalities. Characterization of the B cells using CD11c and CD43 demonstrates a lambda-restricted population with expression of these antigens (*arrows*), consistent with involvement of a B-cell lymphoproliferative disorder (marginal zone lymphoma by morphology). T cells are *green* and kappa- and lambda-expressing B cells are in *blue* and *red*, respectively.

zone lymphoma (Fig. 2), hairy cell leukemia (Fig. 3), Burkitt lymphoma (minimal residual disease (MRD)) (Fig. 4), and follicular lymphoma (Fig. 5) are provided. While individual cases vary, a summary of common immunophenotypes of typical B-cell lymphomas is provided (Table 2). A more comprehensive discussion of immunophenotypes in B-cell NHL can be found in the review by Craig and Foon (2).

The approach for evaluating specimens for B-cell NHL MRD is identical to that used for evaluating samples at disease presentation, with the obvious difference being that the abnormal population post therapy can be very small, sometimes <0.1% of the viable cells. In practice, a series of sequential gates to selectively include

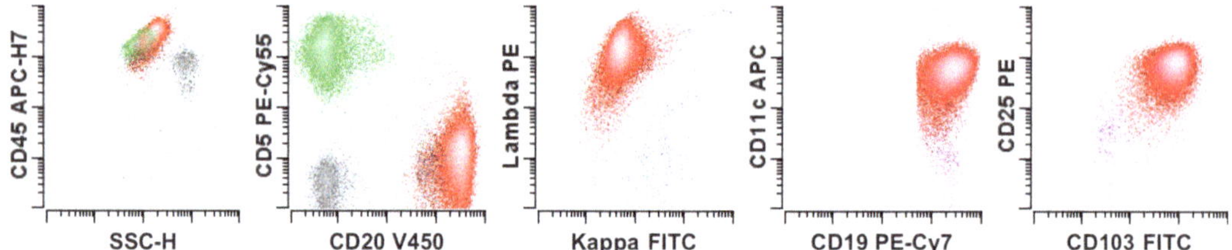

Fig. 3. Immunophenotypic characterization of a case of hairy cell leukemia in the bone marrow. The *first panel* shows all leukocytes, the second all lymphocytes, and the remaining three panels show only B cells. T cells are colored in *green*, kappa-restricted B cells are colored in *blue*, and the lambda-restricted B cells are colored in *red*. The neoplastic hairy cell leukemia population demonstrates increased side light scatter and CD45 expression, bright expression of CD20, lambda light chain expression, and expression of CD11c, CD19, CD25, CD38 (data not shown), and CD103, without CD5 or CD10 (data not shown).

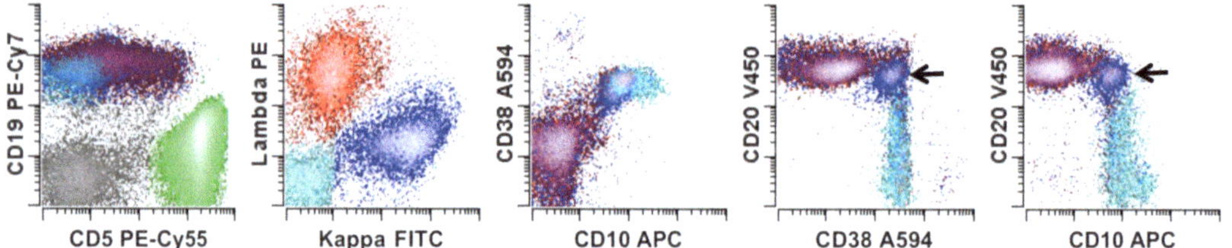

Fig. 4. Minimal residual disease detection of Burkitt lymphoma in the bone marrow. Overall, this bone marrow specimen demonstrates a normal kappa-to-lambda ratio of 1.5. In addition, multiple dot plots do not demonstrate an abnormal B-cell population. However, a small (1% of leukocytes) kappa-restricted B-cell population is noted (*arrows*) with relatively bright expression of CD38 and CD10. This immunophenotype is identical to that identified in the prior presenting mesenteric mass. The first dot plot shows all lymphocytes, while the remaining four panels show B cells. T cells are colored *green*; kappa- and lambda-restricted mature B cells are colored *blue* and *red*, respectively; immature, normal B lymphoblasts are colored *cyan*.

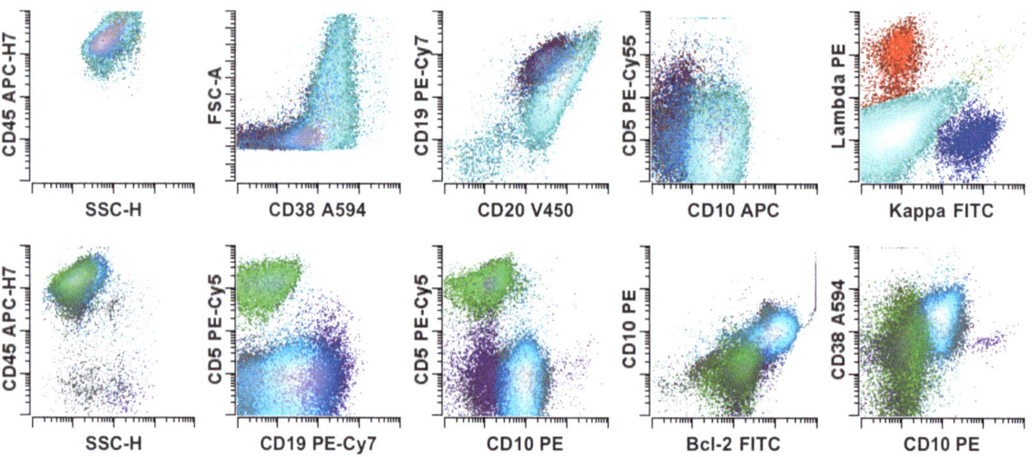

Fig. 5. Immunophenotypic characterization of a case of follicular lymphoma (grade 3). The first dot-plots of both the *top* and *bottom panels* show all leukocytes, the last four in the *top panel* show B cells, and the remaining four plots in the *bottom panel* show lymphocytes. T cells are colored in *green*, kappa-restricted B cells are colored in *blue*, the lambda-restricted B cells are colored in *red*, and neoplastic B cells are colored in *cyan*. The neoplastic cells show expression of CD45, increased forward light scatter, decreased expression of CD19 relatively to the reactive B cells, absent surface light chain restriction, and expression of CD10, CD20, bcl-2 (relative to the reactive T cells), and CD38 (normal to slight decreased relative to the expression of a normal germinal center population (29)) and no expression of CD5.

Table 2
Immunophenotypes of common B-cell NHL

B-cell lymphoma	Immunophenotype
Diffuse large B-cell lymphoma	Increased forward and side light scatter with variable expression of CD19, CD20, and monotypic surface light chain. CD10 may or may not be expressed
Follicular lymphoma	CD45, CD19 (decreased), CD20 (decreased), CD10 (increased), CD38 (decreased), monotypic surface light chain, increased BCL-2 as compared to background T cells
Chronic lymphocytic leukemia/small lymphocytic lymphoma	CD45, CD5, CD20 (dim), mono- or bi-typic dim to absent surface light chain expression, CD23 positive, FMC-7 negative, with variable CD38 and ZAP-70 expression
Mantle zone lymphoma	CD45, CD5, CD20 (normal), monotypic surface light chain restriction, CD23 negative, FMC-7 positive
Burkitt lymphoma	CD45, CD20, CD10, CD38 (increased), monotypic surface light chain expression, no over-expression of BCL2
Hairy cell leukemia	CD20 (bright), CD19, monotypic surface light chain expression, aberrant co-expression of CD11c, CD25, CD103, without CD5 and CD10
Marginal zone lymphoma	CD20, CD19, negative for CD5 and CD10. Occasional cases have expression of CD43

and/or exclude various populations may be required to convincingly isolate the abnormal population and evaluate for definite light chain restriction (Fig. 4). Identifying these small populations is facilitated by knowledge of the original immunophenotype of the abnormal B-cell population at presentation (when the population is usually present at significantly increased proportion) preferably using the identical reagent combination used to immunophenotype the presentation specimen. Other caveats related to the analysis of B cells are described in Notes 4–7.

3.2.2. T-Cell Analysis

In general, the flow cytometric gating strategy for T cells is similar to that for B cells. Compensated data files are first gated to exclude doublet events and then nonviable events (see Note 8). T cells are then identified typically by two gating strategies: (1) forward versus side scatter to identify lymphocytes or (2) CD3 versus side scatter to identify T cells. The former approach helps to identify T cells that have aberrant decreased or absent expression of CD3, while the latter identifies T cells using CD3 to permit identification of potentially larger cells that may have increased side scatter (see also Note 9). While the analysis of the B cells rests on identifying an immunophenotypically abnormal B-cell population that is clonal, evaluation of the T cells involves the recognition of an

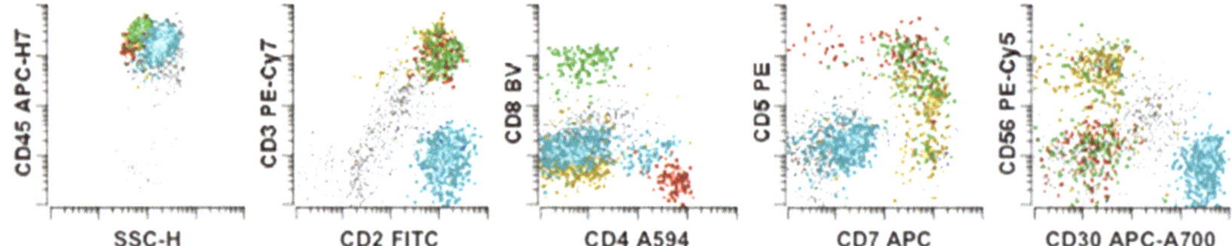

Fig. 6. Example of a CD30+ T-cell lymphoma. An abnormal CD2-positive T-cell population is identified with strong CD30+ expression, CD4 expression on a small subset, and aberrant loss of CD3, CD5, and CD7 without CD8 or CD56 (teal). Reactive CD4, CD8, and CD4/CD8-negative T cells are colored *red*, *green*, and *orange*, respectively. Of note, although the expression of CD2 and possible dim expression of CD4 are not considered lineage specific, subsequent molecular analysis of T-cell receptor gamma gene rearrangement confirms the population is clonal.

immunophenotypically abnormal T-cell population. At some institutions, evaluation of probable clonality of T-cell populations can also be performed using flow cytometry by assessing the T-cell receptor (TCR) V-beta repertoire.

As part of the immunophenotypic evaluation of T cells, it is critical to recognize that numerous reactive T-cell populations may be present in any given sample (see Note 10). These T cells may include memory T cells (decreased CD7 expression), gamma-delta T cells (increased CD3, absent CD4 and CD8 (partial)), and large granular lymphocytes (typically CD8+/CD5 dim/absent) and may be identified in any given patient sample, sometimes in increased proportion (2). Familiarity with this normal spectrum of antigenic variation of reactive T cells is critical to prevent misinterpretation of a reactive population as a malignant clone. On the other hand, it is also possible that a neoplastic, clonal population of T cells may show identical immunophenotypic change as any of these reactive populations. As such, careful clinical and pathologic correlation is required in every case.

With this spectrum of antigenic variation in mind, it is diagnostically fortuitous that the majority of T-cell lymphoproliferative disorders will demonstrate aberrant antigenic expression such that the level of expression of T-cell-associated antigens (CD2, CD3, CD4, CD5, CD7, CD8, and CD45) can be increased, decreased, or completely absent, as compared to the background, normal reactive T cells (6, 7). Multiparametric identification of these antigenic changes consequently permits one to reliably identify an abnormal T-cell population, which in conjunction with histologic findings or clinical data can permit subsequent definite classification. As an example, flow cytometry of a liver core needle sample from a 69-year-old man with multiple liver lesions (Fig. 6) identified an abnormal T-cell population comprising 25.5% of total white cells with increased forward and side scatter and expression of CD2 and CD30 (bright) and with aberrant loss of expression of CD3, CD5, and CD7 without CD8, CD34, or CD56. CD4 may be expressed dimly on a small subset. The immunophenotype of the abnormal

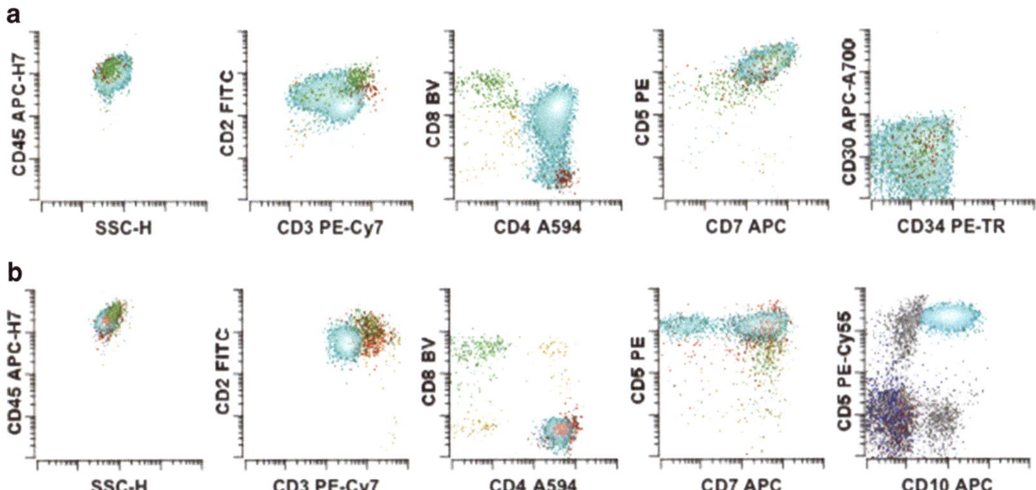

Fig. 7. Examples of other mature T-cell lymphomas. (**a**) T-cell prolymphocytic lymphoma, a population of CD4/CD8+double-positive T cells with decreased expression of CD2, CD3, and CD45, without evidence of CD34 or TdT (data not shown), is identified in a peripheral blood sample of a 73-year-old male with a white blood cell (WBC) count of 398,000 and prior cytogenetic studies showing involvement of a complex karyotype, including inv(14)(q11.2q32). Background reactive CD4+ (*red*) and CD8+ (*green*) T lymphocytes are also shown. (**b**) Angioimmunoblastic T-cell lymphoma, a population of CD4+positive T cells (teal) with aberrant expression of CD2 (decreased), CD3 (decreased), and CD10, with variable CD7 and normal expression of CD5 and CD45, is identified in a 91-year-old female peripheral blood sample, comprising 18.6% of total white cells. Background reactive CD4+ (*red*) and CD8+ (*green*) T lymphocytes are also shown.

CD45, CD30+ cell population was favored to represent a CD30+ peripheral T-cell lymphoma and a clonal TCR gamma gene rearrangement (by PCR) was identified, providing support for T-cell lineage. Other examples of mature T-cell lymphomas, including T-cell prolymphocytic leukemia and probable angioimmunoblastic T-cell lymphoma, are also shown (Fig. 7). While individual cases vary, a summary of common immunophenotypes of typical T-cell lymphomas is provided (Table 3). A more comprehensive discussion of immunophenotypes in T-cell NHL can be found elsewhere (2). See Note 11 for an additional caveat regarding the analysis of T cells.

At presentation, a common question clinically is whether a T-cell population of interest represents a clonal process or alternatively is a reactive oligoclonal or polyclonal expansion of an immunophenotypically distinct subset. Flow cytometric analysis of TCR V-beta repertoire is a methodology that uses fluorescently labeled anti-TCR V-beta antibodies to determine if the identified T-cell populations represent a diverse or a clonal process (8–11). As provided by the manufacturer, the IOTest® Beta Test Mark assay is a set of 24 antibodies that covers approximately 70% of the normal human TCR V-beta repertoire. These V-beta-specific antibodies are each conjugated to one of the three fluorochromes, FITC, PE, or PE, and FITC conjugate. According to the intended original

Table 3
Immunophenotypes of common T-cell NHL

T-cell lymphoma	Immunophenotype
T-PLL	CD4+/–, CD8–/+, variable CD3, CD2, CD5, CD7
AITL	CD4+, CD5+, CD10+/–, CD3–/+
Mycosis fungoides/Sezary syndrome	CD4+, variably decreased CD7, CD3, CD5, CD45
PTCL-NOS	CD4+/–, often with loss of CD5 and/or CD7
ATLL	CD4+, CD25+, CD3+, decreased CD7
ALCL	ALK+/–, CD30+, variable loss of T-cell antigens, TIA1/granzyme B+
T-LGL	CD3+, CD4–, CD8+, with decreased CD5 and or CD7

T-PLL T-cell prolymphocytic leukemia, *AITL* angioimmunoblastic T-cell lymphoma, *PTCL-NOS* peripheral T-cell lymphoma, not otherwise specified, *ATLL* adult T-cell leukemia/lymphoma, *ALCL* anaplastic large cell lymphoma, *T-LGL* T-cell large granular lymphocyte leukemia, *TIA1* T-cell intracellular antigen 1

protocol, the assay is run in eight separate tubes such that three V-beta family-specific antibodies (one labeled with FITC, one labeled with PE, and one labeled with FITC and PE) are present in each tube. Through the use of gating reagents (for example CD3, CD4, or CD8), the T-cell population of interest can be isolated to determine if there is evidence for overrepresentation of a particular TCR V-beta isoform relative to normal, a finding that is strongly suggestive of clonality. Numerous studies have demonstrated the utility of this approach for assessing putative T-cell clonality in both the clinical and research environment (8–11). The reader is advised to consult these publications for performing this approach for TCR V-beta analysis.

At our institution, a simple permutation of the original protocol for TCR V-beta repertoire analysis has been developed and used for several years (manuscript submitted). In this modification, instead of 8-tube analysis as intended by the manufacturer, all of the fluorescently labeled TCR V-beta antibodies are combined into a single tube for combined analysis. This modification permits rapid identification of a putative TCR V-beta-restricted T-cell population with emphasis on identifying aberrant antigenic expression through the use of an increased number of T-cell-specific gating reagents (Table 1). As compared to the standard method, this modified approach is fast, can be adopted in any laboratory currently performing routine TCR V-beta analysis, and minimizes the amount of reagent and sample used for analysis, thus permitting application to specimens that are typically hypocellular in nature, such as skin biopsies and cerebrospinal fluid samples. An example of this modified approach is shown in which an expanded large

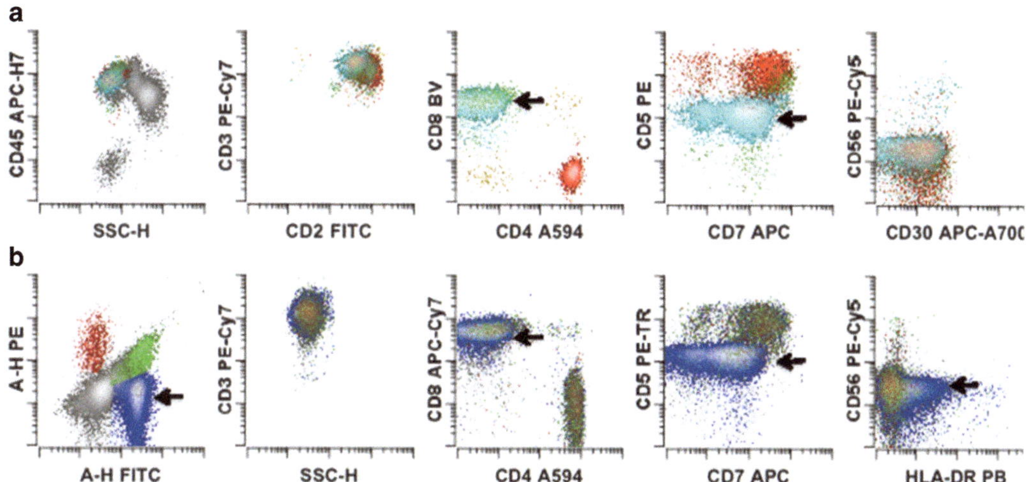

Fig. 8. Example of combined using modified TCR V-beta assay showing clonality of T-LGL population. (**a**) A CD8+ T-cell population (teal, *arrows*) is identified with decreased CD5 and CD7 expression as compared to background CD4+ (*red*) and CD8+ (*green*) T cells. (**b**) Subsequent analysis of TCR V-beta expression using the modified assay in which all TCR V-beta antibodies are combined together shows the T-LGL population (*blue*, *arrows*) with apparent TCR V-beta restriction limited to an FITC-labeled isoform. By comparison, background, reactive T cells show variable expression of PE, FITC, PE-FITC, and unlabeled TCR V-beta isoforms.

granular lymphocyte population is identified with expression of CD8 and decreased expression of CD5 and CD7 (Fig. 8). Subsequent assessment of TCR V-beta repertoire analysis using a modified approach in which all V-beta antibodies are combined together permits identification of probable FITC-fluorochrome-labeled TCR V-beta isoform. If clinically indicated, molecular analysis of TCR gene rearrangement could be subsequently performed to confirm the presumed clonal nature of this population. In addition, if knowledge of the specific TCR V-beta isoform is important, such as for future minimal residual disease monitoring, the standard TCR V-beta repertoire assay could be performed to permit identification of which V-beta isoform is identified by the PE-labeled anti-V beta antibody.

3.2.3. Hodgkin Cell Analysis

CHL is an unusual type of B-cell lymphoma (12–14) in which the neoplastic Hodgkin and Reed–Sternberg (HRS) cells are rare (<1% of the cells in lymph node) (13, 15); the bulk of the cells in an involved lymph node include reactive lymphocytes, eosinophils, plasma cells, and histiocytes (15); and the HRS bind to non-neoplastic T cells, resulting in HRS cell–T-cell rosettes (16–22). Traditionally, CHL has been diagnosed by morphology and immunohistochemistry (HRS cells demonstrate expression of CD15 and CD30 but lack expression of CD20, CD3, and CD45 (12, 15, 23, 24)). Our recent studies, however, have demonstrated that HRS cells can be directly identified by flow cytometry with high

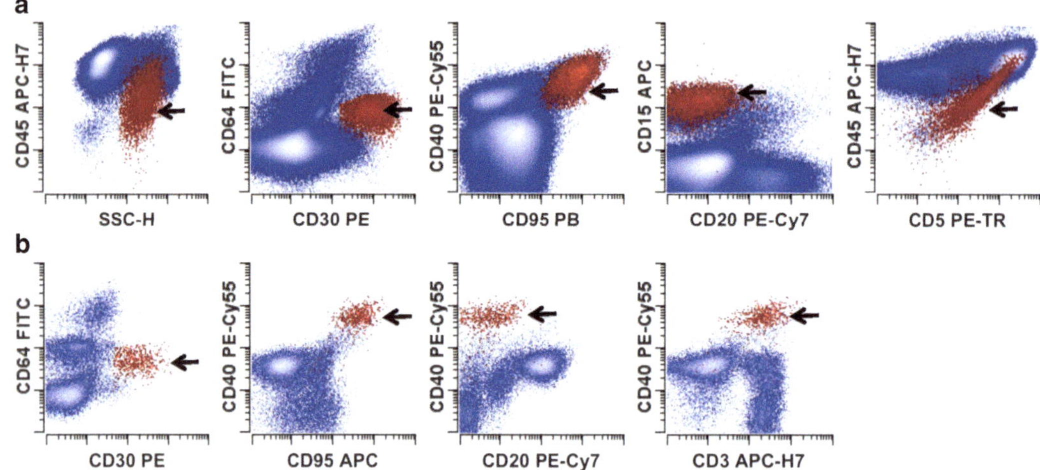

Fig. 9. Representative examples of 9- (**a**) and 6-color (**b**) flow cytometry studies of CHL cases. HRS cells (*arrows*; also shown in *red* and *emphasized*) are identified by their absence of expression of CD64 (position of negative determined by control experiments, not shown), expression of CD30, CD40, CD95, and increased side light scatter (SSC-H) compared to normal lymphocytes; all the remaining viable events are in *blue*. (**a**) The population of neoplastic HRS cells have expression of intermediate CD15, intermediate to bright CD30, intermediate to bright CD40, variable CD71 (data not shown), and intermediate to bright CD95, without expression of CD64 or CD20. The diagonal relationship between CD45 and CD5 is due to the presence of T cells bound to the HRS cells. (**b**) Neoplastic HRS cells have expression of intermediate CD30, bright CD40, and intermediate CD95, without expression of CD64 or CD20. CD3 is expressed suggesting some degree of HRS–T-cell rosetting.

clinical sensitivity and specificity (22), allowing this neoplasm to be diagnosed by this technique (22, 25). The interaction of T cells and HRS cells (rosetting) can be directly detected by flow cytometry, as demonstrated by the observation of a composite immunophenotype of the HRS cells and T cells, that is, cells with expression of both HRS-cell and T-cell antigens (22). T-cell–HRS cell rosetting can be disrupted by unlabeled "blocking" antibodies (antibodies that can compete for the binding of the adhesion molecule binding partner (20, 22)), a practice that is useful for purifying HRS cells (22) but is not necessary for diagnostic flow cytometry (25). Reagent combinations are proposed for either 9-color (25) or 6-color (26) flow cytometry platforms (Table 1). In addition, a reactive T-cell population (CD4+ T-cell population with CD45^{bright}, CD7^{bright}) has been identified in lymph nodes involved by CHL, a finding that can be used to suggest a diagnosis of CHL (see Note 12) (27).

The gating strategy to identify HRS cells is different from that for non-Hodgkin B- and T-cell lymphomas. The first difference is that because of the relatively increased cell size of HRS and rosetted cells, increased side scatter is used to identify these populations (Figs. 9 and 10); HRS cells are then identified by the requisite expression of CD30, CD40, and CD95 with the absence of

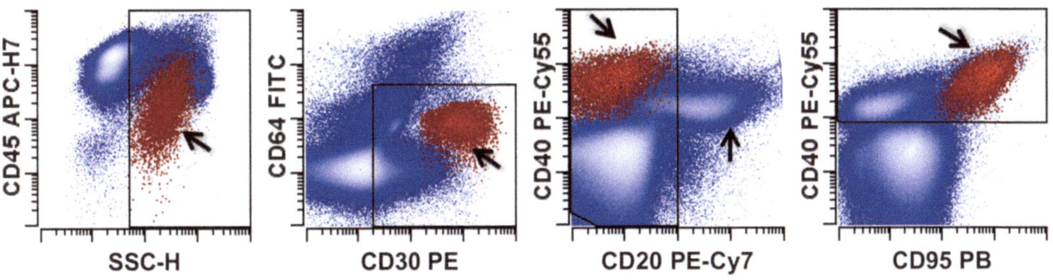

Fig. 10. Gating strategy to identify HRS cells in CHL. HRS cells are in *red* (also identified by *oblique arrows*) and are emphasized, while all other events are in *blue*. The B cells are identified by the *vertical arrow* (CD20, CD40+). Putative HRS cell events must fall in all four gates and form a distinct population. Specifically, the cells must have increased side scatter (1st plot), be CD30+ and have increased auto-fluorescence in the FITC channel compared to small lymphocytes (2nd plot), express no or low CD20 (3rd plot), and express CD40 at or greater intensity than a reactive B cell (4th plot).

Table 4
Criteria for identifying an HRS cell population

1	Express CD30, CD40, and CD95
2	Have increased light scatter properties (compared to background lymphocytes)
3	Lack moderate to bright expression of CD20
4	Lack expression of CD64
5	Represent a discrete cell population in multidimensional projections of the immunophenotypic data with increased autofluorescence as compared to CD30+ reactive immunoblasts

The putative HRS cell population must meet all of the above criteria

strong CD20 expression (25, 26). An HRS cell population must be identifiable in all four gates (Fig. 10) having increased side scatter as compared to background lymphocytes; expression of CD30 and increased autofluorescence, as assessed using the FITC channel; absence of or decreased expression of CD20; and lastly, strong expression of CD40. Of note, the increased autofluorescence seen in HRS cells is critical to distinguishing these from CD30+ reactive immunoblasts, and frequently, CD40 expression on these cells is increased as compared to background cells in the tissue preparation. If properly gated, HRS cells may appear as two subset populations, one with apparent T-cell antigen expression (such as CD3 or CD5) and the second with decreased (but not absent) CD45 but no expression of T-cell antigens due to the presence and absence, respectively, of T-cell rosetting (see Note 13).

A summary of the immunophenotypic criteria (see also Notes 14 and 15) for identifying a putative HRS cell population is provided in Table 4 (22, 25, 26). See Note 16 for a further caveat regarding the analysis of CHL.

4. Notes

1. Some B-cell populations may demonstrate aberrant loss of surface light chains. Evaluation of cytoplasmic light chains may show monotypic restriction in these cases. Aside from hematogones (normal immature B cells in the marrow and rarely in lymphoid tissues) and plasma cells, absence of surface light chain expression usually is considered aberrant antigen expression (28). Note, however, that some normal B-cell populations may exhibit decreased light chain expression (for example, germinal center B cells (1)), and consequently this finding should be interpreted cautiously.
2. Some large cell B-cell lymphomas may be missed if analyzed using the routine gating strategy due to increased light scatter. In every case, it is important to consider flow cytometry-detected events that fall outside the normal expected size range for typical small lymphoid cells based on light scatter properties. Evaluating events with increased scatter properties is helpful to ensure that complete sampling has been achieved.
3. Some B-cell lymphomas may show absent or decreased expression of CD19 (29–31) and/or CD20 (29) and may therefore be inadvertently excluded from analysis using standard gating strategies. To avoid this pitfall, evaluate more than one B-cell antigen and examine their expression on all lymphocytes gated by light scatter.
4. Not all clonal germinal center B-cell populations are neoplastic (32). This not uncommonly occurs in reactive states, particularly Hashimoto's thyroiditis (33). Correlation with other immunophenotypic abnormalities and histologic studies is required to make a determination of malignancy.
5. Not all neoplastic populations that express CD19 represent B-cell lymphomas. B lymphoblastic leukemia/lymphoma (B-ALL) and, rarely, myeloid neoplasms (namely, myeloid neoplasms with t(8;21) (34)) may show expression of CD19. Additional studies are required to completely evaluate these neoplasms (discussion beyond the scope of this paper).
6. Occasionally, dying cells and debris may appear to express B-cell antigens and appear clonal. While the differentiation of debris and authentic abnormal B-cell populations can be challenging, as a general rule, debris binds antibodies nonspecifically, often resulting in a correlated (diagonal) relationship between antigens in multiple 2-dimensional projections of the immunophenotypic data. The exclusion of low forward scatter events, e.g., a viability gate, may significantly reduce apparent nonspecific binding and simplify analysis. However, neoplasms with a high proliferative rate may show

preferential degeneration with loss of forward scatter, e.g., Burkitt lymphoma, so examination of events prior to exclusion by viability gating is important to detect this occurrence.

7. Small clonal B-cell populations that do not represent B-cell lymphoma are relatively common in normal individuals and have been described in the literature (35, 36). Ultimately, flow cytometric evaluation in the context of clinical findings, peripheral blood counts, and tissue morphology is required for the diagnosis of B-cell lymphoma.
8. As with B cells, debris can mimic an authentic abnormal T-cell population. See Note 6 above for B cells for evaluation of this phenomena.
9. Some large T-cell lymphomas may be missed if analyzed using the routine gating strategy due to increased light scatter. In every case, it is important to consider flow cytometry events that fall outside the normal expected size range for typical small lymphoid cells based on light scatter properties. Evaluating events with increased scatter properties is helpful to ensure that complete sampling has been achieved.
10. Not all clonal and immunophenotypically aberrant T-cell populations represent T-cell lymphomas (2). In some situations, clonal populations may arise as part of the normal process of the adaptive immune system, such as in response to cytomegalovirus (37).
11. Some T-cell lymphomas may show aberrant expression of CD19 or CD20 (38, 39). This should not be inadvertently interpreted as a B-cell neoplasm. Correlation with expression of other T-cell antigens (cytoplasmic CD3, CD7, CD4/CD8) and absence of expression of other B-cell antigens (cytoplasmic CD79a, CD19) is recommended.
12. In the absence of specific analysis for HRS cells, the presence of a characteristic, reactive T-cell population may suggest the diagnosis of CHL in the appropriate clinical and histologic context (40, 41). We have noted that increased expression of CD2, CD5, CD7, and CD45 on the CD4+ T cells, increased CD5 and CD45 on the CD8+ T cells, and decreased expression of CD3 on the CD4 and CD8+ T cells is suggestive of CHL (27). The increased expression of CD45 and CD7 on CD4+ T cells (Fig. 11) may be the most useful reactive T-cell finding to suggest involvement of CHL (27).
13. Putative HRS cell populations should be identifiable using the aforementioned gating strategy. To support this diagnosis, one should attempt to identify the presence of a diagonal relationship on a plot of CD45 versus CD3 or CD5. This diagonal relationship occurs due to the presence of varying numbers of T cells rosetting the HRS cells.

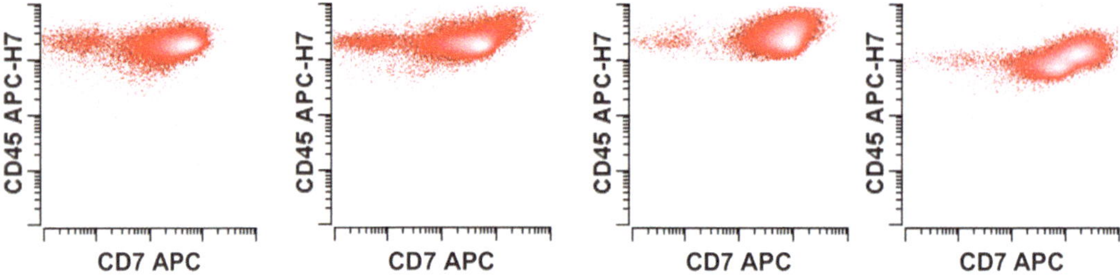

Fig. 11. Reactive T cells in classical Hodgkin lymphoma. CD4+ T cells from CHL demonstrate increased expression of CD45 and CD7 when compared to reactive lymph nodes. 1st dot plot, typical reactive lymph node; 2nd to 4th dot plots, three examples of CHL cases from three different patients.

14. There may be slight variance to the typical HRS cell immunophenotype. For instance, while most HRS cells do not express CD20, low-level expression may occasionally be observed. Further, some HRS cell populations may lack expression of CD15, a finding that correlates with reported immunohistochemical studies (~20% of CHL cases lack expression of CD15 (42–44)).
15. In clinical samples, there may be populations of cells with an apparent immunophenotype that has some resemblance to HRS cells. However, since the immunophenotype does not meet the four basic criteria outlined in Table 4, these events are almost always shown to be a different type of hematologic neoplasm or cellular debris. For example, anaplastic large T-cell lymphoma cells will commonly express CD30 with increased scatter properties. However, its expression of CD40 is usually at a lower level than reactive, background B cells, as also observed in immunohistochemical studies (45, 46). Diffuse large B-cell lymphoma (DLBCL) may occasionally show expression of CD30; however, the level of expression of CD30 in DLBCL (if present) is relatively lower than that seen for a typical HRS cell population. Further, DLBCL cells will have intermediate to bright expression of CD20 (not dim as compared to HRS cells) with a slight increase in side scatter properties (but not as increased as in CHL). Lastly, nodular lymphocyte predominant Hodgkin lymphoma is a type of Hodgkin lymphoma in which the neoplastic cells, referred to now as lymphocyte predominant (LP) cells, express CD20, CD40, and CD45 without CD30 or CD15 (47). Although these cells cannot yet be reliably identified by flow cytometry, the expression of CD20 without CD30 would argue against this population from being a putative HRS cell population.
16. It is critical to note that HRS cells may occasionally be identified by flow cytometry in patients with NHL, such as peripheral T-cell lymphoma (48) and chronic lymphocytic leukemia (49–51).

In these instances, a concurrent abnormal non-Hodgkin B- or T-cell population can also be identified. Accordingly, the determination of the presence of an HRS cell population should be considered in the context of all the flow cytometry data. Ultimately, the diagnosis of lymphoma requires the integration of all the clinical, morphologic, and immunophenotypic data.

Acknowledgement

The authors thank Anju Thomas and the other medical technologists in the Hematopathology laboratory at the University of Washington for their expert technical assistance.

References

1. Jaffe ES, Harris NL, Stein H, Campo E, Pileri SA, Swerdlow SH (2008) Introduction and overview of the classification of lymphoid neoplasms. In: Swerdlow SH, Campos E, Harris NL, Jaffe ES, Pileri SA, Stein H, Thiele J, Vardiman JW (eds) WHO classification of tumours of haematopoietic and lymphoid tissues. IARC Press, Lyon, pp 158–166
2. Craig FE, Foon KA (2008) Flow cytometric immunophenotyping for hematologic neoplasms. Blood 111:3941–3967
3. Wood BL, Borowitz MJ (2007) The flow cytometric evaluation of hematopoietic neoplasia. In: McPherson RA, Pincus MR (eds) Henry's clinical diagnosis and management by laboratory methods. Saunders Elsevier, Philadelphia, PA, pp 599–616
4. Givan AL (2011) Flow cytometry: an introduction. Methods Mol Biol 699:1–29
5. Redelman D (2000) Flow cytometric analyses of cell phenotypes. In: Stewart CC, Nicholson JKA (eds) Immunophenotyping. Wiley-Liss, New York, NY
6. Gorczyca W, Weisberger J, Liu Z, Tsang P, Hossein M, Wu CD, Dong H, Wong JY, Tugulea S, Dee S, Melamed MR, Darzynkiewicz Z (2002) An approach to diagnosis of T-cell lymphoproliferative disorders by flow cytometry. Cytometry 50:177–190
7. Jamal S, Picker LJ, Aquino DB, McKenna RW, Dawson DB, Kroft SH (2001) Immunophenotypic analysis of peripheral T-cell neoplasms. A multiparameter flow cytometric approach. Am J Clin Pathol 116:512–526
8. Beck RC, Stahl S, O'Keefe CL, Maciejewski JP, Theil KS, Hsi ED (2003) Detection of mature T-cell leukemias by flow cytometry using anti-T-cell receptor V beta antibodies. Am J Clin Pathol 120:785–794
9. Langerak AW, van Den Beemd R, Wolvers-Tettero IL, Boor PP, van Lochem EG, Hooijkaas H, van Dongen JJ (2001) Molecular and flow cytometric analysis of the Vbeta repertoire for clonality assessment in mature TCRalphabeta T-cell proliferations. Blood 98:165–173
10. Morice WG, Kimlinger T, Katzmann JA, Lust JA, Heimgartner PJ, Halling KC, Hanson CA (2004) Flow cytometric assessment of TCR-Vbeta expression in the evaluation of peripheral blood involvement by T-cell lymphoproliferative disorders: a comparison with conventional T-cell immunophenotyping and molecular genetic techniques. Am J Clin Pathol 121:373–383
11. Tembhare P, Yuan CM, Xi L, Morris JC, Liewehr D, Venzon D, Janik JE, Raffeld M, Stetler-Stevenson M (2011) Flow cytometric immunophenotypic assessment of T-cell clonality by Vbeta repertoire analysis: detection of T-cell clonality at diagnosis and monitoring of minimal residual disease following therapy. Am J Clin Pathol 135:890–900
12. Chan WC (2001) The Reed-Sternberg cell in classical Hodgkin's disease. Hematol Oncol 19:1–17
13. Stein H, Hummel M (1999) Cellular origin and clonality of classic Hodgkin's lymphoma: immunophenotypic and molecular studies. Semin Hematol 36:233–241
14. Marafioti T, Hummel M, Foss HD, Laumen H, Korbjuhn P, Anagnostopoulos I, Lammert H, Demel G, Theil J, Wirth T, Stein H (2000) Hodgkin and Reed-Sternberg cells represent

an expansion of a single clone originating from a germinal center B-cell with functional immunoglobulin gene rearrangements but defective immunoglobulin transcription. Blood 95: 1443–1450

15. Stein H, Delsol G, Pileri SA, Weiss LM, Poppema S, Jaffe ES (2008) Classical Hodgkin lymphoma, introduction. In: Swerdlow SH, Campos E, Harris NL, Jaffe ES, Pileri SA, Stein H, Thiele J, Vardiman JW (eds) WHO classification of tumours of haematopoietic and lymphoid tissues. IARC Press, Lyon, pp 326–329
16. Dorreen MS, Habeshaw JA, Stansfeld AG, Wrigley PF, Lister TA (1984) Characteristics of Sternberg-Reed, and related cells in Hodgkin's disease: an immunohistological study. Br J Cancer 49:465–476
17. Kadin ME, Newcom SR, Gold SB, Stites DP (1974) Letter: origin of Hodgkin's cell. Lancet 2:167–168
18. Payne SV, Jones DB, Wright DH (1977) Reed-Sternberg-cell/lymphocyte interaction. Lancet 2:768–769
19. Payne SV, Newell DG, Jones DB, Wright DH (1980) The Reed-Sternberg cell/lymphocyte interaction: ultrastructure and characteristics of binding. Am J Pathol 100:7–24
20. Sanders ME, Makgoba MW, Sussman EH, Luce GE, Cossman J, Shaw S (1988) Molecular pathways of adhesion in spontaneous rosetting of T-lymphocytes to the Hodgkin's cell line L428. Cancer Res 48:37–40
21. Stuart AE, Williams AR, Habeshaw JA (1977) Rosetting and other reactions of the Reed-Sternberg cell. J Pathol 122:81–90
22. Fromm JR, Kussick SJ, Wood BL (2006) Identification and purification of classical Hodgkin cells from lymph nodes by flow cytometry and flow cytometric cell sorting. Am J Clin Pathol 126:764–780
23. Harris NL (1999) Hodgkin's disease: classification and differential diagnosis. Mod Pathol 12:159–175
24. Schmitz R, Stanelle J, Hansmann ML, Kuppers R (2009) Pathogenesis of classical and lymphocyte-predominant Hodgkin lymphoma. Annu Rev Pathol 4:151–174
25. Fromm JR, Thomas A, Wood BL (2009) Flow cytometry can diagnose classical Hodgkin lymphoma in lymph nodes with high sensitivity and specificity. Am J Clin Pathol 131:322–332
26. Fromm JR, Wood BL (2010) A six-color flow cytometry tube for immunophenotyping classical Hodgkin lymphoma in lymph nodes. In: ICCS 25th Annual Meeting, Houston, TX, p 395
27. Fromm JR, Thomas A, Wood BL (2010) Increased expression of T cell antigens on T cells in classical Hodgkin lymphoma. Cytometry B Clin Cytom 78:387–388
28. Li S, Eshleman JR, Borowitz MJ (2002) Lack of surface immunoglobulin light chain expression by flow cytometric immunophenotyping can help diagnose peripheral B-cell lymphoma. Am J Clin Pathol 118:229–234
29. Mantei K, Wood BL (2009) Flow cytometric evaluation of CD38 expression assists in distinguishing follicular hyperplasia from follicular lymphoma. Cytometry B Clin Cytom 76: 315–320
30. Yang W, Agrawal N, Patel J, Edinger A, Osei E, Thut D, Powers J, Meyerson H (2005) Diminished expression of CD19 in B-cell lymphomas. Cytometry B Clin Cytom 63:28–35
31. Ray S, Craig FE, Swerdlow SH (2005) Abnormal patterns of antigenic expression in follicular lymphoma: a flow cytometric study. Am J Clin Pathol 124:576–583
32. Kussick SJ, Kalnoski M, Braziel RM, Wood BL (2004) Prominent clonal B-cell populations identified by flow cytometry in histologically reactive lymphoid proliferations. Am J Clin Pathol 121:464–472
33. Chen HI, Akpolat I, Mody DR, Lopez-Terrada D, De Leon AP, Luo Y, Jorgensen J, Schwartz MR, Chang CC (2006) Restricted kappa/lambda light chain ratio by flow cytometry in germinal center B cells in Hashimoto thyroiditis. Am J Clin Pathol 125:42–48
34. Hurwitz CA, Raimondi SC, Head D, Krance R, Mirro J Jr, Kalwinsky DK, Ayers GD, Behm FG (1992) Distinctive immunophenotypic features of t(8;21)(q22;q22) acute myeloblastic leukemia in children. Blood 80:3182–3188
35. Rawstron AC, Green MJ, Kuzmicki A, Kennedy B, Fenton JA, Evans PA, O'Connor SJ, Richards SJ, Morgan GJ, Jack AS, Hillmen P (2002) Monoclonal B lymphocytes with the characteristics of "indolent" chronic lymphocytic leukemia are present in 3.5% of adults with normal blood counts. Blood 100: 635–639
36. Rawstron AC, Shanafelt T, Lanasa MC, Landgren O, Hanson C, Orfao A, Hillmen P, Ghia P (2010) Different biology and clinical outcome according to the absolute numbers of clonal B-cells in monoclonal B-cell lymphocytosis (MBL). Cytometry B Clin Cytom 78(Suppl 1):S19–S23
37. Rodriguez-Caballero A, Garcia-Montero AC, Barcena P, Almeida J, Ruiz-Cabello F, Tabernero MD, Garrido P, Munoz-Criado S, Sandberg Y, Langerak AW, Gonzalez M, Balanzategui A, Orfao A (2008) Expanded

cells in monoclonal TCR-alphabeta+/CD4+/NKa+/CD8-/+dim T-LGL lymphocytosis recognize hCMV antigens. Blood 112:4609–4616

38. Rahemtullah A, Longtine JA, Harris NL, Dorn M, Zembowicz A, Quintanilla-Fend L, Preffer FI, Ferry JA (2008) CD20+ T-cell lymphoma: clinicopathologic analysis of 9 cases and a review of the literature. Am J Surg Pathol 32:1593–1607
39. Rizzo K, Stetler-Stevenson M, Wilson W, Yuan CM (2009) Novel CD19 expression in a peripheral T cell lymphoma: a flow cytometry case report with morphologic correlation. Cytometry B Clin Cytom 76:142–149
40. Seegmiller AC, Karandikar NJ, Kroft SH, McKenna RW, Xu Y (2009) Overexpression of CD7 in classical Hodgkin lymphoma-infiltrating T lymphocytes. Cytometry B Clin Cytom 76:169–174
41. Bosler DS, Douglas-Nikitin VK, Harris VN, Smith MD (2008) Detection of T-regulatory cells has a potential role in the diagnosis of classical Hodgkin lymphoma. Cytometry B Clin Cytom 74:227–235
42. Hsu SM, Jaffe ES (1984) Leu M1 and peanut agglutinin stain the neoplastic cells of Hodgkin's disease. Am J Clin Pathol 82:29–32
43. Stein H, Mason DY, Gerdes J, O'Connor N, Wainscoat J, Pallesen G, Gatter K, Falini B, Delsol G, Lemke H et al (1985) The expression of the Hodgkin's disease associated antigen Ki-1 in reactive and neoplastic lymphoid tissue: evidence that Reed-Sternberg cells and histiocytic malignancies are derived from activated lymphoid cells. Blood 66:848–858
44. Stein H, Uchanska-Ziegler B, Gerdes J, Ziegler A, Wernet P (1982) Hodgkin and Sternberg-Reed cells contain antigens specific to late cells of granulopoiesis. Int J Cancer 29:283–290
45. Carbone A, Gloghini A, Gattei V, Aldinucci D, Degan M, De Paoli P, Zagonel V, Pinto A (1995) Expression of functional CD40 antigen on Reed-Sternberg cells and Hodgkin's disease cell lines. Blood 85:780–789
46. Carbone A, Gloghini A, Pinto A (1996) CD40: a sensitive marker of Reed-Sternberg cells. Blood 87:4918–4919
47. Poppema S, Delsol G, Pileri SA, Stein H, Swerdlow SH, Warnke RA, Jaffe ES (2008) Nodular lymphocyte predominant Hodgkin lymphoma. In: Swerdlow SH, Campos E, Harris NL, Jaffe ES, Pileri SA, Stein H, Thiele J, Vardiman JW (eds) WHO classification of tumours of haematopoietic and lymphoid tissues. IARC Press, Lyon, pp 323–325
48. Quintanilla-Martinez L, Fend F, Moguel LR, Spilove L, Beaty MW, Kingma DW, Raffeld M, Jaffe ES (1999) Peripheral T-cell lymphoma with Reed-Sternberg-like cells of B-cell phenotype and genotype associated with Epstein-Barr virus infection. Am J Surg Pathol 23:1233–1240
49. Mao Z, Quintanilla-Martinez L, Raffeld M, Richter M, Krugmann J, Burek C, Hartmann E, Rudiger T, Jaffe ES, Muller-Hermelink HK, Ott G, Fend F, Rosenwald A (2007) IgVH mutational status and clonality analysis of Richter's transformation: diffuse large B-cell lymphoma and Hodgkin lymphoma in association with B-cell chronic lymphocytic leukemia (B-CLL) represent 2 different pathways of disease evolution. Am J Surg Pathol 31: 1605–1614
50. Momose H, Jaffe ES, Shin SS, Chen YY, Weiss LM (1992) Chronic lymphocytic leukemia/small lymphocytic lymphoma with Reed-Sternberg-like cells and possible transformation to Hodgkin's disease. Mediation by Epstein-Barr virus. Am J Surg Pathol 16:859–867
51. Ohno T, Smir BN, Weisenburger DD, Gascoyne RD, Hinrichs SD, Chan WC (1998) Origin of the Hodgkin/Reed-Sternberg cells in chronic lymphocytic leukemia with "Hodgkin's transformation". Blood 91:1757–1761
52. Crespo M, Bosch F, Villamor N, Bellosillo B, Colomer D, Rozman M, Marce S, Lopez-Guillermo A, Campo E, Montserrat E (2003) ZAP-70 expression as a surrogate for immunoglobulin-variable-region mutations in chronic lymphocytic leukemia. N Engl J Med 348:1764–1775
53. Damle RN, Wasil T, Fais F, Ghiotto F, Valetto A, Allen SL, Buchbinder A, Budman D, Dittmar K, Kolitz J, Lichtman SM, Schulman P, Vinciguerra VP, Rai KR, Ferrarini M, Chiorazzi N (1999) Ig V gene mutation status and CD38 expression as novel prognostic indicators in chronic lymphocytic leukemia. Blood 94:1840–1847

Chapter 3

Laser-Based Microdissection of Single Cells from Tissue Sections and PCR Analysis of Rearranged Immunoglobulin Genes from Isolated Normal and Malignant Human B Cells

Ralf Küppers, Markus Schneider, and Martin-Leo Hansmann

Abstract

Normal and malignant B cells carry rearranged immunoglobulin (Ig) variable region genes, which due to their practically limitless diversity represent ideal clonal markers for these cells. We describe here an approach to isolate single cells from frozen tissue sections by microdissection using a laser-based method. From the isolated cells rearranged IgH and Igκ genes are amplified in a semi-nested PCR approach, using a collection of V gene family-specific primers recognizing nearly all V gene segments together with primers for the J gene segments. By sequence analysis of V genes from distinct cells, the clonal relationship of the B lineage cells can unequivocally be determined and related to the histological distribution of the cells. The approach is also useful to determine V, D, and J gene usage. Moreover, the presence and pattern of somatic Ig V gene mutations give valuable insight into the stage of differentiation of the B cells.

Key words: B cells, B cell lymphoma, Clonality, Hodgkin's lymphoma, Micromanipulation, V gene recombination, Single-cell PCR, Somatic hypermutation, Taq DNA polymerase errors

1. Introduction

To generate a functional B cell antigen receptor (BCR), gene segments coding for the variable region of the immunoglobulin (Ig) heavy and light chains have to be assembled by a somatic recombination process during early B cell development in the bone marrow. Heavy-chain V region genes are composed of three gene segments, called variable (V_H), diversity (D_H), and joining (J_H) (Fig. 1), whereas V regions of light chains are composed of V and J segments. There are two types of light chains, κ and λ. The human IgH locus on chromosome 14 contains about 50 functional V_H genes (depending on the haplotype), 27 D_H genes, and 6 J_H genes (1–3). As each V_H segment can recombine with each D_H

Ralf Küppers (ed.), *Lymphoma: Methods and Protocols*, Methods in Molecular Biology, vol. 971,
DOI 10.1007/978-1-62703-269-8_3, © Springer Science+Business Media, LLC 2013

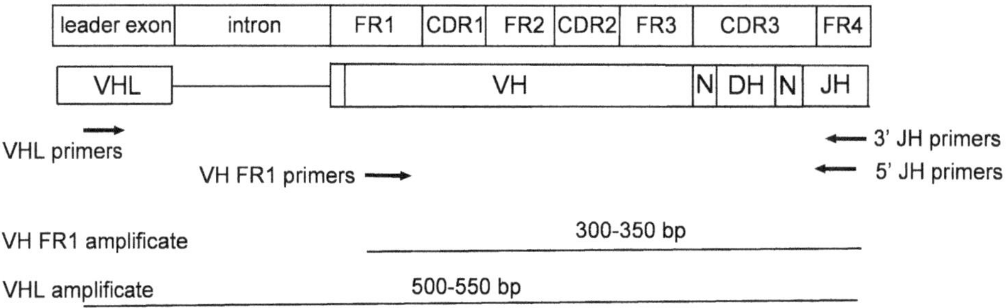

Fig. 1. Structure of the V_H region gene and location of PCR primers. The V_H region gene consists of two exons. The first encodes the most part of the V_H leader peptide, the second exon encodes the rest of the leader exon, and the mature V_H region. The latter is composed of the V_H, D_H and J_H gene segments. At the joining sites between V_H–D_H and D_H–J_H, one usually finds several bases of non-germline-encoded N nucleotides. The V_H region is subdivided into four relatively conserved framework regions (FR) and three diverse complementarity determining regions (CDR). CDR1 and 2 are encoded by the V_H gene segments, CDR3 is generated in the rearrangement process at the V–D–J joining sites. The approximate location of the binding sites for V_H FR1, V_H leader, and 5′ and 3′ J_H primer mixes is indicated. With the V_H FR1 and J_H primers, amplificates of about 300–350 bp are obtained, the amplificates with the V_H leader and J_H primers are about 500–550 bp long. The length of the substructures of the V_H region gene as shown here are not to scale.

and each J_H segment, a large diversity of heavy-chain V gene rearrangements is generated. This diversity is further increased by loss of nucleotides at the ends of the rearranging gene segments, by addition of non-germline-encoded nucleotides (N nucleotides) between the gene segments, and the variable opening of a hairpin structure generated as a recombination intermediate (P nucleotides). The diversity is more limited for light-chain gene rearrangements, as D segments are missing. Nevertheless, a large number of functional V and J gene segments is also available for V–J recombination: The human Igκ locus on chromosome 2 encompassed 30–35 functional V_κ genes, and 5 J_κ genes, and the Igλ locus on chromosome 22 consists of 30–37 functional V_λ genes and 4 J_λ genes (4–7). Moreover, loss of nucleotides and the addition of N nucleotides and P nucleotides between the V and J segments occur also during light-chain gene rearrangements. Hence, a large diversity can also be generated for light-chain V region genes. Further diversity can be generated at later stages of B cell differentiation, when B cells are antigen-activated and become involved in T cell-dependent immune responses. During such immune responses, B cells establish histological structures called germinal centers (GC) in secondary lymphoid organs (e.g., lymph nodes). In proliferating GC B cells, V region genes acquire somatic mutations at a high rate through a process called somatic hypermutation (8, 9).

Due to the nearly limitless diversity of V region genes, each mature B cell is equipped with a unique BCR, which is inherited to the daughter cells when B cells proliferate. Consequently, V region genes represent ideal markers for members of a B cell clone. This can be used to study B cell developmental processes and the clonal relationship of B lineage cells, such as the clonal composition and intraclonal diversification in GC B cells (8). An analysis of rearranged Ig V genes can also be used to study malignant B cells. For example, we showed that the malignant Hodgkin and Reed–Sternberg (HRS) cells of Hodgkin's lymphoma harbor clonal Ig gene rearrangements (10, 11). This proved the B cell origin of these cells and demonstrated that HRS cells represent monoclonal cell populations, a hallmark of tumor cells.

In many situations, PCR amplification of rearranged Ig V genes from cell populations is sufficient to identify clonal B cell populations and obtain the Ig rearrangements of dominant normal or malignant B cell clones. Such approaches are often applied to study B cell non-Hodgkin lymphomas, where usually most of the B cells in the tissue belong to the lymphoma clone (12–14). However, for a characterization of rare cells, such as the HRS cells in Hodgkin's lymphoma, that usually account for only about 1% of cells in the tissue, analysis of single HRS cells is mandatory for a reliable characterization of the cells. Moreover, for a detailed analysis of B cells that are defined by their location in the histological environment, a single cell isolation and PCR analysis is needed.

We established a PCR protocol to amplify rearranged human Ig V genes from genomic DNA of single B cells using primers binding to the 5′ end of the V gene segments in framework region (FR)1 combined with primers binding to the J gene segments in FR4 (8). As the human V_H and V_L genes are grouped into families based on at least 80% DNA sequence homology for family members, V gene family-specific primers were designed that bind to all or at least most members of a given V gene family. Starting from a single target molecule in a single cell, one round of PCR is not sufficient to obtain sufficient amplificate for direct sequence analysis. Therefore, the PCR protocol includes a second round of PCR with a small aliquot of the first-round amplificate using the same V gene primers, but J primer mixes that are located 5′ (i.e., internal) of the J primers used in the first round of PCR (semi-nested PCR) (Fig. 2). With this strategy, PCR products of about 300–350 bp length are obtained that cover most of the V region gene. Hence, besides the determination of the clonal relationship of distinct cells, the sequence analysis of the PCR products also allows one to identify the V, (D), and J gene segments used in the rearrangements and analyze the sequences for somatic mutations.

We used this single-cell PCR approach for the analysis of single cells isolated from tissue sections by Laser-assisted micromanipulation, as described in this chapter. However, the PCR protocol can also be used to amplify rearranged Ig V genes from cells isolated by flow cytometry (15–17). This is an updated version of a manuscript that described a combination of hydraulic micromanipulation with single-cell Ig V gene PCR (18).

a

```
                              FR4                                        intron
Cons. GCT GAC TAC TTC GAC TTC TGG GGC CAA GGG ACC CTG GTC ACC GTC TCC TCA GGT
J1    --- --A --- --- --G CA- --- --- --G --C --- --- --- --- --- --- --- ---
J2    TAC TGG --- --- --T C-- --- --- -GT --C --- --- --- --T --- --- --- ---
J3a       --T GCT --T --T G-- --- --- --- --- --A A-- --- --- --- --T --- ---
J3b       --T GCT --T --T A-- --- --- --- --- --A A-- --- --- --- --T --- ---
J4a       --- --- --T --- -A- --- --- --- --A --- --- --- --- --- --- --- ---
J4b       --- --- --T --- -A- --- --- --G --A --- --- --- --- --- --- --- ---
J5a    AC A-- -GG --- --- -C- --- --- --- --A --- --- --- --- --- --- --- ---
J5b    AA A-- -GG --- --- CC- --- --- --G --A --- --- --- --- --- --- --- ---
J6a (TAC)4T-- GGT A-G --- G-- --- --G --- --- --- AC- --- --- --- --- --- --- ---
J6b (TAC)4T-- GGT A-G --- G-- --- --- --- --- --- AC- --- --- --- --- --- --- ---
J6c (TAC)4T--     A-G --- G-- --- --- A-- --- --- AC- --- --- --- --- --- --- ---

5' primers
JH1,4,5                           --- --G --M --- --- --- --- ---
JH2                               --- -GT --C --- --- --- --T ---
JH3                               --- --- --- --A A-- --- --- ---
JH6                               --- M-- --- --- AC- --- --- ---

3' primers
JH1,2,4,5                                     --- --- --- --- --- --- --- ---
JH3                                           --A A-- --- --- --- --T --- ---
JH6                                           --- AC- --- --- --- --- --- ---


M: A+C
```

b

```
                FR4                                         intron
J1    G TGG ACG TTC GGC CAA GGG ACC AAG GTG GAA ATC AAA CGT GAGTAG
J2    - -AC --T --T --- --G --- --- --- C-- --G --- --- --- A----C
J3    A -TC --T --- --- -CT --- --- --A --- --T --- --- --- A----C
J4    - CTC --T --- --- GG- --- --- --- --- --G --- --- --- A---GC
J5    - ATC --C --- --- --- --- --A CGA C-- --G --T --- --- A--C-T

3' primers
Jk1,2,4                       -- --- --- S-- --R --- --- --- ----
Jk3                           -- --- --A --- --T --- --- --- A----C
Jk5                           -- --A CGA C-- --G --T --- --- A--C


5' primers
Jk1,2                 -- --R --- --- --- S-- --R --- --
Jk3                   -- -CT --- --- --A --- --T --- --
Jk4                   -- GG- --- --- --- --- --G --- --
Jk5                   -- --- --- --A CGA C-- --G --T --


S: G+C; R: A+G
```

Fig. 2. Sequences of human J_H and J_κ genes and of the J primer sequences. Shown are the sequences of the human J_H (**a**) and J_κ (**b**) genes and the location of the first- and second-round PCR J primers. The primer sequences are the reverse complement of the J sequences shown here (see Table 1). For J_H3, 4, 5, and 6, several alleles exist (denoted as a, b, c). The beginning of framework region 4 (FR4) and of the intron 3′ of the J segments are indicated. The bases upstream of FR4 belong to CDR3. The numbers in the primer designations indicate the J segments for which the primers were designed. Note, however, that the primers may also amplify other J segments of the respective Ig loci.

2. Materials

2.1. Histological Stainings

1. Aceton.
2. Antibodies: For example CD30 (Ber-H2, mouse anti-human, DAKO, Hamburg, Germany), 1:100 diluted, or CD20 (L26, mouse anti-human, DAKO), 1:100 diluted.

3. Biotinylated $F(ab')_2$ fragment of rabbit anti-mouse immunoglobulins (DAKO), 1:200 diluted.
4. SuperSensitive™ Link-Label IHC Detection System, MultiLink®—alkaline phosphatase (concentrated) (Biogenex, Fremont, CA, USA).
5. Permanent Red Substrate-Chromogen, liquid (DAKO).
6. Tris-buffer saline (TBS): Dissolve 8 g of NaCl, 0.2 g of KCl, and 3 g of Tris base in ca. 800 ml of H_2O, adjust pH to 7.4 with HCl, and add H_2O to 1 l.
7. TBS/BSA: TBS supplemented with 1% or 5% (w/v) bovine serum albumin.
8. Mayer's hemalaun.
9. Membrane covered slides (Zeiss, Jena, Germany).

2.2. Laser-Assisted Microdissection

1. PALM MicroBeam (Zeiss).
2. Axiovert 200 microscope (Zeiss).
3. PCR tubes.
4. PCR buffer (without $MgCl_2$) from Expand High Fidelity PCR system (Roche, Mannheim, Germany).

2.3. Proteinase K Digestion

1. Proteinase K, PCR grade (Roche).
2. 10 mM Tris–HCl, pH 7.6.
3. Mineral oil.

2.4. First Round of PCR

1. High fidelity polymerase mixture, 3.5 units/μl (Roche) with high fidelity PCR buffer and $MgCl_2$.
2. dNTPs.
3. V_H/Vκ primer mix: Mix same volumes of 2.5 μM working dilutions (V_H1–6, 3′ J_H primers, Vκ1–6 and 3′ Jκ primers) (Table 1).

2.5. Second Round of PCR

2.5.1. V_H PCR

1. Taq DNA polymerase (5 units/μl) with PCR buffer and $MgCl_2$.
2. dNTPs.
3. Mineral oil.
4. V_H primers: See first round of PCR, 2.5 μM working dilution.
5. J_H primer mix: Mix same volumes of 2.5 μM working dilutions of the four 5′ J_H primers (0.625 μM per 5′ J_H primer) (Table 2).

2.5.2. $V_κ$ PCR

1. Taq DNA polymerase (5 units/μl) with PCR buffer and $MgCl_2$.
2. dNTPs.

Table 1
Primers for the first round of PCR

Name	Sequence (5′–3′)
VH1	CAG-TCT-GGG-GCT-GAG-GTG-AAG-A
VH2	GTC-CTR-CGC-TGG-TGA-AAC-CCA-CAC-A
VH3	GGG-GTC-CCT-GAG-ACT-CTC-CTG-TGC-AG
VH4	GAC-CCT-GTC-CCT-CAC-CTG-CRC-TGT-C
VH5	AAA-AAG-CCC-GGG-GAG-TCT-CTG-ARG-A
VH6	ACC-TGT-GCC-ATC-TCC-GGG-GAC-AGT-G
3′JH1.2.4.5	ACC-TGA-GGA-GAC-GGT-GAC-CAG-GGT
3′JH3	ACC-TGA-AGA-GAC-GGT-GAC-CAT-TGT
3′JH6	ACC-TGA-GGA-GAC-GGT-GAC-CGT-GGT
Vk1	GAC-ATC-CRG-WTG-ACC-CAG-TCT-CCW-TC
Vk2	CAG-WCT-CCA-CTC-TCC-CTG-YCC-GTC-A
Vk3	TTG-TGW-TGA-CRC-AGT-CTC-CAG-SCA-CC
Vk4	AGA-CTC-CCT-GGC-TGT-GTC-TCT-GGG-C
Vk5	CAG-TCT-CCA-GCA-TTC-ATG-TCA-GCG-A
Vk6	TTT-CAG-TCT-GTG-ACT-CCA-AAG-GAG-AA
3′Jk1.2.4	ACT-CAC-GTT-TGA-TYT-CCA-SCT-TGG-TCC
3′Jk3	GTA-CTT-ACG-TTT-GAT-ATC-CAC-TTT-GGT-CC
3′Jk5	GCT-TAC-GTT-TAA-TCT-CCA-GTC-GTG-TCC

R: A&G; W: A&T; Y: C&T; S: G&C
The numbers in the V primer designations indicate the respective V gene families. The numbers in the J primers indicate that J segments that are preferentially amplified by these primers

Table 2
Sequences of J_H and Jκ primers for the second round of PCR

Name	Sequence (5′–3′)
5′JH1.4.5	GAC-GGT-GAC-CAG-GGT-KCC-CTG-GCC
5′JH2	GAC-AGT-GAC-CAG-GGT-GCC-ACG-GCC
5′JH3	GAC-GGT-GAC-CAT-TGT-CCC-TTG-GCC
5′JH6	GAC-GGT-GAC-CGT-GGT-CCC-TTK-GCC
5′Jk1.2	TTG-ATY-TCC-ASC-TTG-GTC-CCY-TGG-C
5′Jk3	TTG-ATA-TCC-ACT-TTG-GTC-CCA-GGG-C
5′Jk4	TTG-ATC-TCC-ACC-TTG-GTC-CCT-CCG-C
5′Jk5	TTA-ATC-TCC-AGT-CGT-GTC-CCT-TGG-C

K: G&T; Y: C&T; S: G&C

3. Mineral oil.
4. Vκ primers: See first round of PCR, 2.5 μM working dilution.
5. Jκ primer mix: 2.5 μM per 5′ Jκ primer (Table 2).

2.6. PCR Product Purification and Direct Sequence Analysis

1. Agarose.
2. GelRed Nucleic Acid stain, 10,000× in water (Biotium, Hayward, CA, USA).
3. 3 M Na-acetate, pH 5–5.5.
4. Ethanol 100%, 70%.
5. TAE: 40 mM Tris–acetate, 1 mM EDTA.
6. Qiaex gel II extraction kit (Qiagen, Hilden, Germany).
7. DNA mass ladder.
8. Big Dye Sequencing Kit (PE Biosystems, Weiterstadt, Germany).
9. Na-acetate/dextranblue: 3 M Na-acetate (pH 5–5.5), 12.5 mg/ml dextranblue.

3. Methods

3.1. Histological Staining

Dry frozen sections overnight on membrane-covered slides. Fix in acetone for 10 min. Dry sections for 30–60 min. Incubate for 15 min with TBS/3% BSA. Dilute primary antibody in TBS/3% BSA. Incubate sections with primary antibody for 30 min at room temperature. Wash three times with TBS followed by 5-min washing with TBS on a rocker. Dilute MultiLink antibody (a biotinylated secondary bridge-antibody) 1:20 in TBS/3% BSA. Incubate for 20 min at room temperature. Wash three times with TBS followed by 5-min washing with TBS on a rocker. Dilute peroxidase-labeled streptavidin 1:20 in TBS/3% BSA. Incubate for 20 min at room temperature. Wash three times with TBS followed by 5-min washing with TBS on a rocker. Incubate for 1–15 min with Permanent Red prepared according to the instructions of the supplier until staining intensity controlled under the microscope is as desired. Stop reaction by washing with distilled water. Counterstain with Mayer's hemalaun. Dry sections for 2 h on a heating plate at 37°C (see Note 1).

3.2. Microdissection

1. The stained tissue sections are mounted on the PALM Stage II of the PALM Robot MicroBeam system.
2. Navigation through the tissue section is done with the PALM RoboSoftware which displays the microscope view onto the monitor screen. A magnification of 40× is used for microdissection.
3. The single cells or group of cells are selected and marked for microdissection by using the pencil tool of the software. With

this pencil the cell or area to be microdissected is surrounded and thereby marked (see Note 2).

4. For microdissection and catapulting of the cells 20 μl of 1× Expand High Fidelity PCR buffer (without $MgCl_2$) is added into the cap of a thin-walled PCR tube and this is loaded into the tube holder of the CapMover, which moves the cap into the appropriate position for microdissection right above the tissue section (see Note 3).
5. Select RoboLPC as cutting mode and start the laser-based microdissection. In RoboLPC the cell is first cut out along the drawn path and then catapulted into the cap of the mounted tube.
6. The PCR tubes are centrifuged at full speed for 1 min to spin down the microdissected cells and are stored at −20°C or −80°C.
7. Before and after micromanipulation a photograph is taken to allow the exact localization of the microdissected cell within the histological microenvironment.
8. In order to monitor possible contamination by DNA from fragmented neighboring cells, every 5th to 10th cell should be followed by one sample which contains only membrane from the slide but no parts of the tissue section (these samples are used as negative controls for microdissection). Microdissection can be carried out for about 24 h without a detectable decrease of PCR efficiency.

3.3. Proteinase K Digestion

Dilute proteinase K 1:1 with 10 mM Tris–HCl, pH 7.6. Add 0.5 μl of this enzyme dilution to the 20 μl PCR buffer containing the single cell and overlay with mineral oil. Incubate for 2 h at 50°C. Inactivate the enzyme by incubation at 95°C for 10 min. Cool down to 4°C.

3.4. First Round of PCR

Prepare a master mix consisting of 2.5 μl dNTP solution (2 mM), 3 μl 10× PCR buffer high fidelity, 5 μl primer mix, 5 μl 25 mM $MgCl_2$, and 13.5 μl H_2O per reaction. Add 29 μl master mix to the reaction mixture from the proteinase K digestion (20.5 μl). Add 0.5 μl Expand High Fidelity enzyme mix after the first denaturation (see below), and carry out PCR. The PCR program consists of 95°C 2 min, 65°C pause (add enzyme), 72°C 1 min, 34× (95°C 50 s, 59°C 30 s, 72°C 60 s), 72°C 5 min, and 10°C pause (see Notes 4–7).

3.5. Second Round of PCR

3.5.1. V_H PCR

Prepare on ice six master mixes (one for each of the six V_H family-specific primers), each consisting of 5 μl dNTPs (2 mM), 5 μl 10× PCR buffer, 2.5 μl of the respective V_H primer, 2.5 μl 5′ J_H primer mix, 3 μl 25 mM $MgCl_2$ (5 μl for V_H1), 30.7 μl H_2O (28.7 μl for V_H1), and 0.3 μl Taq DNA polymerase. Pipette 49 μl master mix in fresh PCR tube. Overlay with mineral oil. To master mix add 1 μl of the first-round reaction. Start PCR program but put PCR tubes from ice to heating block only after block has reached 95°C. The program consists of 95°C 1 min, 45× (95°C 50 s, 61°C 30 s, 72°C 60 s), 72°C 5 min, and 10°C pause.

3.5.2. V_κ PCR

Prepare on ice six master mixes (one for each of the six V_κ family-specific primers), each consisting of 5 μl dNTPs (2 mM), 5 μl 10× PCR buffer, 2.5 μl of the respective V_κ primer, 2.5 μl J_κ primer mix, 5 μl 25 mM $MgCl_2$, 28.7 μl H_2O, and 0.3 μl Taq DNA polymerase. Pipette 49 μl master mix in fresh PCR tube. Overlay with mineral oil. To master mix add 1 μl of the first-round reaction. Start PCR program but put PCR tubes from ice to heating block only after block has reached 95°C. The program consists of 95°C 1 min, 45× (95°C 50 s, 61°C 30 s, 72°C 60 s), 72°C 5 min, and 10°C pause (see Note 8).

3.6. PCR Product Purification and Direct Sequence Analysis

Since primer dimers and additional bands are often observed when PCR products are analyzed by agarose gel electrophoresis, it is recommended to purify the PCR products from agarose gels before sequencing. The amount of purified product is estimated from an agarose gel using a DNA mass ladder to estimate the DNA concentration (see Note 9).

1. PCR products are first analyzed on a 2% analytical agarose gel containing GelRed Nucleic Acid stain and visualized under UV light.
2. For samples with a PCR product, the second round of PCR is repeated with 2–3 reactions in 50 μl or 100 μl volume. These reactions are then mixed and precipitated by adding 1/10 volume of 3 M Na-acetate and 2 volumes 100% ethanol. After centrifugation at ca. 12,000 × *g* for 30 min and washing of the pellet with 70% ethanol, the pellet is dissolved in 20 μl TE and loaded on a 2% preparative agarose gel.
3. The PCR product is cut out of the gel under UV light and isolated with the Qiaex II gel extraction kit, as recommended by the supplier.
4. One-tenth of the DNA is run on a 2% agarose gel together with a DNA mass ladder. By comparing the strength of the PCR product bands to the strength of the DNA fragments of the mass ladder the DNA concentration is estimated.
5. Sequence reaction: Mix the following components on ice: 30–40 ng PCR product, 1.5 μl primer (2.5 μM; usually the V primer), 2 μl Big Dye reaction mix plus × μl H_2O to obtain a final volume of 10 μl. Overlay with one drop of mineral oil. Run PCR: 96°C 2 min, 34 times (96°C 30 s, 50°C 15 s, 60°C 4 min), and cool to 10°C.

6. Precipitation (at room temperature): Add 40 μl H_2O, remove oil, transfer reaction mixture to fresh tube, and add 5 μl 3 mM Na-acetate/dextranblue and 125 μl 100% ethanol. Mix by vortexing. Directly centrifuge at ca. 12,000×*g* for 30 min. Remove supernatant and wash the pellet with 400 μl 70% ethanol. Centrifuge at ca. 12,000×*g* for 5 min, remove supernatant, and let the pellet air-dry. Store at 4°C; if not run on a sequencing gel on the same day.
7. Analyze sequence reaction on an ABI sequencer.

3.7. Sequence Evaluation

V gene sequences can be evaluated for the identity of the gene segments used in the rearrangement and the presence of somatic mutations by comparing to GenBank entries (www.ncbi.nlm.nih.gov/blast/). However, a more convenient way is to use the IMGT database and the DNA plot software (http://www.imgt.org/IMGT_vquest/share/textes/ or http://vbase.mrc-cpe.cam.ac.uk/index.php?&MMN_position=1:1) or the Igblast tool (http://www.ncbi.nlm.nih.gov/igblast/). With these tools, rearranged human Ig genes are analyzed for sequences homologous to V, D, and J gene segments and the most homologous germline gene sequences are given (see Note 10).

4. Notes

1. While the protocol described here uses alkaline phosphatase for immunostaining, the procedure can also be performed using horseradish peroxidase as enzyme for the color reaction. Moreover, PCR works also efficiently with cells labeled by in situ hybridization, e.g., for EBER transcripts to detect Epstein–Barr virus-infected B cells (19).
2. The software allows the marking/selection of multiple objects before microdissecting them, also on different glass slides mounted on the Stage II in parallel. However, sometimes the slides slip a bit by navigating through the section and then all drawings are not surrounding the cells anymore and have to be adjusted again. So do not mark too many cells before microdissection.
3. In general about 200–300 cells can be microdissected in one cap in one turn. However, microdissection should not take too long as the fluid in the cap vaporizes after some time. If you need more cells it is possible to pick about 300 cells, centrifuge down the fluid, add another 20 μl of buffer into the cap, and microdissect an additional 200–300 cells. This can be repeated several times. Microdissection can also be performed dry without adding PCR buffer. Then, adhesive caps should be used. Here up to 400 cells can be picked as a cell conglomerate

forms in the cap which prevents proper sticking of the additional cells to the adhesive cap.

4. The protocol described here uses V_H and Vκ family-specific primers binding to the FR1 of the respective genes together with combinations of J segment primers (Fig. 1). We also established PCR strategies for the co-amplification of V_H and Vλ gene rearrangements from single cells (20), and a strategy to co-amplify V_H, Vκ, and Vλ gene rearrangements (20). Also in these approaches, V gene family-specific primers binding to sequences in the FR are used together with primers for the J gene segments. For the amplification of V_H gene rearrangements, V_H family-specific primers binding to sequences in the leader exon may be used instead of the FR1 primers (21) (Table 3). With the V_H leader primers, amplificates of about 500–550 bp length are obtained (Fig. 1). There are usually less somatic mutations in the leader exon than in FR1. Therefore, usage of the leader primers is advantageous when highly mutated B cells are studied, because there is less chance for leader primers than for FR1 primers that they cannot bind efficiently to the V genes due to somatic mutations at their binding sites. A further advantage when using V_H leader primers is that the complete V_H gene sequence is obtained. However, when the DNA quality is not optimal, the longer amplificates with V_H leader primers may be amplified with lower efficiency. We also established a protocol for the amplification of Vκ gene rearrangements from single cells using Vκ family-specific leader primers (22). For a detailed characterization of the IgH and Igκ loci of human B lineage cells, there are also protocols available to amplify D_HJ_H joints or fragments indicative of germline configuration of the IgH locus (23), and to detect rearrangements of the kappa-deleting element (KDE) (24).

Table 3
Family-specific VH leader exon primers

Name	Sequence (5′–3′)
VH1L	CTC-ACC-ATG-GAC-TGG-ACC-TGG-AG
VH2L	TGC-TCC-ACR-CTC-CTG-CTR-CTG-A
VH3L	ACC-ATG-GAG-TTT-GGG-CTG-AGC-TG
VH3.2 L	ACC-ATG-GAA-CTG-GGG-CTC-CGC-TG
VH4L	CTC-CTG-GTG-GCA-GCT-CCC-AGA-T
VH5L	ATC-ATG-GGG-TCA-ACC-GCC-ATC-CT

Two primers are used to cover the members of the V_H3 family. No V_H leader primer was designed for the single-gene family V_H6

5. Since 60% of human B cells use kappa light chains, the V_H and Vκ primer combination is the preferred one to analyze human B cells for clonality and the presence of somatic mutations. However, one has to keep in mind that most λ light chain expressing B cells carry VκJκ joints, and that these joints are usually inactivated by rearrangements of a KDE that is located downstream of the Cκ gene (24–26). Rearrangement of the KDE to a sequence in the Jκ–Cκ intron deletes not only the Cκ gene but also the two Igκ enhancers. However, VκJκ joints remain on the chromosome and can be amplified. As the Ig enhancers are needed for somatic hypermutation activity, such VκJκ joints remain unmutated when the other Ig loci in the cell acquire mutations (17).
6. The protocol described here allows one to co-amplify V_H and Vκ gene rearrangements from single human cells. By performing first a whole genome amplification with the DNA of a single cell, it is possible to analyze the cells in parallel for multiple genes (27). Using this strategy, single cells can be analyzed in parallel for IgH, Igκ, and Igλ gene rearrangements and also for other features (28), such as mutations in oncogenes or tumor-suppressor genes, or infection by viruses (19, 29). A similar approach to combine whole genome amplification with PCR for rearranged human Ig genes has also been described by Brezinschek and colleagues (30, 31).
7. PCR contamination is a major concern when performing single-cell PCR. Therefore, a number of precautions should be followed to obtain reliable results: Pre- and post-PCR work should be performed in separate rooms. Gloves should be worn and frequently changed. Separate pipettes, reaction tubes, aerosol-resistant pipette tips, and other equipment should be dedicated only for the pre-PCR work of the first round of PCR. Tubes containing first-round reaction products should not be opened in the room where the pre-PCR work is done.
8. Since Vκ genes of the Vκ5 and Vκ6 families are only rarely used, the Vκ analysis may be restricted to families 1–4.
9. The PCR products amplified with the V gene primers binding to the FR1 of human V_H and Vκ genes and the corresponding J segment primers are usually 300–350 bp long. However, occasionally also longer or shorter PCR products are obtained. These should not be disregarded, as they also may represent V gene rearrangements. On the one hand, one has to keep in mind that the J gene segments are located close to each other in the IgH and Igκ loci (200–600 bp distances; refs. (3, 4)), and that J primer mixes are used that bind to all J_H and Jκ segments. Hence, longer PCR products may represent amplificates

ranging from the V primer-binding site to a J gene segment downstream of the one that is actually rearranged. If a somatic mutation in the rearranged J gene segments prevents successful amplification from this J segment, the product from the next downstream J gene segment may be the predominant one. On the other hand, somatic hypermutation introduces not only point mutations but also deletions or duplications. These events account for ca. 4% of somatic mutations in non-selected (e.g., out-of-frame) rearrangements (15). Overall, about 40% of mutated out-of-frame rearrangements from human B cells have been found to carry deletions and/or duplications, which may range in length from single nucleotides to several hundred basepairs (15).

10. While the single cell approach is technically demanding and laborious, it has several advantages compared with PCR analysis of cell populations: (1) PCR products can be directly sequenced without cloning. This is not only fast but also has the important advantage that errors due to Taq DNA polymerase mistakes do not pose a problem for the reliable identification of somatic mutations, as a "consensus" sequence is obtained. Even if a Taq DNA polymerase error happens in the first cycle of amplification, it should principally be present only in a quarter of the final molecules of the PCR (assuming that both strands of the target molecule are replicated in that cycle and that the mutated sequence does not have an amplification advantage in the further cycles). (2) PCR hybrid artifacts, which may represent a considerable fraction of sequences if PCR products obtained from DNA of cell populations are cloned before sequencing (32), do not play a role in single-cell PCR. (3) V_H and V_L genes can be coamplified from a single cell so that they can be assigned to distinct cells, which is not possible if heavy- and light-chain genes are amplified from cell populations. (4) The sizes of B cell clones can be estimated when single cells from defined histological locations are analyzed. If PCR is performed with DNA from cell populations, it is difficult to discern whether identical sequences derive from a single cell or from an expanded clone.

Acknowledgements

This work was supported by the Deutsche Forschungsgemeinschaft and the Deutsche Krebshilfe. We are grateful to all present and previous members of the group and coworkers that were involved in the establishment of the protocols described here.

References

1. Cook GP, Tomlinson IM (1995) The human immunoglobulin VH repertoire. Immunol Today 16:237–242
2. Corbett SJ, Tomlinson IM, Sonnhammer EL, Buck D, Winter G (1997) Sequence of the human immunoglobulin diversity (D) segment locus: a systematic analysis provides no evidence for the use of DIR segments, inverted D segments, "minor" D segments or D-D recombination. J Mol Biol 270:587–597
3. Ravetch JV, Siebenlist U, Korsmeyer S, Waldmann T, Leder P (1981) Structure of the human immunoglobulin mu locus: characterization of embryonic and rearranged J and D genes. Cell 27:583–591
4. Hieter PA, Maizel JV Jr, Leder P (1982) Evolution of human immunoglobulin kappa J region genes. J Biol Chem 257:1516–1522
5. Kawasaki K, Minoshima S, Nakato E, Shibuya K, Shintani A, Schmeits JL, Wang J, Shimizu N (1997) One-megabase sequence analysis of the human immunoglobulin lambda gene locus. Genome Res 7:250–261
6. Schäble KF, Zachau HG (1993) The variable genes of the human immunoglobulin kappa locus. Biol Chem Hoppe Seyler 374: 1001–1022
7. Vasicek TJ, Leder P (1990) Structure and expression of the human immunoglobulin lambda genes. J Exp Med 172:609–620
8. Küppers R, Zhao M, Hansmann ML, Rajewsky K (1993) Tracing B cell development in human germinal centres by molecular analysis of single cells picked from histological sections. EMBO J 12:4955–4967
9. Rajewsky K (1996) Clonal selection and learning in the antibody system. Nature 381: 751–758
10. Kanzler H, Küppers R, Hansmann ML, Rajewsky K (1996) Hodgkin and Reed-Sternberg cells in Hodgkin's disease represent the outgrowth of a dominant tumor clone derived from (crippled) germinal center B cells. J Exp Med 184:1495–1505
11. Küppers R, Rajewsky K, Zhao M, Simons G, Laumann R, Fischer R, Hansmann ML (1994) Hodgkin disease: Hodgkin and Reed-Sternberg cells picked from histological sections show clonal immunoglobulin gene rearrangements and appear to be derived from B cells at various stages of development. Proc Natl Acad Sci U S A 91:10962–10966
12. Küppers R, Zhao M, Rajewsky K, Hansmann ML (1993) Detection of clonal B cell populations in paraffin-embedded tissues by polymerase chain reaction. Am J Pathol 143:230–239
13. McCarthy KP, Sloane JP, Wiedemann LM (1990) Rapid method for distinguishing clonal from polyclonal B cell populations in surgical biopsy specimens. J Clin Pathol 43:429–432
14. Trainor KJ, Brisco MJ, Wan JH, Neoh S, Grist S, Morley AA (1991) Gene rearrangement in B- and T-lymphoproliferative disease detected by the polymerase chain reaction. Blood 78:192–196
15. Goossens T, Klein U, Küppers R (1998) Frequent occurrence of deletions and duplications during somatic hypermutation: Implications for oncogene translocations and heavy chain disease. Proc Natl Acad Sci U S A 95:2463–2468
16. Klein U, Rajewsky K, Küppers R (1998) Human immunoglobulin (Ig)M+IgD+ peripheral blood B cells expressing the CD27 cell surface antigen carry somatically mutated variable region genes: CD27 as a general marker for somatically mutated (memory) B cells. J Exp Med 188:1679–1689
17. Goossens T, Bräuninger A, Klein U, Küppers R, Rajewsky K (2001) Receptor revision plays no major role in shaping the receptor repertoire of human memory B cells after the onset of somatic hypermutation. Eur J Immunol 31:3638–3648
18. Küppers R (2004) Molecular single-cell PCR analysis of rearranged immunoglobulin genes as a tool to determine the clonal composition of normal and malignant human B cells. Methods Mol Biol 271:225–238
19. Kurth J, Spieker T, Wustrow J, Strickler GJ, Hansmann LM, Rajewsky K, Kuppers R (2000) EBV-infected B cells in infectious mononucleosis: viral strategies for spreading in the B cell compartment and establishing latency. Immunity 13:485–495
20. Bräuninger A, Küppers R, Spieker T, Siebert R, Strickler JG, Schlegelberger B, Rajewsky K, Hansmann ML (1999) Molecular analysis of single B cells from T cell-rich B-cell lymphoma shows the derivation of the tumor cells from mutating germinal center B cells and exemplifies means by which immunoglobulin genes are modified in germinal center B cells. Blood 93:2679–2687
21. Braeuninger A, Küppers R, Strickler JG, Wacker HH, Rajewsky K, Hansmann ML (1997) Hodgkin and Reed-Sternberg cells in lymphocyte predominant Hodgkin disease represent clonal populations of germinal

center-derived tumor B cells. Proc Natl Acad Sci U S A 94:9337–9342

22. Müschen M, Küppers R, Spieker T, Bräuninger A, Rajewsky K, Hansmann ML (2001) Molecular single-cell analysis of Hodgkin- and Reed-Sternberg cells harboring unmutated immunoglobulin variable region genes. Lab Invest 81:289–295
23. Küppers R, Bräuninger A, Müschen M, Distler V, Hansmann ML, Rajewsky K (2001) Evidence that Hodgkin and Reed-Sternberg cells in Hodgkin disease do not represent cell fusions. Blood 97:818–821
24. Bräuninger A, Goossens T, Rajewsky K, Küppers R (2001) Regulation of immunoglobulin light chain gene rearrangements during early B cell development in the human. Eur J Immunol 31:3631–3637
25. Hieter PA, Korsmeyer SJ, Waldmann TA, Leder P (1981) Human immunoglobulin kappa light-chain genes are deleted or rearranged in lambda-producing B cells. Nature 290:368–372
26. Korsmeyer SJ, Hieter PA, Sharrow SO, Goldman CK, Leder P, Waldmann TA (1982) Normal human B cells display ordered light chain gene rearrangements and deletions. J Exp Med 156:975–985
27. Zhang L, Cui X, Schmitt K, Hubert R, Navidi W, Arnheim N (1992) Whole genome amplification from a single cell: implications for genetic analysis. Proc Natl Acad Sci U S A 89:5847–5851
28. Kanzler H, Küppers R, Helmes S, Wacker HH, Chott A, Hansmann ML, Rajewsky K (2000) Hodgkin and Reed-Sternberg-like cells in B-cell chronic lymphocytic leukemia represent the outgrowth of single germinal-center B-cell-derived clones: potential precursors of Hodgkin and Reed-Sternberg cells in Hodgkin's disease. Blood 95:1023–1031
29. Müschen M, Re D, Bräuninger A, Wolf J, Hansmann ML, Diehl V, Küppers R, Rajewsky K (2000) Somatic mutations of the CD95 gene in Hodgkin and Reed-Sternberg cells. Cancer Res 60:5640–5643
30. Brezinschek HP, Brezinschek RI, Lipsky PE (1995) Analysis of the heavy chain repertoire of human peripheral B cells using single-cell polymerase chain reaction. J Immunol 155: 190–202
31. Brezinschek HP, Foster SJ, Dörner T, Brezinschek RI, Lipsky PE (1998) Pairing of variable heavy and variable kappa chains in individual naive and memory B cells. J Immunol 160:4762–4767
32. Küppers R (1997) Ongoing somatic mutation in mantle cell lymphomas questioned. Br J Haematol 97:932–934

Chapter 4

PCR-Based Analysis of Rearranged Immunoglobulin or T-Cell Receptor Genes by GeneScan Analysis or Heteroduplex Analysis for Clonality Assessment in Lymphoma Diagnostics

Elke Boone, Brenda Verhaaf, and Anton W. Langerak

Abstract

The assessment of the presence of clonal lymphoproliferations via polymerase chain reaction (PCR)-based analysis of rearranged immunoglobulin (Ig) or T-cell receptor (TCR) genes is a valuable technique in the diagnosis of suspect lymphoproliferative disorders. Furthermore this technique is more and more used to evaluate dissemination of non-Hodgkin lymphoma and/or the presence of (minimal) residual disease. In this chapter we describe an integrated approach to assess clonality via analysis of Ig heavy chain (IGH), Ig kappa (IGK), TCR beta (TCRB), and TCR gamma (TCRG) gene rearrangements. The described PCR protocol is based on the standardized multiplex PCRs as developed by the European BIOMED-2 collaborative study (Concerted Action BMH4-CT98-3936). Furthermore it also includes the pre-analytical DNA isolation step from various tissues (formalin fixed paraffin-embedded tissue, fresh tissues, body fluids, peripheral blood and bone marrow), GeneScan analysis of labeled PCR products on a genetic analyzer, heteroduplex analysis of unlabeled PCR products, and post-analytical guidelines for the interpretation of the obtained "molecular morphology" patterns.

Key words: BIOMED-2, Immunoglobulin, T-cell receptor, Gene rearrangements, Clonality, Heteroduplex analysis, GeneScan analysis, DNA isolation, Polyacrylamide electrophoresis, Capillary electrophoresis

1. Introduction

Cancer cells have the unique feature that they all originate from a single transformed cell and are therefore clonally related. Mature lymphoid B- or T-cells have several unique DNA sequences coding for unique antigen-receptor molecules (1, 2). This feature is very useful for demonstrating the clonal expansion of B- or T-cells in

On behalf of the EuroClonality/BIOMED-2 Collaborative Study Group

Ralf Küppers (ed.), *Lymphoma: Methods and Protocols*, Methods in Molecular Biology, vol. 971,
DOI 10.1007/978-1-62703-269-8_4, © Springer Science+Business Media, LLC 2013

lymphoid malignancies. B-cells are characterized by unique rearranged DNA sequences in the IGH, IGK, and/or IGL loci, whereas in T-cells rearrangements of the TCRD, TCRG, TCRB, and/or TCRA loci can occur. The unique DNA sequences are formed during lymphocyte development by a stepwise rearrangement of DNA genes (3). In case of the IGH, TCRB, and TCRD loci variable (V), diversity (D), and joining (J) genes are joint to form a functional V(D)J exon (3). IGK, IGL, TCRA, and TCRG gene rearrangements are the result of joining of one specific V gene to a J gene (3). Based on sequence homology, the V and D genes, though each slightly different, can be grouped into families. The number of families is strongly dependent on the individual locus. For instance, the 44 functional V genes of the IGH locus cluster into seven families, whereas the 47 functional TCRBV genes can only be grouped into 23 families, of which 12 families contain just 1 member. Depending on the locus, a different number of V, D, and J genes are available for the recombination process. Therefore the so-called combinatorial repertoire of these genes can range from $\sim 5 \times 10^3$ to $\sim 3 \times 10^6$ possible rearrangements. Besides functional genes, also pseudogenes are found in the locus, which cannot be rearranged or do not result in a functional receptor upon rearrangement. During the joining process random insertion and/or deletion of nucleotides occurs, resulting in highly diverse (both in size and nucleotide sequence), unique junctional regions per cell. However, among the different Ig and TCR genes the level of junctional diversity is not identical. Particularly, IGK and IGL rearranged genes show a restricted junctional repertoire as compared to the other loci. Based on the combinatorial and junctional diversity the total—polyclonal—repertoire of Ig and TCR molecules in healthy individuals is estimated to be $>10^{12}$.

During early lymphoid differentiation the loci are rearranged in a sequential order. Pre-B-cells first rearrange the IGH locus, followed by the IGK locus (4). In case the rearranged IGK gene is not functional, the gene is deleted and IGL rearrangement occurs (4). This deletion of the IGK locus is mediated via a special rearrangement in the IGK locus involving the so-called kappa deleting element (Kde) that can rearrange to any of the Vκ genes or to an isolated recombination signal sequence (RSS) in the J-Cκ intron (4). Eventually, B-cells thus express either IgH/κ or IGH/λ receptors. In case of T-cells, either expression of TCRαβ or TCRγδ receptors is found on the cell surface. First the TCRD locus rearranges, followed by the TCRG locus (5). In case of nonfunctionality, TCRB gene rearrangement will occur followed by deletion of TCRD and rearrangement of the TCRA locus (5). In a normal mature B- or T-cell repertoire both monoallelic and biallelic gene rearrangements can be found. In the latter case one of the rearrangements resulted in a nonfunctional receptor gene, whereas lymphoid cells with two functional rearrangements will not fully differentiate into mature cells. Of note, malignant, lymphoid cells sometimes escape differentiation control checkpoints, resulting in unusual rearrangement patterns within one cell.

To discriminate between a polyclonal, reactive, lymphoid repertoire, and a monoclonal lymphoproliferation by polymerase chain reaction (PCR) analysis, the primers in the PCR reaction should be able to amplify all possible rearranged V(D)J exons and should flank the junctional region. This way, a polyclonal cell population will result in differently sized PCR products, corresponding to the presence of different V(D)J rearrangements which show a Gaussian distribution with respect to the amount of inserted or deleted nucleotides in the junctional region. In case of a monoclonal cell expansion only PCR products of the same size and sequence will be generated.

In the past, a large number of different PCR protocols were developed, all however with their sometimes limited analytical sensitivity to detect clonality in the suspect lymphoproliferation. Due to the known sequence diversity among the different V, D, and J genes some of the developed family and/or consensus primers showed an improper annealing, resulting in false-negative data. Furthermore, in case of clonal expansions of germinal center or post-germinal center B-cells, improper annealing might occur due to the presence of mutations that occur through the process of somatic hypermutation (such somatic mutations are mostly single-nucleotide mutations, but also deletions and insertions are seen) in the rearranged Ig genes, potentially also resulting in false-negativity.

Therefore, the BIOMED-2 Concerted Action BMH4 CT98-3936 was initiated to develop standardized PCR protocols which would maximally cover all possible rearrangements (including somatically mutated Ig rearrangements) for all the lymphoid gene targets. A total of 47 institutes from 7 European countries collaborated in this project and used their expertise on Ig and TCR rearrangement processes to develop a standardized protocol for each target gene (3), with the exception of TCRA due to its high level of complexity. A total of 97 new primers were designed, covering the majority of functional gene segments and representing 418 single PCR reactions. After initial evaluation of all separate 418 PCR tests, careful combination of the primers resulted in only 14 Ig/TCR multiplex PCR tubes: three for complete IGH (VH-JH) to circumvent possible mismatching due to somatic mutations in the V region, two for incomplete IGH (DH-JH), two for IGK (Vκ-Jκ and Kde rearrangements), one for IGL (Vλ-Jλ), two for complete TCRB (Vβ-Jβ), one for incomplete TCRB (Dβ-Jβ), two for TCRG (Vγ-Jγ), and one for all TCRD gene rearrangements (3).

A second major technical pitfall is the occurrence of false-positive results due to a lack of optimal discrimination between a polyclonal, oligoclonal, or monoclonal PCR product by the techniques used to visualize the obtained DNA molecules. If the PCR

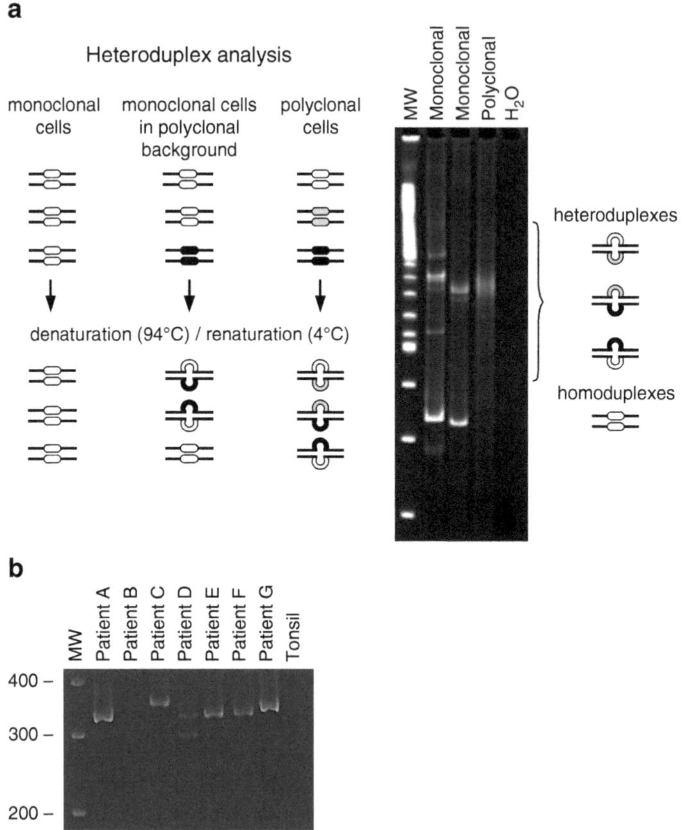

Fig. 1. Heteroduplex analysis. (**a**) Scheme of heteroduplex analysis technique, showing migration effect of homo- and heteroduplexes upon gel electrophoresis. (**b**) Example of polyacrylamide gel showing clonal homoduplexes in patients *A*, *B* (biallelic), *C*, and *E–G*. In patient *B* and in tonsil no clonal PCR products are observed.

products obtained are visualized by normal agarose or acrylamide gel electrophoresis, both polyclonal and monoclonal PCR products will result in a band of apparently the same size and intensity. Therefore either heteroduplex analysis or GeneScan fragment analysis is required to identify the differences in size and/or sequence of the obtained PCR products (3). Heteroduplex analysis uses double-stranded PCR products and takes advantage of the sequence diversity in the junctional regions to discriminate between a polyclonal or monoclonal population (3). After PCR, the double-stranded PCR products are denatured, followed by a very slow renaturation of the single strands at 4°C. In polyclonal reactions renaturation will also occur between PCR products with similar V and J genes, but different junctional sequences forming typical heteroduplexes. In case of clonal PCR products perfect double-stranded homoduplexes will be present after the denaturation/renaturation process. Heteroduplexes run on a polyacrylamide gel will result in a smear, whereas the double-stranded homoduplexes will migrate faster as a single band and at a predictable, expected size range (Fig. 1).

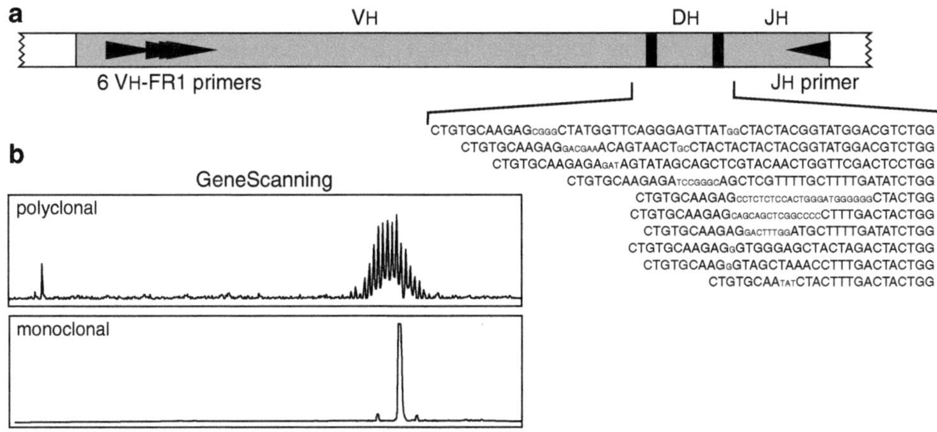

Fig. 2. GeneScan fragment analysis. (**a**) IGH rearrangements showing junctional region diversity, reflecting different levels of deletion and insertion (lower cases) of nucleotides. (**b**) Upon PCR amplification with VH and JH primers (see (**a**)), polyclonal PCR products give rise to a normal Gaussian size distribution in GeneScan fragment analysis, whereas in case of monoclonal PCR product one clear peak of identically sized products is seen.

GeneScan fragment analysis requires PCR products which are fluorescently labeled (3). Therefore the reverse primers in each BIOMED-2 multiplex PCR contain a 5′ coupled fluorochrome which enables detection of the PCR products with automated sequencing equipment. PCR products are denatured, followed by size separation of the single-stranded products in denaturing polyacrylamide gels or capillary polymer. The labeled PCR products are visualized by automated laser scanning of the migrating molecules and capturing of the fluorescent signals by a CCD camera. Polyclonal proliferations will generate a Gaussian distribution of multiple peaks, corresponding to the differently sized PCR products. Pure monoclonal proliferations will result in a single peak due to the presence of only one type of PCR product in the reaction (Fig. 2).

Heteroduplex analysis is more laborious and less sensitive than the automated GeneScan fragment analysis. However, as heteroduplex analysis uses both the size and sequence diversity of the junctional region, this technique is particularly useful for loci with restricted junctional size diversity. On the other hand, due to superior sensitivity of GeneScan fragment analysis and the exact identification of the size of the clonal PCR product, this method can be used for demonstrating the presence of a similar clonal proliferation in case of, e.g., suspected bone marrow invasion by lymphoma cells or for monitoring the presence of the clonal proliferation during followup of the patient.

Three large-scale BIOMED-2 studies demonstrated that by using these PCR protocols, the false-negative rate for detecting clonal lymphoproliferations in five categories of B-cell malignancies (mantle cell lymphoma, MCL; chronic lymphocytic leukemia, CLL; follicular lymphoma, FL; marginal zone lymphoma, MZL; diffuse large B-cell lymphoma, DLBCL) could be reduced to less than 1% by using a combined strategy of the different IGH and

IGK tubes (6). The PCR for IGL rearrangements did not show any additional information and can therefore be omitted in most screening strategies. Likewise, the combination of the 3 TCRB tubes and 2 TCRG tubes resulted in an overall clonality detection rate of 99% in T-cell prolymphocytic leukemia (T-PLL), T-cell large granular leukemia (T-LGL) peripheral T-cell lymphoma, unspecified (PTCL-U), angioimmunoblastic T-cell lymphoma (AITL), and a rate of 94% when also the category of anaplastic large cell lymphoma (ALCL) was included (7). The TCRD tube can be omitted, except in cases of suspected TCRγδ or immature T-cells. In the vast majority of reactive samples Ig or TCR gene rearrangements appeared to be polyclonal, whereas occasionally monoclonality was seen that prompted further diagnostic work-up (8).

Finally, it is extremely important to realize the limitation of this technique with regard to the assignment of a malignant status and/or a specific cell lineage status to the analyzed sample (9). Clinical benign clonal TCRαβ$^+$ proliferations can regularly be found in peripheral blood (PB) of elderly individuals. PB or lymph nodes of patients with acute Epstein–Barr virus (EBV) or cytomegalovirus (CMV) infections can show a restricted TCR repertoire, often resulting in an (oligo)clonal peak pattern, especially in immunodeficient patients. Benign monoclonal gammopathies often show a clonal IGH or IGK pattern. Furthermore, TCR gene rearrangements occur in 10–20% of B-cell malignancies and Ig gene rearrangements occur in 5–10% of T-cell malignancies, mostly however in a single locus (6, 7). Therefore, the identification of a clonal pattern in a single locus cannot be used as a marker for B/T-lineage assignment.

In this chapter we present a complete overview of the different technical steps for clonality analysis in suspected lymphoproliferations. Also the technical interpretation and reporting of the patterns obtained will be addressed briefly. However, a good knowledge of the immunobiology of Ig and TCR rearrangements is mandatory to be able to correctly interpret the different molecular patterns. Furthermore, it is of utmost importance that molecular clonality results are eventually interpreted in the context of available clinical, morphological, and immunophenotypic data.

2. Materials

2.1. DNA Extraction

1. Xylene (molecular biology quality grade).
2. 99% ethanol, absolute pro analyze (molecular biology quality grade).
3. Tissue grinder system, e.g., Medimachine (Becton Dickinson) including CellWASH (Becton Dickinson) and Medicon 35 μm filter (Becton Dickinson).
4. Proteinase K.

5. ATL buffer (QIAGEN, 19076).
6. Genomic DNA isolation kit, e.g., QIAamp DNA blood mini kit (QIAGEN) or GenElute mammalian genomic DNA miniprep kit (Sigma) (see Note 1); contains all necessary reagents for DNA extraction from PB, bone marrow (BM) or cell suspensions.
7. Centrifuge, e.g., Biofuge Pico (Heraeus) or equivalent equipment.

2.2. PCR Setup

1. PCR-grade H_2O.
2. dNTPs (10 mM): Combine equal amounts of 100 mM dATP, dTTP, dCTP, and dGTP (e.g., Amersham Biosciences) and dilute this dNTP mix 1/10 in PCR-grade H_2O.
3. Ampli*Taq* Gold 5 U/μl supplied with 25 mM $MgCl_2$ and 10× ABI buffer II or 10× GOLD (Applied Biosystems).
4. TE-buffer for primers (1 mM Tris–HCl pH 8.0, 0.01 mM EDTA): combine 10 μl of 1 M Tris–HCl pH 8.0, 1 ml of 0.1 mM EDTA, and 8.99 ml PCR-grade H_2O.
5. Highly purified, lyophilized primers to be dissolved in TE-buffer at 100 pmol/μl (100 μM).
6. Primer mix for B-cell targets (IGH, IGK): combine 100 μl of each primer (100 μM) and adjust, when necessary, to 800 μl with TE-buffer (Table 1); alternatively, IGH and IGK unlabeled (InVivoScribe, 2-101-0010, 2-101-0020, 2-101-0030, 2-102-0010, 2-102-0020) or labeled (InVivoScribe, 2-101-0011, 2-101-0101, 2-101-0031, 2-102-0011, 2-102-0021) master mixes.
7. Primer mix for TCRB targets: combine 100 μl of each primer (100 μM) and adjust, when necessary, to 3,200 μl with TE-buffer (Table 2); alternatively, TCRB unlabeled (InVivoScribe, 2-205-0010, 2-205-0020, 2-205-0030) or labeled (InVivoScribe, 2-205-0011, 2-205-0021, 2-205-0031) master mixes.
8. Primer mix for TCRG targets: Combine 100 μl of each primer (100 μM) (Table 3) or TCRG unlabeled (InVivoScribe, 2-207-0030, 2-207-0040) or labeled (InVivoScribe, 2-207-0031, 2-207-0041) master mixes.
9. Thermo cyclers, e.g., ABI 9700, 2700, 2720, Verity (Applied Biosystems), PTC-0200 (MJ Research), or equivalent equipment (see Note 2).

2.3. GeneScan Fragment Analysis

1. Genetic analyzer, e.g., 3,100/3,130×l/3,700 (Applied Biosystems) or equivalent systems (see Note 3).
2. Capillary array, 36-cm (Applied Biosystems) or equivalent capillary arrays.
3. MicroAmp Optical 96-well reaction plate (Applied Biosystems) or equivalent plates.

4. MicroAmp Caps (8 caps/strip, Applied Biosystems) or equivalent caps/seals.
5. Size standard, e.g., preferably GeneScan 500 ROX size standard (Applied Biosystems).
6. Hi-Di formamide (e.g., Applied Biosystems).
7. Performance Optimized Polymer 7 (POP7, Applied Biosystems) or equivalent polymers.
8. PCR-grade H_2O.
9. Centrifuge, e.g., Megafuge 1.0 with plate holder (Heraeus) or equivalent equipment.

2.4. Heteroduplex Analysis

1. Vertical polyacrylamide gel electrophoresis (PAGE) system (e.g., Hoefer SE600 Ruby).
2. 30% acrylamide/bisacrylamide (29:1).
3. 10× TBE buffer.
4. TEMED.
5. Ammonium persulfate (APS).
6. Ethidium bromide 10 mg/ml.

3. Methods

3.1. Samples and Quality Controls

Clonality analysis can be performed on DNA extracted from any human tissue. The specific extraction procedure for isolating DNA depends on the sample which is received. We here describe the column-based extraction procedure of QIAGEN for formalin-fixed paraffin-embedded (FFPE) tissue, fresh tissues, cell suspensions, PB and BM; equivalent isolation systems are also possible (see Note 1). Quality controls are crucial for correct evaluation of the clonality results (see Note 4). To monitor PCR reaction performance three control samples are crucial (see Note 4):

- No template control.
- Polyclonal control: DNA extracted from tonsil or mononuclear PB cells.
- Clonal control: DNA extracted from cell lines or clonal patient samples.

3.2. DNA Extraction

3.2.1. Extraction from Formalin-Fixed Paraffin-Embedded Tissue

1. Place two 10–20 μm sections of FFPE tissue into a 1.7 ml tube (see Note 5).
2. Add 1,200 μl xylene and vortex thoroughly for 10 s (work in a protective cabinet).
3. Centrifuge 5 min, 16,200 ×*g*.

4. Remove the supernatant carefully and dispense in specific waste containers.
5. Add 1,200 μl 99% ethanol.
6. Centrifuge 5 min, 16,200 × *g*.
7. Remove carefully the ethanol.
8. Add again 1,200 μl 99% ethanol and mix carefully by inverting the tube.
9. Centrifuge 5 min, 16,200 × *g*.
10. Remove all ethanol.
11. Dry the remaining tissue by leaving the tube open and incubate for 10–15 min at 37°C in a thermo block.
12. Resuspend the dried tissue in 180 μl ATL buffer + 20 μl proteinase K and vortex thoroughly for 10 s (see Note 6).
13. Incubate overnight at 54–56°C in a thermo block (see Note 6).
14. Continue the procedure as described in Subheading 3.2.3 from step 3 onwards.

3.2.2. Extraction from Fresh Tissues or Human Fluids

1. Make a cell suspension of the fresh tissue by using a Medimachine mixer (see Note 7).
2. If necessary, first cut the biopsy in small pieces with a sterile blade so that tissue pieces of maximum 10 mm^3 are obtained free of fat or necrotizing tissue.
3. Place the tissue together with 1 ml of CellWASH in the Medicon holder.
4. Place the Medicon holder in the Medimachine and start the machine.
5. Depending on the tissue type different run times should be used: 2 × 30 s for firm/solid biopsies and 1 × 45 s for skin biopsies. If necessary, an extra 1 ml of CellWASH can be added to the Medicon holder to obtain a cell suspension and determine the cell concentration.
6. Stop the machine and remove the Medicon holder from the machine.
7. Carefully remove the cover of the Medicon holder and use a small syringe to recuperate the obtained cell suspension.
8. Determine the cell concentration of the suspension.
9. If the suspension contains less than 625 cells/μl, concentrate the cell suspension by centrifugation for 1 min at 16,200 × *g* and dissolve the pellet in a smaller volume of CellWASH. Likewise, if the concentration is >40,000 cells/μl dilute the suspension (see Note 8).
10. Put 200 μl of the cell suspension in a 1.7 ml tube.

Table 1
Composition of multiplex primer mixes for rearranged Ig genes

Primer mix	Primer sequence (5′→3′)
IGH tube A VH FR1-JH	
VH1/7-FR1	GGCCTCAGTGAAGGTCTCCTGCAAG
VH2-FR1	GTCTGGTCCTACGCTGGTGAAACCC
VH3-FR1	CTGGGGGGTCCCTGAGACTCTCCTG
VH4-FR1	CTTCGGAGACCCTGTCCCTCACCTG
VH5-FR1	CGGGGAGTCTCTGAAGATCTCCTGT
VH6-FR1	TCGCAGACCCTCTCACTCACCTGTG
JH consensus (5′ FAM)	CTTACCTGAGGAGACGGTGACC
IGH tube B VH FR2-JH	
VH1-FR2	CTGGGTGCGACAGGCCCCTGGACAA
VH2-FR2	TGGATCCGTCAGCCCCCAGGGAAGG
VH3-FR2	GGTCCGCCAGGCTCCAGGGAA
VH4-FR2	TGGATCCGCCAGCCCCCAGGGAAGG
VH5-FR2	GGGTGCGCCAGATGCCCGGGAAAGG
VH6-FR2	TGGATCAGGCAGTCCCCATCGAGAG
VH7-FR2	TTGGGTGCGACAGGCCCCTGGACAA
JH consensus (5′ FAM)	CTTACCTGAGGAGACGGTGACC
IGH tube C VH FR3-JH	
VH1-FR3	TGGAGCTGAGCAGCCTGAGATCTGA
VH2-FR3	CAATGACCAACATGGACCCTGTGGA
VH3-FR3	TCTGCAAATGAACAGCCTGAGAGCC
VH4-FR3	GAGCTCTGTGACCGCCGCGGACACG
VH5-FR3	CAGCACCGCCTACCTGCAGTGGAGC
VH6-FR3	GTTCTCCCTGCAGCTGAACTCTGTG
VH7-FR3	CAGCACGGCATATCTGCAGATCAG
JH consensus (5′ FAM)	CTTACCTGAGGAGACGGTGACC
IGH tube D DH-JH	
DH1	GGCGGAATGTGTGCAGGC
DH2	GCACTGGGCTCAGAGTCCTCT
DH3	GTGGCCCTGGGAATATAAAA

(continued)

Table1 (continued)

Primer mix	Primer sequence (5′→3′)
DH4	AGATCCCCAGGACGCAGCA
DH5	CAGGGGGACACTGTGCATGT
DH6	TGACCCCAGCAAGGGAAGG
JH consensus (5′ FAM)	CTTACCTGAGGAGACGGTGACC
IGK tube A Vκ-Jκ	
Vκ1f/6	TCAAGGTTCAGCGGCAGTGGATCTG
Vκ2f	GGCCTCCATCTCCTGCAGGTCTAGTC
Vκ3f	CCCAGGCTCCTCATCTATGATGCATCC
Vκ4	CAACTGCAAGTCCAGCCAGAGTGTTTT
Vκ5	CCTGCAAAGCCAGCCAAGACATTGAT
Vκ7	GACCGATTTCACCCTCACAATTAATCC
Jκ1–4 (5′ FAM)	CTTACGTTTGATCTCCACCTTGGTCCC
Jκ5 (5′ FAM)	CTTACGTTTAATCTCCAGTCGTGTCCC
IGK tube B Vκ/intron-Kde	
Vκ1f/6	TCAAGGTTCAGCGGCAGTGGATCTG
Vκ2f	GGCCTCCATCTCCTGCAGGTCTAGTC
Vκ3f	CCCAGGCTCCTCATCTATGATGCATCC
Vκ4	CAACTGCAAGTCCAGCCAGAGTGTTTT
Vκ5	CCTGCAAAGCCAGCCAAGACATTGAT
Vκ7	GACCGATTTCACCCTCACAATTAATCC
INTR	CGTGGCACCGCGAGCTGTAGAC
Kde (5′ FAM)	CCTCAGAGGTCAGAGCAGGTTGTCCTA

11. Continue the procedure as described in Subheading 3.2.3 from step 3 onwards.

3.2.3. Extraction from Peripheral Blood or Bone Marrow

1. Place 200 μl fresh or frozen EDTA-PB or EDTA-BM in a 1.7 ml tube (see Note 9).
2. Add 20 μl Qiagen protease directly into the PB or BM (see Note 10).
3. Add 200 μl AL buffer.
4. Vortexing the tube for at least 15 s at high speed (see Note 10).
5. Incubate 10 min at 56°C in a thermo block.

Table 2
Composition of multiplex primer mixes for rearranged TCRB genes

Primer name	Primer sequence (5′→3′)	TCRB tube A Vβ-Jβ	TCRB tube B Vβ-Jβ	TCRB tube C Dβ-Jβ
Vβ2	AACTATGTTTTGGTATCGTCA	+	+	–
Vβ4	CACGATGTTCTGGTACCGTCAGCA	+	+	–
Vβ5/1	CAGTGTGTCCTGGTACCAACAG	+	+	–
Vβ6a/11	AACCCTTTATTGGTACCGACA	+	+	–
Vβ6b/25	ATCCCTTTTTTGGTACCAACAG	+	+	–
Vβ6c	AACCCTTTATTGGTATCAACAG	+	+	–
Vβ7	CGCTATGTATTGGTACAAGCA	+	+	–
Vβ8a	CTCCCGTTTTCTGGTACAGACAGAC	+	+	–
Vβ9	CGCTATGTATTGGTATAAACAG	+	+	–
Vβ10	TTATGTTTACTGGTATCGTAAGAAGC	+	+	–
Vβ11	CAAAATGTACTGGTATCAACAA	+	+	–
Vβ12a/3/13a/15	ATACATGTACTGGTATCGACAAGAC	+	+	–
Vβ13b	GGCCATGTACTGGTATAGACAAG	+	+	–
Vβ13c/12b/14	GTATATGTCCTGGTATCGACAAGA	+	+	–
Vβ16	TAACCTTTATTGGTATCGACGTGT	+	+	–
Vβ17	GGCCATGTACTGGTACCGACA	+	+	–
Vβ18	TCATGTTTACTGGTATCGGCAG	+	+	–
Vβ19	TTATGTTTATTGGTATCAACAGAATCA	+	+	–
Vβ20	CAACCTATACTGGTACCGACA	+	+	–

Vβ21	TACCCTTTACTGGTACCGGCAG	+	+	–
Vβ22	ATACTTCTATTGGTACAGACAAATCT	+	+	–
Vβ23/8b	CACGGTCTACTGGTACCAGCA	+	+	–
Vβ24	CGTCATGTACTGGTACCAGCA	+	+	–
Jβ1.1 (5′ HEX)	CTTACCTACAACTGTGAATCTGGTG	+	–	+
Jβ1.2 (5′ HEX)	CTTACCTACAACGGTTAACCTGGTC	+	–	+
Jβ1.3 (5′ HEX)	CTTACCTACAACAGTGAGCCAACTT	+	–	+
Jβ1.4 (5′ HEX)	CATACCCAAGACAGAGAGCTGGGTTC	+	–	+
Jβ1.5 (5′ HEX)	CTTACCTAGGATGGAGAGTCGAGTC	+	–	+
Jβ1.6 (5′ HEX)	CATACCTGTCACAGTGAGCCTG	+	–	+
Jβ2.1 (5′ FAM)	CCTTCTTACCTAGCACGGTGA	–	+	+
Jβ2.2 (5′ FAM)	CTTACCCAGTACGGTCAGCCT	+	–	+
Jβ2.3 (5′ FAM)	CCCGCTTACCGAGCACTGTCA	–	+	+
Jβ2.4 (5′ FAM)	CCAGCTTACCCAGCACTGAGA	–	+	+
Jβ2.5 (5′ FAM)	CGCGCACACCGAGCAC	–	+	+
Jβ2.6 (5′ FAM)	CTCGCCCAGCACGGTCAGCCT	+	–	+
Jβ2.7 (5′ FAM)	CTTACCTGTAACCGTGAGCCTG	+	–	+
Dβ1	GCCAAACAGCCTTACAAAGAC	–	–	+
Dβ2	TTTCCAAGCCCCACACAGTC	–	–	+

Table 3
Composition of multiplex primer mixes for rearranged TCRG genes

Primer mix	Primer sequence (5′→3′)
TCRG tube A Vγ-Jγ	
Vγfl	GGAAGGCCCCACAGCRTCTT
Vγ10	AGCATGGGTAAGACAAGCAA
Jγ1.1/2.1 (JP1/2) (5′ FAM)	TTACCAGGCGAAGTTACTATGAGC
Jγ1.3/2.3 (J1/2) (5′ HEX)	GTGTTGTTCCACTGCCAAAGAG
TCRG tube B Vγ-Jγ	
Vγ9	CGGCACTGTCAGAAAGGAATC
Vγ11	CTTCCACTTCCACTTTGAAA
Jγ1.1/2.1 (JP1/2) (5′ FAM)	TTACCAGGCGAAGTTACTATGAGC
Jγ1.3/2.3 (J1/2) (5′ HEX)	GTGTTGTTCCACTGCCAAAGAG

6. Shortly spin the tube before proceeding (see Note 10).
7. Add 200 μl 99% ethanol.
8. Vortex the tube for at least 15 s at maximum rpm.
9. Shortly spin the tube before proceeding.
10. Label de QIAamp columns and put them in the 2 ml collection tubes.
11. Carefully apply the lysed sample onto the column (see Note 10).
12. Centrifuge 1 min, 6,200 × *g*.
13. Discard the collection tube with the flow through and place the column in a new collection tube.
14. Put 500 μl AW1 buffer on the column.
15. Centrifuge 1 min, 6,200 × *g*.
16. Discard the collection tube with the flow through and place the column in a new collection tube.
17. Put 500 μl AW2 buffer on the column.
18. Centrifuge 3 min, 16,200 × *g*.
19. Discard the collection tube with the flow through and place the column in a new collection tube.
20. Centrifuge 1 min, 16,200 × *g* (see Note 11).
21. Label new 1.7 ml tubes.
22. Place the column in a labeled 1.7 ml tube.
23. Put 200 μl AE buffer on the middle of the column.

24. Incubate 5 min at room temperature (18–25°C).
25. Centrifuge 1 min, 6,200 × *g*.
26. Discard the column and keep the 1.7 ml tube containing the DNA solution.
27. Determine the DNA concentration and dilute if necessary to 50 ng/μl with AE buffer (see Note 12).

3.3. PCR Setup

1. PCR setup will be dependent on the available information accompanying the sample. In most cases either pathology results or flow cytometry results will indicate whether the sample is suspicious for the presence of abnormal B- or T-cell proliferations. In case the origin is not known, a combination of the targets mentioned below should be analyzed (Fig. 3) (see Note 13).
2. Analysis should be performed in duplicate for each target, to be able to distinguish between a true clonal proliferation of cells or an inconsistent clonal pattern due to the presence of few B- or T-lymphocytes or, in case of FFPE tissue, degraded DNA (see Note 14).
3. A standard protocol is used for all PCR reactions: 95°C 10 min, (94°C 30 s, 60°C 30 s, 72°C 1 min) × 35, /72°C 10 min, /4°C hold (see Note 2).
4. We here describe a PCR setup of primer mixes that consist of individually ordered primers. An alternative PCR setup would make use of commercially available master mixes that include primers, dNTPs, buffer, and $MgCl_2$.

3.3.1. Suspect B-Cell Proliferations

1. Make primer mix for the following targets: IGH tube A VH FR1-JH, IGH tube B VH FR2-JH, IGH tube C VH FR3-JH, IGH tube D DH-JH, IGK tube A Vκ-Jκ, and IGK tube B Vκ/intron-Kde.
2. For each target, combine 32 μl PCR grade H_2O, 5 μl 10× buffer GOLD, 3 μl 25 mM $MgCl_2$, 4 μl 10 mM dNTPs, and 0.8 μl primer mix (see Note 15). Vortex thoroughly.
3. Add 0.2 μl Ampli*Taq* Gold and mix by pipetting (do not vortex!).
4. Pipet 45 μl master mix in a PCR tube or PCR plate (see Note 16).
5. Add 5 μl of the 50 ng/μl DNA solution (see Note 12).
6. Start the PCR reaction.

3.3.2. Suspect T-Cell Proliferations

1. Make primer mix for the following targets: TCRB tube A Vβ-Jβ, TCRB tube B Vβ-Jβ, TCRB tube C Dβ-Jβ, TCRG tube A Vγ-Jγ, and TCRG tube B Vγ-Jγ.

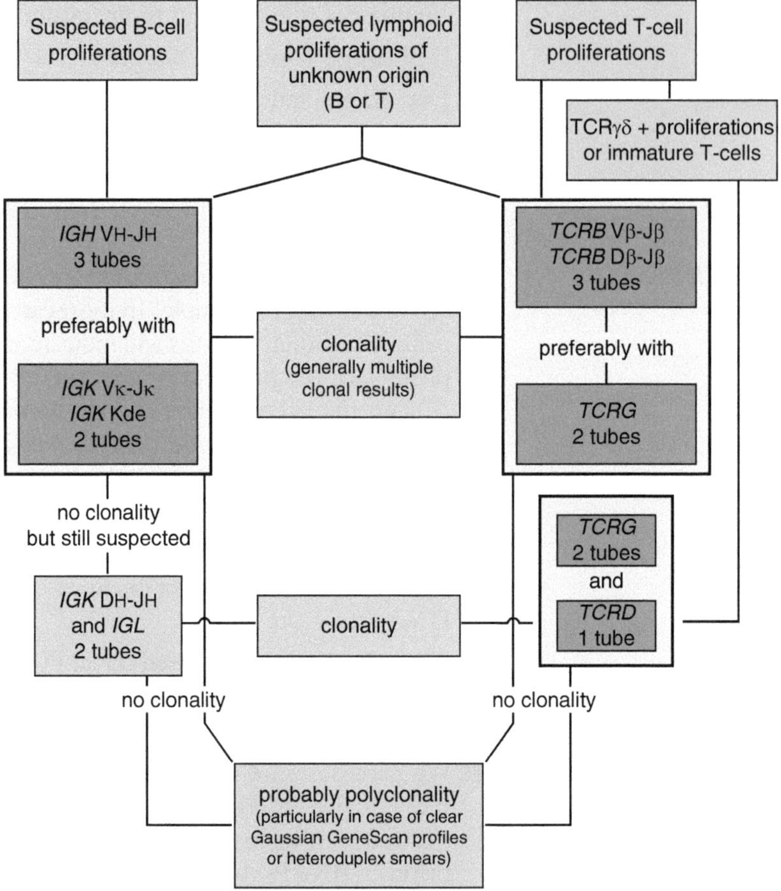

Fig. 3. Proposed Ig/TCR clonality testing algorithm.

2. For TCRB tube A and B, combine 26.4 μl PCR-grade H_2O, 5 μl 10× buffer II, 6 μl 25 mM $MgCl_2$, 4 μl 10 mM dNTPs, and 3.2 μl primer mix. Vortex thoroughly. Add 0.4 μl Ampli*Taq* Gold and mix by pipetting (do not vortex!).
3. For TCRB tube C, combine 29.6 μl PCR-grade H_2O, 5 μl 10× buffer II, 3 μl 25 mM $MgCl_2$, 4 μl 10 mM dNTPs, and 3.2 μl primer mix. Vortex thoroughly. Add 0.2 μl Ampli*Taq* Gold and mix by pipetting (do not vortex!).
4. For TCRG tube A and B, combine 32.4 μl PCR-grade H_2O, 5 μl 10× buffer II, 3 μl 25 mM $MgCl_2$, 4 μl 10 mM dNTPs, and 0.4 μl primer mix (see Note 15). Vortex thoroughly. Add 0.2 μl Ampli*Taq* Gold and mix by pipetting (do not vortex!).
5. Pipet 45 μl master mix in a PCR tube or PCR plate (see Note 16).
6. Add 5 μl of the 50 ng/μl DNA solution (see Note 12).
7. Start the PCR reaction.

3.4. GeneScan Fragment Analysis

We here describe a GeneScan analysis procedure using a genetic analyzer from Applied Biosystems/Life Technologies.

1. First denature the PCR product in Hi-Di formamide or H_2O together with (preferably) the GeneScan 500 ROX size standard. Add together 9.25 μl Hi-Di formamide (or H_2O), 0.25 μl ROX standard and 1 μl PCR product in the specific 96-well plate for the genetic analyzer (see Note 17).
2. Close the wells with the 8-cap strips or seals.
3. Shortly spin the plate in the centrifuge (e.g., Megafuge 1.0).
4. Put the plate in a thermo cycler and denature the samples for 2–5 min at 95°C, followed by a minimal incubation of 2 min at 4°C.
5. Mount the plate on the genetic analyzer and analyze on a 36 cm capillary filled with POP7 with the following instrument protocol parameters: Dye Set D; Oven_Temperature 60°C; Poly_Fill_Vol 6,500 steps; Current_Stability 5.0 μA; PreRun_Voltage 15.0 kV; Pre_Run_Time 180 s; Injection_ Voltage 1.2 kV; Injection_Time 23 s; Voltage_Number_Of_ Steps 20 nk; Voltage_Step_Interval 15 s; Data_Delay_Time 60 s; Run_Voltage 15.0 kV; Run_Time 800 s.
6. Analyze the obtained .fsa files with Genemapper software v3.7 or freely available PeakScanner software by using the default microsatellite analysis parameters with one exception: use the advanced peak detection algorithm and change the baseline window from 51 to 251 data points (see Note 18).
7. Resolve any indicated analysis problems by reviewing all samples for which the software indicated a bad sizing quality. Make sure to set the sample plot settings so that off-scale peaks are identified in the plots. Due to overloading of the CCD camera during data collection, off-scale peaks (broad, capped peaks) appear, which sometimes might give the impression of >1 peak. Also these off-scale peaks can give problems during size standard analysis as they result in extra peaks in the ROX color due to incomplete color compensation.
8. To display and review the analyzed data, it could be helpful to use a specific sample plot setting for Ig or TCR genes. The Ig targets contain only FAM-labeled PCR products, whereas in the TCR targets both FAM and HEX labels are present (with the exception of TCRB tube B). Therefore a sample plot for Ig should only demonstrate FAM and ROX labeled peaks, whereas for TCR also the HEX color should be displayed. For the *X*-axis always display the full base pair range (50–500 bp) to be able to identify peaks outside the expected size range. Off-scale peaks should be shown, and the *Y*-axis (corresponding to the

signal intensity) is set at 4,000 for Ig targets and 8,000 for TCR targets (see Note 19).

9. Verify whether the control samples give the expected pattern (see Note 20).
10. The molecular interpretation of the patterns obtained of each sample should be performed in several steps:
    (a) First, give a technical interpretation of the obtained pattern per tube. Identify possible nonspecific peaks and assign a technical description to the pattern (see Note 21). Some examples are given in Fig. 4.
    (b) Secondly, interpret the different tubes of the same gene together and evaluate whether the number of clonal peaks is in agreement with the presence of one clonal population (see Note 22).
    (c) Finally, the clonality result obtained needs to be integrated with the clinical, morphological, and immunophenotypic data to make the final diagnosis and classification (see Note 23).

3.5. Heteroduplex Analysis

1. Heat 20 μl PCR product for 5 min at 94°C in a thermal cycler.
2. Immediately cool the products to 4°C and incubate for 60 min to reanneal the PCR products.
3. Assemble the glass plates in the holder (use clean plates, appropriate clamps, and spacers).
4. Prepare a 6% non-denaturing polyacrylamide gel (40 ml) by mixing 8 ml 30% acrylamide/bisacrylamide (29:1) and 2 ml 10× TBE add MilliQ water to a final volume of 40 ml (see Note 24).
5. Add 240 μl 10% APS.
6. Add 60 μl TEMED and mix well.
7. Pour or pipette the mixture between the glass plates.
8. Place the comb.
9. Let the gel polymerize for approximately 1 h.
10. Remove the comb.
11. Assemble the plates in the vertical PAGE unit.
12. Fill the upper and lower compartments with 0.5 × TBE.
13. Add 20 μl (denatured + reannealed) PCR product to one the wells of a 96-well plate containing 4 μl non-denaturing bromophenol blue loading buffer (ratio 1:6).
14. Immediately load the samples on the 6% non-denaturing polyacrylamide gel (see Note 25).
15. Run at 40–50 V overnight, or alternatively 1 h at 110 V followed by several hours at 180 V (see Note 24).

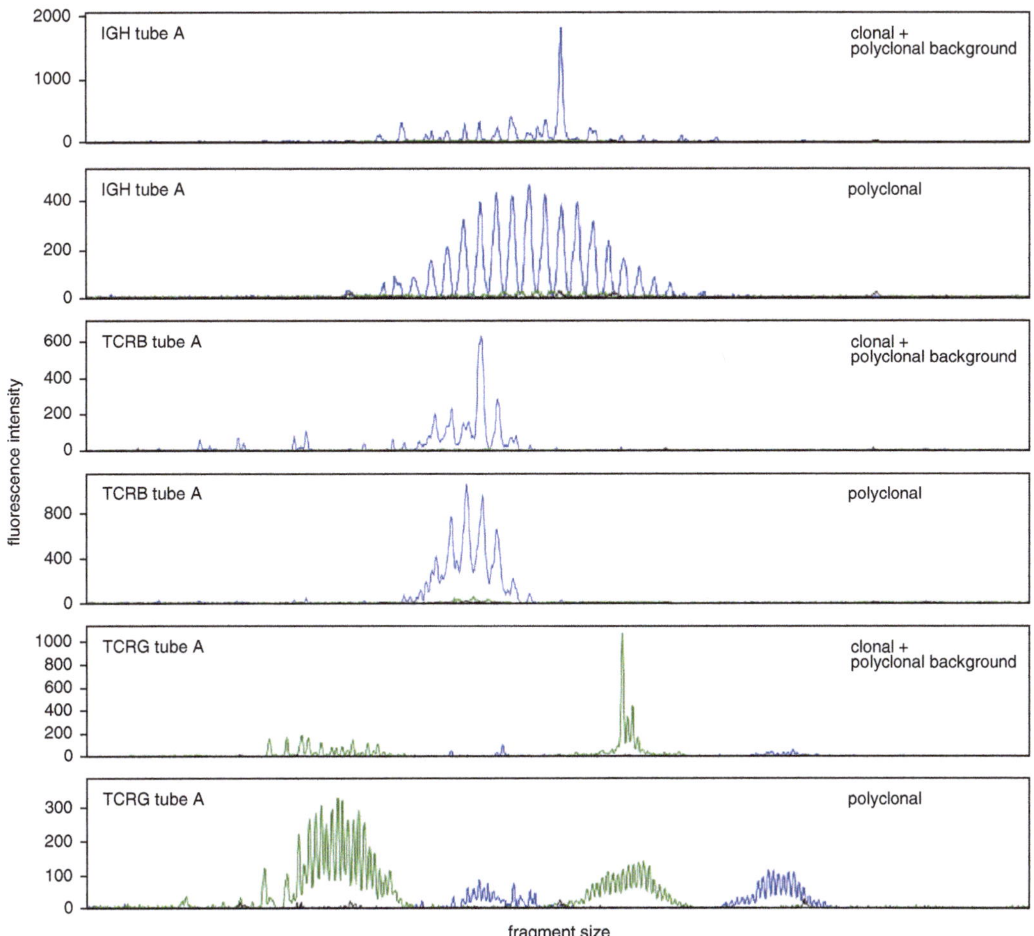

Fig. 4. Examples of Ig/TCR profiles in GeneScan fragment analysis. From top to bottom the following profiles are shown: (1) IGH FR1 (tube A) multiplex PCR with clear monoclonal product on a polyclonal background; (2) IGH FR1 (tube A) multiplex PCR showing normal Gaussian distribution of PCR products reflecting polyclonality; (3) TCRB tube A multiplex PCR with monoclonal product on a polyclonal background; (4) TCRB tube A multiplex PCR showing polyclonality with normal Gaussian distribution of PCR products; (5) TCRG tube A multiplex PCR with monoclonal product on a background of polyclonality; (6) TCRG tube A multiplex PCR showing typical polyclonal profile that actually consists of four Gaussian size distributions due to variation in V and J primer positions.

16. Disassemble the electrophoresis unit and carefully remove one of the glass plates.
17. Incubate the gel (loosely attached to the other glass plate) for 2–3 min in 250 ml water with 12.5 μl ethidium bromide 10 mg/ml (or use an alternative stain).
18. Discard the staining solution appropriately.
19. Remove access ethidium bromide by washing the gel for 2 min with water.
20. Visualize the DNA fragments by placing the gel on a transilluminator and take a picture.

4. Notes

1. Any DNA extraction technique (column-based or other) might be used, as long as the resulting DNA solution is free of proteins and RNA. Extraction methods which extract both DNA and RNA in parallel are not suitable. The RNA present in the solution negatively influences the PCR reaction resulting in an abnormal, disturbed polyclonal pattern.
2. The BIOMED-2 PCR protocols were validated by analysis of the same samples in different laboratories using different types of thermo cyclers. This demonstrates the robustness of the developed protocols with regard to the used thermo cycler and PCR cycling conditions. However, it was shown that with older PCR equipment better results are obtained with slightly different cycling conditions. Therefore, upon implementation of the technique it is worthwhile to evaluate different cycling conditions (3).
3. Different automated sequencing systems can be used for GeneScan analysis. It is important to know that the molecular peak profiles obtained are influenced by the polymer used to separate the single-stranded PCR products and by the settings of the software used to identify a peak.
4. A negative control sample (e.g., H_2O or buffer) is used to monitor possible contaminations throughout the entire procedure. Contamination during DNA extraction is sometimes underestimated; especially in samples with few lymphoid cells a contamination during extraction with a strong clonal sample can cause a false-positive result. To this end buffer (or H_2O) can be used as negative control. Additionally, an extra control reaction containing only master mix is useful to monitor contamination during PCR reaction setup. The polyclonal control sample is crucial as it allows to evaluate whether the PCR reaction was successful and to identify the polyclonal range of each tube. To obtain a complete, polyclonal TCR pattern, DNA from mononuclear PB cells of healthy individuals is most optimal, although DNA from tonsil can be used as well. For Ig polyclonal pattern tonsil DNA is preferred because of higher B-cell numbers. Although derived from alleged healthy individuals it might be, especially for TCR targets, that minor clones are seen. Therefore each new batch of polyclonal control needs to be validated up-front. The clonal control sample might be omitted during routine PCR setup as it mainly serves to monitor correct base pair calling by the GeneScan software.
5. The exact tissue fixation protocol will have an effect on the quality of the extracted DNA. It is important that neutral, buffered, formalin is used. Tissues fixated in Bouin or picric acid cannot be used for DNA analysis. Regarding the fixation

time no standardized protocols are available. However, prolonged fixation has been shown to induce too much cross-linking between DNA and other biomolecules resulting in a bad DNA quality for PCR analysis. Paraffin embedding should optimally be done with undiluted, fresh reagents and beeswax, or high-melt temperature paraffin should be avoided.

6. ATL buffer contains sodium dodecyl sulfate (SDS) which at lower temperatures might form a precipitate. If precipitation is seen, first incubate the buffer at 56°C to dissolve the SDS. If the amount of tissue is too large, 180 μl of ATL buffer will not be sufficient to lyse the tissue entirely. In this case a double volume of buffer and proteinase K can be used. All subsequent volumes of buffers should then also be doubled until application of the obtained lysate on the column. The total lysate can be put on the column by two centrifugation steps. An overnight incubation is sufficient, but longer incubations (e.g., over the weekend) will not hamper the DNA quality.

7. Often fresh tissue samples will also be tested by flow cytometry to determine the presence of clonal B- or T-cell populations. In this case, it is advised to use the same cell suspension for both molecular and flow analysis so that an optimal correlation between the results can be made. The cell suspensions should be made in neutral, buffered solutions such as PBS or CellWASH. All tissue grinding procedures will do (e.g., simple grinding of the tissue pieces in a Petri plate with the top of a sterile 15 ml Falcon tube might work as well). Flow cytometry fixatives should be avoided, or at least tested up-front, as the obtained DNA might no longer be usable for PCR analysis due to irreversible cross-linking.

8. The QIAGEN extraction method described here is able to extract high-quality DNA from cell suspensions ranging from 625 cells/μl until 40,000 cells/μl as determined by the BIOMED-2 control gene primer set (Subheading 10 in ref. (3)). Suspensions below 625 cells/μl will result in lower DNA concentration (<10 ng/μl), with accordingly weaker PCR amplification results. Therefore it is better to work with concentrated cell suspensions above 625 cells/μl. Concentrations >20,000 cells/μl should be diluted as they might result in a bad DNA quality due to the presence of proteins which are not completely removed. Also, the used columns get saturated and therefore no gain in DNA concentration is seen with higher cell numbers.

9. Preferably use EDTA or citrate treated PB/BM. Heparin stabilized PB or BM is less useful for PCR analysis as the heparin will remain in the DNA solution and will inhibit downstream PCR reactions.

10. To obtain high-quality DNA several points should be taken into consideration. In case the sample volume is <200 μl, add CellWASH before starting the procedure to maintain the

buffer proportions in the subsequent steps. It is important to avoid direct contact between QIAGEN protease and AL buffer as this will result in reduced protease activity. Therefore always put the protease into the PB or BM directly. Another crucial step for obtaining high-quality DNA is the 15 s vortexing of each sample with protease and AL buffer at high speed. To prevent contamination each vortexing step should be followed by a short spin before opening the tubes.

11. Be sure that all washing buffer is removed before proceeding as this buffer is an PCR inhibitor.
12. In the subsequent PCR reactions 5 μl of the DNA solution will be used, corresponding to 250 ng DNA input. The original BIOMED-2 PCR protocol was validated using 100 ng of DNA in a 50 μl PCR. However, a DNA input between 100 and 500 ng can be used without problem. Higher amounts of input DNA (or contaminating RNA) might lead to inhibition of the PCR reaction, but lower inputs might still be successful to determine the presence of a clonal cell population.
13. In a second series of reports the standardized PCR protocols described here were used for clonality analysis of various disease entities. From these studies a guideline was developed for implementing clonality analysis in routine lymphoma diagnostics (Fig. 3) (10). In suspect B-cell proliferations it is advised to use at least the three IGH VH-JH tubes combined with analysis of the two IGK (Vκ-Jκ and Kde rearrangements) tubes, to avoid delay in the diagnostic process. Subsequently, IGH DH1-6-JH PCR analysis (potentially combined with IGL analyses) can be performed in those cases where the previous PCR analysis failed to demonstrate clonality and no clear polyclonality was demonstrated, as this might indicate a false-negative result due to primer mismatching. In suspect T-cell proliferations it was shown that combining the three TCRB tubes with the two TCRG tubes is the preferred strategy. TCRD analysis should be avoided in routine practice because of its complexity of interpretation. Only in case of suspected TCRγδ$^+$ T-cell proliferations this tube provides complementary information. In this report we therefore restricted description to the protocols for IGH, IGK, TCRB, and TCRG PCR analysis.
14. When few B- or T-cells are present in a sample, or in case of degraded DNA, the pattern obtained might look clonal as only few amplifiable targets are present, resulting in a peak pattern instead of a polyclonal pattern. By performing the analysis in duplicate a true clonal pattern will give the same result in both PCR reactions, whereas a "false" clonal pattern will give differently sized peaks. Some labs use different amounts of input DNA to evaluate possible artifacts due to few amplifiable

DNA targets (FFPE tissues or PCR inhibitors). Depending on the kind of tissues analyzed (many or few lymphoid cells, FFPE or fresh) and the experience, one might decide not to perform a duplicate analysis up-front, but rather in second line for those samples in which no clear results are obtained. The latter strategy should however be used with great caution, as false clonality is an important factor, especially in the more sensitive GeneScan analysis, leading to a wrong interpretation of the results if not taken into account.

15. The master mixes can also be obtained commercially, ready for use, in both heteroduplex or GeneScan fragment analysis format via InVivoScribe Technologies (Carlsbad, CA, USA, www.invivoscribe.com). By this way the time-consuming and difficult validation and quality control of the BIOMED-2 multiplex PCR tubes can be avoided. Master mix can be made up-front if stored at <–70°C. Instead of 50 μl reactions, 25 μl reactions are also feasible, when using 100 ng of input DNA (3). Verify whether higher amounts of DNA also work before changing the protocol to 25 μl.
16. The protocols described make use of the hot-start Ampli*Taq* Gold protocols allowing PCR setup at room temperature.
17. The optimal amount of PCR product used for GeneScan analysis is dependent on the amount of input DNA, the strength of the PCR reaction, and/or the type of instrument used. It is therefore recommended to validate the GeneScan setup before performing routine analysis. If too much PCR product is present, the CCD camera recording the fluorescent signals might get over-loaded, leading to difficult pattern interpretation. In this case, a true clonal pattern might look as an oligoclonal pattern due to the detection of minor PCR amplification products which are the result of some normal lymphoid cells still present in the sample. This can be solved by adapting the fluorescence intensity scale (*Y*-axis) to the intensity of the dominant clonal peak(s).
18. The used analysis parameters of the Genemapper software influence the peak pattern and the identification of a fluorescent signal as a peak. The newer Genemapper software uses a slightly different algorithm as compared to the older software used during the BIOMED-2 project. It was observed that when using a baseline window of 51 data points an irregular polyclonal pattern, due to few lymphoid cells in the samples, might look like an oligoclonal, or even clonal pattern. Also the signal intensity of the peaks in the pattern is smaller as more background is subtracted. By using 251 or 501 data points, less background subtraction occurs and the polyclonal, Gaussian, pattern is clearer defined. Therefore this setting is better to identify true clonal populations.

19. The intensity of the peaks is dependent on a variety of parameters: lymphoid cell repertoire, quality of the obtained DNA, quality of the primers and/or PCR reaction, dilution factor of the PCR products, and genetic analyzer settings. Therefore, the proposed sample plot settings are only a guideline. It might be necessary to review a specific sample with different plot settings to be able to evaluate the obtained pattern correctly.
20. The expected polyclonal patterns and possible nonspecific products are listed in Table 4. Some of the mentioned nonspecific products will only be seen in samples with few lymphoid cells. Also, in IGH tube D DH-JH, in case of few B-cells, the irregular polyclonal pattern will look like a pattern with multiple peaks instead of a disturbed Gaussian distribution. No amplification products should be seen in the NTC sample and the peak size of the PCR products of the clonal control sample should be as formerly determined.
21. It is not possible within the frame of this chapter to describe all details and pitfalls of the molecular interpretation of GeneScan patterns. These items have however already been described in several special articles:
    (a) To make a correct interpretation of each PCR tube, an excellent knowledge of the analyzed gene rearrangements per tube is imperative. Details can be found in the different sections of the report of the BIOMED-2 Concerted Action BMH4 CT98-3936 (3).
    (b) Most clonal PCR products will have a number of nucleotides within the polyclonal size range. However, occasionally, specific clonal products can be found outside the expected range due to, e.g., insertions or deletions in the rearranged genes. Therefore, all isolated peaks, different from the known nonspecific peaks, should be taken into account when interpreting the obtained pattern (11).
    (c) When no specific PCR products are seen it is crucial that the DNA quality of the sample is evaluated by the BIOMED-2 control gene primer set (Subheading 10 in ref. (3)). Only in case of optimal DNA quality, the absence of specific PCR products can be attributed to an absence of cells with rearranged Ig or TCR genes.
    (d) To give a technical interpretation of the obtained pattern per tube, a standardized universal description of clonality data has been developed by the EuroClonality group (13). This scoring system is based on the fact that the type of the obtained pattern is determined by the underlying immunobiological features of the sample.
    (e) To avoid a biased interpretation it is advised to interpret the pattern first without any extra information on the lymphoid

Table 4
Size ranges, nonspecific bands, and method of analysis for Ig/TCR multiplex PCRs (updated in EuroClonality consortium in December 2009)

Tube	Polyclonal size range (bp)	Nonspecific products (bp)	Preferred method of analysis
IGH tube A VH FR1-JH	301–370	~85[a]	GS and HD both suitable
IGH tube B VH FR2-JH	242–288	~85[a] and 228[a]	idem
IGH tube C VH FR3-JH	96–167	211[a]	idem
IGH tube D DH-JH	110–290 (DH1/2/4/5/6-JH) 390–420 (DH3-JH)	345[b]	HD slightly preferred over GS (amplicon variation hampers GS)
IGK tube A Vκ-Jκ	120–160 (Vκ1f/6/Vκ7-Jκ) 190–210 (Vκ3f-Jκ) 260–300 (Vκ2f/Vκ4/Vκ5-Jκ)	None	HD slightly preferred over GS(less junctional size diversity + amplicon variation hampers GS)
IGK tube B Vκ/intron-Kde	210–250 (Vκ1f/6/Vκ7-Kde) 270–300 (Vκ3f/intron-Kde) 350–390 (Vκ2f/Vκ4/Vκ5-Kde)	404[a]	idem
TCRB tube A Vβ-Jβ	240–285	~213[a,c] and 273[a,c]	GS and HD both suitable
TCRB tube B Vβ-Jβ	240–285	~93, ~126 and 221[a,c]	idem
TCRB tube C Vβ-Jβ	170–210 (Dβ2) 285–325 (Dβ1)	~128 and 337[a,c]	idem
TCRG tube A Vγ-Jγ	145–175 (Vγ10-Jγ1.3/2.3) 175–195 (Vγ10-Jγ1.1/2.1) 195–230 (Vγfl-Jγ1.3/2.3) 230–255 (Vγfl-Jγ1.1/2.1)	None	GS and HD both suitable
TCRG tube B Vγ-Jγ	80–110 (Vγ11-Jγ1.3/2.3) 110–140 (Vγ11-Jγ1.1/2.1) 160–195 (Vγ9-Jγ1.3/2.3) 195–220 (Vγ9-Jγ1.1/2.1)	None	idem

[a]Intensity and/or presence of the nonspecific product will depend on the lymphoid repertoire present in the sample. Few lymphoid cells will result in stronger nonspecific products

[b]Nonspecific band as a result of cross-annealing of the DH2 primer to a sequence upstream of JH4. In GeneScan analysis this nonspecific band does not comigrate with D-J products.

[c]Presence of these nonspecific bands also depends on primer quality (primer batch dependent)

cell repertoire in the sample. However, immunophenotypic data from cytomorphology, pathology, and/or flow cytometry are very useful tools to further interpret the overall obtained molecular clonality results.

(f) In case of followup samples, the size of the clonal peaks is useful to determine whether the observed peaks correspond to the original malignant clone.

(g) In case of request for BM invasion by lymphoma, especially in elderly people, it is important to correlate possible clonal peaks in the bone marrow, or peripheral blood, with the original clonal population found in the lymph node. As people tend to live longer dual pathologies can be seen at the same time, e.g., CLL and lymphoma. Both will result in clonal patterns, however, with differently sized peaks.

22. The number of clonal peaks still corresponding to the presence of one single clonal cell expansion is gene target dependent (12).

23. In case clonality results are being reported outside a multidisciplinary setting a clear remark on the report should be made indicating that the presence of a clonal cell proliferation is not equivalent to the presence of a malignant process. Also, the specific background of the request for clonality analysis should be taken into account before making a final conclusion.

24. Percentage polyacrylamide gel and electrophoresis time are dependent on the expected size of the PCR products.

25. For convenient loading use special tips for loading vertical electrophoresis gels, e.g., Prot/Elec tips, Bio-Rad, 223-9915.

References

1. Tonegawa S (1983) Somatic generation of antibody diversity. Nature 302:575–581
2. Davis MM, Bjorkman PJ (1988) T-cell antigen receptor genes and T-cell recognition. Nature 334:395–402
3. Van Dongen JJM, Langerak AW, Brüggemann M, Evans PA, Hummel M, Lavender FL, Delabesse E, Davi F, Schuuring E, García-Sanz R, Van Krieken JHJM, Droese J, González D, Bastard C, White HE, Spaargaren M, González M, Parreira A, Smith JL, Morgan GJ, Kneba M, Macintyre EA (2003) Design and standardization of PCR primers and protocols for detection of clonal immunoglobulin and T-cell receptor gene recombinations in suspect lymphoproliferations: report of the BIOMED-2 Concerted Action BMH4-CT98-3936. Leukemia 17:2257–2317
4. Van Zelm MC, Van der Burg M, De Ridder D, Barendregt BH, De Haas EF, Reinders MJ, Lankester AC, Révész T, Staal FJT, Van Dongen JJM (2005) Ig gene rearrangement steps are initiated in early human precursor B cell subsets and correlate with specific transcription factor expression. J Immunol 175:5912–5922
5. Dik WA, Pike-Overzet K, Weerkamp F, De Ridder D, De Haas EF, Baert MRM, Van der Spek P, Koster EEL, Reinders MJT, Van Dongen JJM, Langerak AW, Staal FJT (2005) New insights on human T-cell development by quantitative T-cell receptor gene rearrangement studies and gene expression profiling. J Exp Med 201:1715–1723
6. Evans PA, Pott C, Groenen PJ, Salles G, Davi F, Berger F, Garcia JF, Van Krieken JHJM,

Pals ST, Kluin PM, Schuuring E, Spaargaren M, Boone E, González D, Martinez B, Villuendas R, Gameiro P, Diss TC, Mills K, Morgan GJ, Carter GI, Milner BJ, Pearson D, Hummel M, Jung W, Ott M, Canioni D, Beldjord K, Bastard C, Delfau-Larue MH, Van Dongen JJM, Molina TJ, Cabeçadas J (2007) Significantly improved PCR-based clonality testing in B-cell malignancies by use of multiple immunoglobulin gene targets. Report of the BIOMED-2 Concerted Action BHM4-CT98-3936. Leukemia 21:207–214

7. Brüggemann M, White H, Gaulard P, Garcia-Sanz R, Gameiro P, Oeschger S, Jasani B, Ott M, Delsol G, Orfao A, Tiemann M, Herbst H, Langerak AW, Spaargaren M, Moreau E, Groenen PJTA, Sambade C, Foroni L, Carter GI, Hummel M, Bastard C, Davi F, Delfau MH, Kneba M, Van Dongen JJM, Macintyre EA, Molina TJ (2007) Powerful strategy for PCR-based clonality assessment in T-cell malignancies. Report of the BIOMED-2 Concerted Action BHM4-CT98-3936. Leukemia 21:215–221
8. Langerak AW, Molina TJ, Lavender FL, Pearson D, Flohr T, Sambade C, Schuuring E, Al ST, Van Dongen JJM, Van Krieken JHJM (2007) PCR-based clonality testing in tissue samples with reactive lymphoproliferations: usefulness and pitfalls. A study from the BIOMED-2 Concerted Action BMH4-CT98-3936. Leukemia 21:222–229
9. Diss TC, Molina TJ, Cabeçadas J, Langerak AW (2012) Molecular diagnostics in lymphoma: why, when and how to apply. Diagn Histopathol 18:53–63
10. Van Krieken JHJM, Langerak AW, Macintyre EA, Kneba M, Smith JL, Garcia SR, Morgan GJ, Parreira A, Molina T, Cabeçadas J, Gaulard P, Jasani B, Garcia JF, Ott M, Hannsmann ML, Berger F, Hummel M, Davi F, Brüggemann M, Lavender FL, Schuuring EMD, Evans PAS, White H, Salles G, Groenen PJTA, Gameiro P, Pott C, Van Dongen JJM (2007) Improved reliability of lymphoma diagnostics via PCR-based clonality testing. Report of the BIOMED-2 Concerted Action BHM4-CT98-3936. Leukemia 21:201–206
11. Rothberg PG, Langerak AW, Verhaaf B, Van Dongen JJM, Burack WR, Johnson MD, Slate D, Laughlin TS, Payne K, Figueiredo L, Bandoh BN, Yan Q, Bacon CM, Wright P, Bench A, Du MQ, Liu H (2012) Clonal antigen receptor gene PCR products outside the expected size range. J Hematopathol 5:57–67
12. Langerak AW, Van Dongen JJM (2012) Multiple clonal Ig/TCR products: implications for interpretation of clonality findings. J Hematopathol 5:35–43
13. Langerak AW, Groenen PJTA, Brüggemann M, Beldjord K, Bellan C, Bonello L, Boone E, Carter I, Catherwood M, Davi F, Delfau-Larue MH, Diss T, Evans PAS, Gameiro P, Garcia Sanz R, Gonzalez D, Grand D, Håkansson A, Hummel M, Liu H, Lombardia L, Macintyre EA, Milner B, Montes-Moreno S, Schuuring E, Spaargaren M, Hodges E, Van Dongen JJM (2012) EuroClonality / BIOMED-2 guidelines for interpretation and reporting of Ig/TCR clonality testing in suspected lymphoproliferations. Leukemia 26:2159–2172

Chapter 5

Expression Cloning of Human B Cell Immunoglobulins

Hedda Wardemann and Juliane Kofer

Abstract

The majority of lymphomas originate from B cells at the germinal center stage or beyond. Preferential selection of B cell clones by a limited set of antigens has been suggested to drive lymphoma development. However, little is known about the specificity of the antibodies expressed by lymphoma cells, and the role of antibody-specificity in lymphomagenesis remains elusive. Here, we describe a strategy to characterize the antibody reactivity of human B cells. The approach allows the unbiased characterization of the human antibody repertoire on a single cell level through the generation of recombinant monoclonal antibodies from single primary human B cells of defined origin. This protocol offers a detailed description of the method starting from the flow cytometric isolation of single human B cells, to the RT-PCR-based amplification of the expressed *Igh*, *Igκ*, and *Igλ* chain genes, and Ig gene expression vector cloning for the in vitro production of monoclonal antibodies. The strategy may be used to obtain information on the clonal evolution of B cell lymphomas by single cell Ig gene sequencing and on the antibody reactivity of human lymphoma B cells.

Key words: Antibody, Expression cloning, Immunoglobulin, Repertoire, Single cell RT-PCR

1. Introduction

The virtually unlimited diversity of the antibody repertoire efficiently protects from infection. Antibody diversity is primarily established early in lymphocyte development through combinatorial joining of variable (V), diversity (D), and joining (J) gene segments (1). A drawback of the random Ig gene recombination is the formation of self-reactive immunoglobulins (2). In healthy individuals, a series of checkpoints purge self-reactive B cells from the repertoire, both centrally in the bone marrow during B cell development and in peripheral lymphoid tissues (3–6). Additional diversity in the B cell repertoire is generated by random somatic Ig gene mutations in antigen-activated B cells primarily during T cell-dependent immune responses in germinal centers (GC). Somatically mutated

Ralf Küppers (ed.), *Lymphoma: Methods and Protocols*, Methods in Molecular Biology, vol. 971,
DOI 10.1007/978-1-62703-269-8_5, © Springer Science+Business Media, LLC 2013

GC B cells with high affinity for their cognate antigen are positively selected and differentiate into humoral effector cells, i.e., memory B cells and plasma cells (7–9). The random somatic mutation process may again lead to the development of self-reactive antibodies, but how newly arising self-reactive GC B cells are regulated to avoid autoimmunity is unknown.

The majority of B cell-associated lymphomas arise from B cells at the GC stage or beyond, indicating that malignant transformation often occurs in GC B cells (10, 11). The predominance of GC-derived lymphomas may be due to the special microenvironment provided in the GC, where Ig gene modification processes, including DNA breaks due to isotype class switch recombination, and intense cellular proliferation occur (12). B cell receptor (BCR) expression and antigen recognition are essential in B cell lymphoma development and pathogenesis (10, 12, 13). Stereotyped antigen binding sites and biased V gene usage suggest a limited set of antigens as driving force for B cell activation (14–21). However, Ig gene sequence analysis alone does not permit predictions on antibody reactivity. Thus, it remains unclear whether antibody-specificity plays a role in lymphoma development.

Human monoclonal antibodies can be produced by different methods such as immortalization of B cells with Epstein–Barr virus (22, 23) or the production of B-cell hybridomas (24). However, immortalization and fusion efficiencies are often low, largely depend on the maturation status of the B cell, and may require B cell stimulation. Phage display technology on the other hand relies on the random pairing of human Ig heavy and light chains in vitro and therefore the antibodies that are produced may not mimic the natural situation (25).

Here, we describe an unbiased and highly efficient strategy to characterize the Ig genes and antibody reactivity of primary human B cells on a single cell level (4, 26). This protocol offers a detailed guide comprising the single cell sort of B cells by flow cytometry, the subsequent unbiased reverse transcription polymerase chain reaction (RT-PCR) amplification of human *Igh*, *Igκ*, and *Igλ* chain genes and Ig gene expression vector cloning for in vitro antibody production (Fig. 1). The sequence data obtained provide detailed information on Ig gene rearrangement, V gene somatic mutations, composition of the antigen-binding region CDR3, and clonal relatedness of individual cells. Importantly, the recombinant monoclonal antibodies can be used to determine their antigen-reactivity

Fig. 1. (continued) or Igλ constant region. Nested second PCRs are performed with forward primer mixes specific for FWR1 and respective nested reverse primers specific for the IgH, Igκ, or Igλ constant regions. Second PCR products are sequenced. For expression vector cloning, sequence information is used to design a variable region specific PCR with V and J gene element specific primers containing restriction sites. The first PCR product is used as template. Amplicons obtained by gene-specific PCR are cloned into the respective Igγ1, Igκ, or Igλ expression vectors. Igγ1 and corresponding Igκ or Igλ vectors are co-transfected into HEK 293 T cells to produce monoclonal antibodies in vitro. Recombinant antibodies are purified from cell culture supernatants to be tested in antibody reactivity assays.

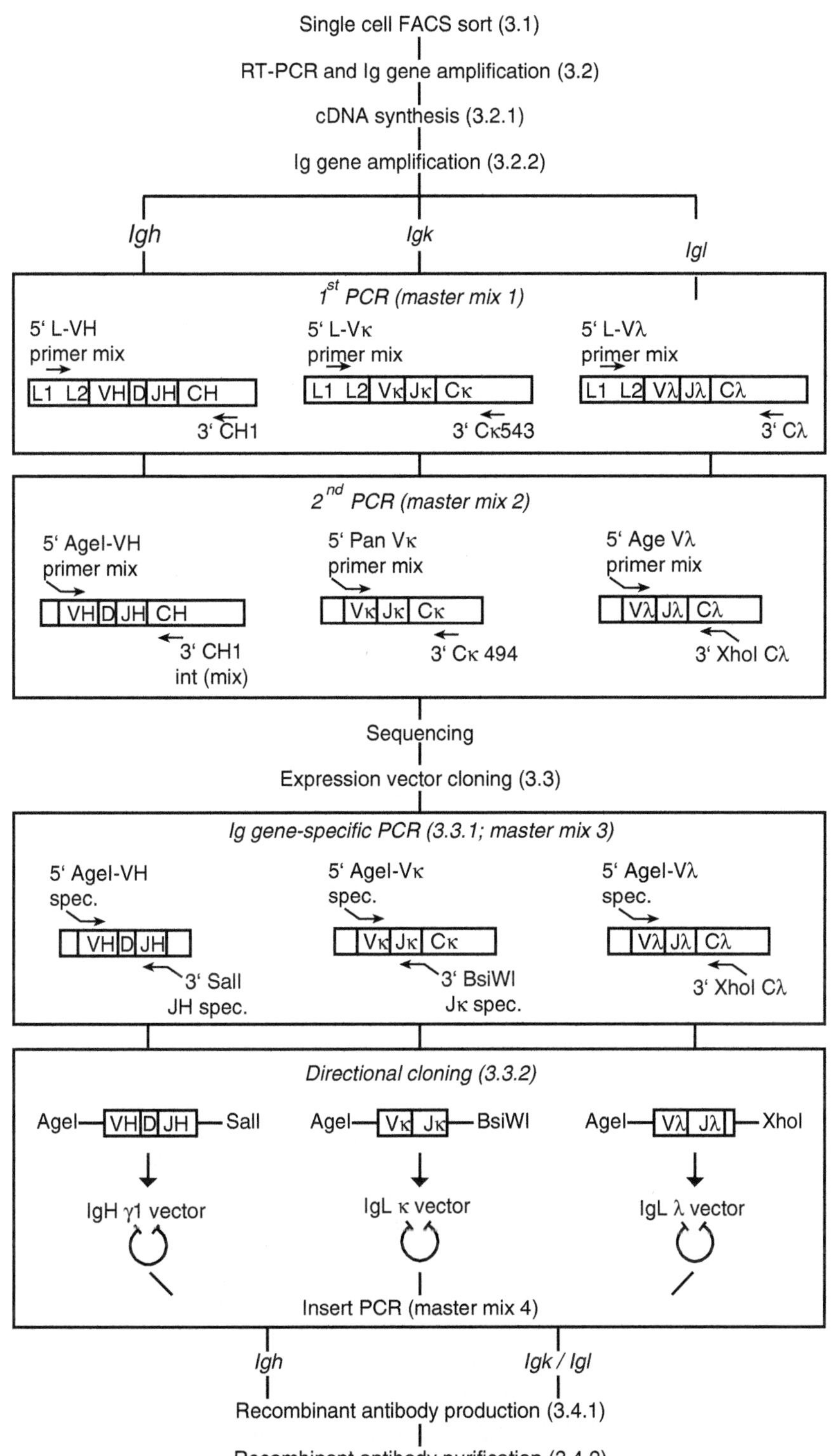

Fig. 1. Strategy to amplify human *IgH*, *Igκ*, and *Igλ* chain transcripts for the production of human monoclonal antibodies. Single human B cells are isolated by fluorescence activated cell sorting (FACS) and cDNA is synthesized by amplification with random hexamers. *IgH*, *Igκ*, and *Igλ* chain genes are amplified by nested RT-PCR from single cell cDNA. First PCRs are performed with forward primer mixes specific for the leader region and reverse primers specific for the respective IgH, Igκ,

and epitope specificity (4, 16, 17, 26–29). A first characterization of recombinant antibodies from B cell chronic lymphocytic leukemia (CLL) cells using this approach revealed the basis of the clinically observed difference between cases of *IGVH*-mutated CLL and *IGVH*-unmutated CLL cases, and demonstrated that antigen binding and BCR stimulation may contribute to clonal selection and expansion of neoplastic B cell clones (16, 17).

2. Materials

2.1. Isolation of Single Human B Cells by Flow Cytometry

1. Mononuclear cells (see Note 1).
2. Staining solution: Cold PBS (without Ca^{2+} or Mg^{2+}) with 2% heat-inactivated fetal calf serum (FCS) and anti-human antibodies or antigen at the final concentration (see Notes 2–4).
3. PBS/FCS: Cold PBS (without Ca^{2+} or Mg^{2+}) with 2% heat-inactivated FCS, sterile; store at 4°C.
4. 7-Aminoactinomycin D (7-AAD) solution: Prepare a 2 μg/ml solution; store protected from light at 4°C (see Note 5).
5. Lysis solution: 0.5× PBS (without Ca^{2+} or Mg^{2+}, molecular biology grade), 10 mM dithiothreitol (DTT), 8 U RNAsin® (Promega, Mannheim, Germany). Keep on ice until transferred to a 96-well PCR plate (see Note 6).
6. 96-well PCR plate, skirted (see Note 7).
7. Adhesive aluminum foil (see Note 8).

2.2. RT-PCR and Ig Gene Amplification

1. PCR cycler compatible with 96-well PCR plates.
2. Nuclease-free PCR water.
3. Random hexamer primer (pd(N)6; Roche Applied Science, Mannheim, Germany): prepare a 300 ng/μl solution; store in small aliquots at −20°C.
4. Igepal CA-630, molecular biology grade: prepare a 10% solution; store in aliquots at −20°C.
5. RNAsin® (40 U/μl; Promega).
6. SuperScript® III Reverse Transcriptase (200 U/μl; Invitrogen, Karlsruhe, Germany) with 5× RT buffer.
7. DTT, molecular biology grade: prepare a 100 mM solution; store in small aliquots at −20°C.
8. dNTP solution, prepare a mix with 25 mM of each nucleotide; store in small aliquots at −20°C.
9. HotStarTaq® (5 U/μl; Qiagen, Hilden, Germany) with 10× PCR buffer.

10. 5′ primer or primer mix: prepare a working dilution of 50 μM (50 pmol/μl) for each primer. For primer mixes, combine equal volumes of 50 μM working dilutions of each primer (Table 1); store in small aliquots at −20°C.

11. 3′ primer: Prepare a working dilution of 50 μM for each primer. For primer mixes, combine equal volumes of 50 μM working dilutions of each primer (Table 1); store in small aliquots at −20°C.

Table 1
Primer sequences

PCR step	Primer name	5′–3′ Sequence
Igh first PCR		
	5′ L-VH 1	ACAGGTGCCCACTCCCAGGTGCAG
	5′ L-VH 3	AAGGTGTCCAGTGTGARGTGCAG
	5′ L-VH 4/6	CCCAGATGGGTCCTGTCCCAGGTGCAG
	5′ L-VH 5	CAAGGAGTCTGTTCCGAGGTGCAG
	3′ Cμ CH outer	GGAAGGAAGTCCTGTGCGAGGC
	3′ Cγ CH1	GGAAGGTGTGCACGCCGCTGGTC
	3′ Cα CH1	TGGGAAGTTTCTGGCGGTCACG
Igh second PCR		
	5′ *Age*I VH1	CTGCAACCGGTGTACATTCCCAGGTGCAGCTGGTGCAG
	5′ *Age*I VH1/5	CTGCAACCGGTGTACATTCCGAGGTGCAGCTGGTGCAG
	5′ *Age*I VH3	CTGCAACCGGTGTACATTCTGAGGTGCAGCTGGTGGAG
	5′ *Age*I VH3–23	CTGCAACCGGTGTACATTCTGAGGTGCAGCTGTTGGAG
	5′ *Age*I VH4	CTGCAACCGGTGTACATTCCCAGGTGCAGCTGCAGGAG
	5′ *Age*I VH 4–34	CTGCAACCGGTGTACATTCCCAGGTGCAGCTACAGCAGTG
	3′ Cμ CH1	GGGAATTCTCACAGGAGACGA
	3′ IgG (internal)	GTTCGGGGAAGTAGTCCTTGAC
	3′ Cα CH1–2	GTCCGCTTTCGCTCCAGGTCACACT
Igh-specific PCR		
	5′ *Age*I VH1/4/6 specific	CTGCAACCGGTGTACATTCC[FWR1] (see Note 9)
	5′ *Age*I VH3/7 specific	CTGCAACCGGTGTACATTCT[FWR1] (see Note 9)
	3′ *Sal*I JH 1/2/4/5	TGCGAAGTCGACGCTGAGGAGACGGTGACCAG
	3′ *Sal*I JH 3	TGCGAAGTCGACGCTGAAGAGACGGTGACCATTG
	3′ *Sal*I JH 6	TGCGAAGTCGACGCTGAGGAGACGGTGACCGTG
Igk first PCR		
	5′ L Vκ 1/2	ATGAGGSTCCCYGCTCAGCTGCTGG
	5′ L Vκ 3	CTCTTCCTCCTGCTACTCTGGCTCCCAG
	5′ L Vκ 4	ATTTCTCTGTTGCTCTGGATCTCTG
	3′ Cκ 543	GTTTCTCGTAGTCTGCTTTGCTCA
Igk second PCR		
	5′ Pan Vκ	ATGACCCAGWCTCCABYCWCCCTG
	3′ Cκ 494	GTGCTGTCCTTGCTGTCCTGCT

(continued)

Table 1 (continued)

PCR step	Primer name	5′–3′ Sequence
Igk-specific PCR		
	5′ *Age*I Vκ specific	CTGCAACCGGTGTACAT[FWR1] (see Note 9)
	3′ *Bsi*WI Jκ 1/4	GCCACCGTACGTTTGATYTCCACCTTGGTC
	3′ *Bsi*WI Jκ 2	GCCACCGTACGTTTGATCTCCAGCTTGGTC
	3′ *Bsi*WI Jκ 3	GCCACCGTACGTTTGATATCCACTTTGGTC
	3′ *Bsi*WI Jκ 5	GCCACCGTACGTTTAATCTCCAGTCGTGTC
Igl first PCR		
	5′ L Vλ 1	GGTCCTGGGCCCAGTCTGTGCTG
	5′ L Vλ 2	GGTCCTGGGCCCAGTCTGCCCTG
	5′ L Vλ 3	GCTCTGTGACCTCCTATGAGCTG
	5′ L Vλ 4/5	GGTCTCTCTCSCAGCYTGTGCTG
	5′ L Vλ 6	GTTCTTGGGCCAATTTTATGCTG
	5′ L Vλ 7	GGTCCAATTCYCAGGCTGTGGTG
	5′ L Vλ 8	GAGTGGATTCTCAGACTGTGGTG
	3′ Cλ	CACCAGTGTGGCCTTGTTGGCTTG
Igl second PCR		
	5′ *Age*I Vλ 1	CTGCTACCGGTTCCTGGGCCCAGTCTGTGCTGACKCAG
	5′ *Age*I Vλ 2	CTGCTACCGGTTCCTGGGCCCAGTCTGCCCTGACTCAG
	5′ *Age*I Vλ 3	CTGCTACCGGTTCTGTGACCTCCTATGAGCTGACWCAG
	5′ *Age*I Vλ 4/5	CTGCTACCGGTTCTCTCTCSCAGCYTGTGCTGACTCA
	5′ *Age*I Vλ 6	CTGCTACCGGTTCTTGGGCCAATTTTATGCTGACTCAG
	5′ *Age*I Vλ 7/8	CTGCTACCGGTTCCAATTCYCAGRCTGTGGTGACYCAG
	3′ *Xho*I Cλ	CTCCTCACTCGAGGGYGGGAACAGAGTG
Insert check		
	5′ Ab sense	GCTTCGTTAGAACGCGGCTAC
	3′ IgG (internal)	GTTCGGGGAAGTAGTCCTTGAC
	3′ Cκ 494	GTGCTGTCCTTGCTGTCCTGCT
	3′ Cλ	CACCAGTGTGGCCTTGTTGGCTTG

Restriction sites are underlined

12. Adhesive seal, PCR suitable.
13. Agarose.
14. TAE buffer: 40 mM Tris–acetate, 1 mM ethylenediamine tetra-acetic acid (pH 8.0).
15. 10 mg/ml Ethidium bromide solution: dilute 1:2,000 in TAE buffer.
16. Loading dye for DNA gels: 60% w/v sucrose, 1 mM Cresol Red.
17. 100 bp DNA ladder.

2.3. Expression Vector Cloning

1. PCR cycler compatible with 96-well PCR plates.
2. 5′ primer: Prepare a working dilution of 5 μM for each primer (Table 1, see Note 9).
3. 3′ primer: Prepare a working dilution of 5 μM for each primer (Table 1, see Note 9).
4. Adhesive seal, PCR suitable.
5. Nuclease-free PCR water.
6. HotStarTaq® (5 U/μl; Qiagen) with 10× PCR buffer.
7. dNTP solution, prepare a mix with 25 mM of each nucleotide; store in small aliquots at −20°C.
8. PCR purification kit, e.g., Qiaquick® 96 PCR Purification Kit (Qiagen) or NucleoSpin® 96 Extract II (Macherey and Nagel).
9. *Age*I-HF (20 U/μl; New England Biolabs (NEB), Frankfurt/Main, Germany) with NEBuffer 4 (10×) and BSA (100×).
10. *Sal*I-HF (20 U/μl; NEB) with NEBuffer 4 (10×).
11. *Bsi*WI (10 U/μl; NEB) with NEBuffer 3 (10×).
12. *Xho*I (20 U/μl; NEB) with NEBuffer 4 (10×) and BSA (100×).
13. NEBuffer 1 (NEB).
14. Linearized human IgH γ1 expression vector with cloning sites for *Age*I and *Sal*I, and ampicillin resistance gene (50 ng/μl; see Note 10).
15. Linearized human IgL κ1 expression vector with cloning sites for *Age*I and *Bsi*WI, and ampicillin resistance gene (50 ng/μl; see Note 10).
16. Linearized human IgL λ2 expression vector with cloning sites for *Age*I and *Xho*I, and ampicillin resistance gene (50 ng/μl; see Note 10).
17. T4 DNA ligase (400 U/μl) with 10× T4 DNA ligase reaction buffer.
18. Chemical competent *Escherichia coli* (*E. coli*) DH10B cells (Invitrogen).
19. LB medium: 10 g bacto-tryptone, 5 g bacto yeast extract, 10 g NaCl, adjust to pH 7 with NaOH and fill to 1 l with distilled H_2O; store at room temperature.
20. LB agar plates containing 100 μg/ml ampicillin.
21. Taq DNA polymerase (5 U/μl) with 10× PCR buffer.
22. 5′ Ab sense (Table 1): Prepare a working dilution of 50 μM; store in small aliquots at −20°C.
23. 3′ primer (3′ IgG (internal), 3′ Cκ 494, 3′ Cλ; Table 1): Prepare a working dilution of 50 μM; store in small aliquots at −20°C.

24. Terrific Broth containing 75 μg/ml ampicillin; prewarmed to room temperature.
25. Round bottom tube with snap cap, 13 ml.
26. Plasmid purification kit, e.g., QIAprep® Spin MiniPrep Kit (Qiagen) or NucleoSpin® Plasmid (Macherey-Nagel, Düren, Germany).

2.4. Expression of Recombinant Monoclonal Antibodies

1. Human embryonic kidney (HEK) 293T cells (ATCC, No. CRL-11268).
2. 150 mm Cell culture plates.
3. D-MEM growth medium: Dulbecco's Modified Eagle's Medium (D-MEM; GibcoBRL, Darmstadt, Germany) supplemented with 10% heat-inactivated FCS (Invitrogen), 100 μg/ml streptomycin, 100 U/ml penicillin G, and 0.25 μg/ml amphotericin B (all GibcoBRL).
4. D-MEM wash medium: D-MEM (GibcoBRL) supplemented with 100 μg/ml streptomycin, 100 U/ml penicillin G, and 0.25 μg/ml amphotericin B (all GibcoBRL).
5. D-MEM nutridoma medium: D-MEM (GibcoBRL) supplemented with 1% Nutridoma-SP (Roche Applied Sciences, Mannheim, Germany), 100 μg/ml streptomycin, 100 U/ml penicillin G, and 0.25 μg/ml amphotericin B (all GibcoBRL).
6. 150 mM NaCl, sterile.
7. Vector DNA: Equal amounts (10–15 μg) of Igγ1 expression vector DNA and Igκ/Igλ expression vector DNA.
8. Polyethylenimine, branched (PEI; Sigma): Prepare a PEI solution of 1 μg/ml in deionized water, filter sterilize. Store at 4°C for up to 6 months.
9. Sodium azide (NaN_3) solution.
10. Protein G Sepharose™ (GE Healthcare, Munich, Germany).
11. PBS, sterile.
12. Chromatography spin columns (Bio-Rad).
13. 0.1 M glycine, pH 3.0.
14. 1 M Tris, pH 8.0, supplemented with 0.5% NaN_3.

3. Methods

3.1. Isolation of Single Human B Cells by Flow Cytometry

Single primary human B cells can be isolated from various sources (e.g., blood, bone marrow, tissue) based on their phenotype or antigen-reactivity by flow cytometry (see Note 1). Throughout all steps it is important to work under RNAse-free conditions and on ice. When fluorochromes are used, incubation should be carried out in the dark.

1. Incubate purified mononuclear cells with 50–100 μl staining solution for 30 min (see Notes 1–6, 11, and 12).
2. Add 1 ml PBS/FCS, centrifuge for 4 min at 4,000 × *g*, 4°C. Carefully discard supernatant.
3. Repeat washing step.
4. Resuspend cell pellet in 100 μl of 7-AAD solution, incubate for 15 min in the dark.
5. Add 1 ml PBS/FCS, centrifuge cells 4 min at 4,000 × *g*, 4°C. Carefully discard supernatant.
6. Repeat washing step.
7. Pre-wet 35 μm cell strainer cap with PBS/FCS (see Note 13).
8. Resuspend cell pellet in 400 μl PBS/FCS. Filter through strainer cap into round bottom tube, centrifuge for 1 min at 8 × *g*.
9. Keep samples on ice until loaded onto the sorter.
10. Set up the flow cytometer for 96-well sorting: sort 100 beads into a test 96-well plate. A single droplet should be visible. Make sure that the bead droplet is located in the center of the bottom of each well (see Notes 14–17).
11. Sort the cells of interest into 96-well PCR plates containing 4 μl of lysis solution per well (see Notes 6, 18, and 20).
12. Immediately seal sort plates with Microseal F foil and put on dry ice before storing at −80°C (see Note 19).

3.2. RT-PCR and Ig Gene Amplification

3.2.1. cDNA Synthesis

cDNA is synthesized in the original 96-well sort plate. It is important throughout all steps to work under RNAse-free conditions and on ice. A master mix is prepared with the indicated surplus, but unless otherwise noted all reaction mixes are specified per well.

1. Thaw a plate of single sorted cells on ice and spin down (see Notes 6 and 20).
2. Prepare the RHP mix consisting of 2.35 μl nuclease-free water, 0.5 μl RHP, 0.5 μl Igepal CA-630, and 0.15 μl RNAsin (see Notes 6 and 21).
3. Prepare the RT mix consisting of 3 μl RT buffer, 2.05 μl nuclease-free water, 1 μl DTT, 0.5 μl dNTP solution, 0.2 μl RNAsin, and 0.25 μl SuperScript III (see Notes 6 and 22).
4. Carefully remove the Microseal foil of the sort plate, add 3.5 μl RHP mix to each well. Mix carefully and seal plate (see Notes 6, 18, and 20).
5. Incubate sort plate for 1 min at 68°C, transfer back onto ice.

6. Add 7 µl RT mix to each well. Mix carefully and seal plate (see Notes 18 and 20).
7. Perform reverse transcription at 42°C for 5 min, 25°C for 10 min, 50°C for 60 min, 94°C for 5 min.
8. Store cDNA at −20°C until further use or use cDNA plate directly to set up single cell PCRs.

3.2.2. Ig Gene Amplification

Human *Igh* and *Igκ* or *Igλ* gene transcripts from the same cell are amplified separately by nested PCRs. All steps are conducted on ice while wearing gloves. A master mix is prepared with the indicated surplus, but unless otherwise noted all reaction mixes are specified per well.

1. Prepare PCR master mix 1 consisting of 32.16 µl water, 4 µl PCR buffer, 0.13 µl 5′ primer mix, 0.13 µl 3′ primer, 0.4 µl dNTP solution, and 0.18 µl HotStarTaq® (see Notes 6 and 22).
2. Transfer PCR master mix 1 into a new 96-well PCR plate (37 µl per well; see Note 18).
3. Carefully transfer 3 µl cDNA from the cDNA plate into the first PCR plate with PCR master mix 1, mix carefully, and seal plate with PCR film (see Notes 18 and 20).
4. Run first PCR at 94°C for 15 min; 50 cycles at 94°C for 30 s, 58°C (IgH and Igκ) or 60°C (Igλ) for 30 s, 72°C for 55 s; 72°C for 10 min.
5. Prepare PCR master mix 2 consisting of 31.66 µl water, 4 µl PCR buffer, 0.13 µl 5′ primer mix, 0.13 µl 3′ primer, 0.4 µl dNTP solution, and 0.18 µl HotStar® Taq DNA polymerase (see Notes 6, 22, and 23).
6. Transfer PCR master mix 2 into new 96-well PCR plate (36.5 µl per well; see Note 18).
7. Add 3.5 µl of the first PCR product, mix carefully, and seal plate (see Notes 18 and 20).
8. Run second PCR at 94°C for 15 min; 50 cycles at 94°C for 30 s, 58°C (IgH and Igκ) or 60°C (Igλ) for 30 s, 72°C for 45 s; 72°C for 10 min.
9. Load 3 µl of second PCR product mixed with 3 µl of loading dye onto a 2% analytical agarose gel in TAE buffer containing ethidium bromide and run for 25 min at 120 V. Visualize DNA bands under ultraviolet (UV) light.
10. Sequence second PCR products of the expected size (450 bp for IgH, 510 bp for Igκ, and 405 bp for Igλ) with the respective 3′ primer.
11. Analyze Ig gene sequences by IgBLAST comparison with GenBank (http://www.ncbi.nlm.nih.gov/igblast/; see Notes 24–28).

3.3. Expression Vector Cloning

All steps are conducted on ice while wearing gloves. A master mix is prepared with the indicated surplus, but unless otherwise noted all reaction mixes are specified per well.

3.3.1. Ig Gene-Specific PCR

All primers used in Ig gene-specific PCRs contain restriction sites (*Age*I and *Sal*I for *Igh*, *Age*I and *Bsi*WI for *Ig*κ, and *Age*I and *Xho*I for *Ig*λ), which allows direct cloning into expression vectors (see Note 29) (4, 26).

1. Prepare PCR master mix 3 consisting of 29.2 μl water, 4 μl PCR buffer, 0.4 μl dNTP solution, and 0.2 μl HotStar® Taq (see Note 22).
2. Transfer PCR master mix 3 into new 96-well PCR plate (33.8 μl into each well; see Note 18).
3. Deposit 2 μl each of 5′ and 3′ gene-specific primer into the corresponding well of the 96-well PCR plate containing PCR master mix 3 (see Note 9).
4. Add 4.2 μl first PCR product to the corresponding well of the 96-well PCR plate, mix carefully, and seal plate (see Note 20).
5. Run second PCR program (see Subheading 3.2.2, step 8).
6. Run second PCR products on a 2% agarose gel to verify amplification.

3.3.2. Directional Cloning into Expression Vectors

1. Purify specific PCR products using a PCR purification kit, elute in 60 μl water (see Notes 29 and 30).
2. Use 31.5 μl purified PCR product, add 3.5 μl NEBuffer 4 for IgH and Igλ chain V gene PCR products and 3.5 μl NEBuffer 1 for Igκ chain V gene PCR products, respectively (see Notes 31 and 32).
3. Prepare enzyme mix 1 for IgV gene PCR products according to Table 2. Add 5 μl of enzyme mix 1 to purified PCR product plate, mix carefully, and seal with PCR film. Incubate for 2 h at 37°C. For Igκ V gene PCR products, prepare enzyme mix 2,

Table 2
Pipetting scheme for restriction digest

	IgH V gene PCR products	Igκ V gene PCR products	Igλ V gene PCR products
Enzyme mix 1 (see Note 22)	4.0 μl water 0.5 μl NEBuffer 4 0.4 μl BSA 0.05 μl *Age*I 0.05 μl *Sal*I	4.05 μl water 0.5 μl NEBuffer 1 0.4 μl BSA 0.05 μl *Age*I	4.0 μl water 0.5 μl NEBuffer 4 0.4 μl BSA 0.05 μl *Age*I 0.05 μl *Xho*I
Enzyme mix 2 (see Note 22)		1.7 μl water 0.2 μl NEBuffer 1 0.1 μl *Bsi*WI	

transfer 2 μl to *Age*I-digested Igκ V gene PCR products, mix carefully, and seal with PCR film. Incubate for 2 h at 55°C (see Notes 22 and 33).

4. Purify digested PCR products using the PCR purification kit, elute in 60 μl water (see Note 30).
5. Ligate 8 μl digested and purified gene-specific PCR product into 0.5 μl expression vector using 1 μl T4 DNA ligase reaction buffer and 0.5 μl T4 DNA ligase (see Note 34). Incubate for 2 h at room temperature or overnight at 16°C.
6. Thaw competent bacteria on ice. Add 3 μl ligation product to 5 μl competent DH10B bacteria, mix carefully. Incubate for 30 min. Perform a heat shock for 45 s at 42°C, put bacteria/ligation mix back on ice.
7. Add 200 μl LB medium and incubate at 37°C for 30 min under moderate shaking.
8. Plate bacteria on ampicillin containing LB agar plate. Incubate upside down over night at 37°C. Label colonies used for insert PCR to screen bacterial colonies for the presence of appropriately sized inserts.
9. Prepare PCR master mix 4 consisting of 19.1 μl water, 2.5 μl PCR buffer, 2.5 μl dNTP solution, 0.2 μl 5′ Ab sense primer, 0.2 μl 3′ primer, and 0.5 μl Taq DNA polymerase (see Notes 22 and 35).
10. Add PCR master mix 4 to new 96-well PCR plate, place on ice, inoculate with bacterial colony (see Notes 35 and 36).
11. Run insert PCR at 94°C for 15 min; 27 cycles at 94°C for 30 s, 58°C for 30 s, 72°C for 60 s; 72°C for 10 min.
12. Run PCR products on a 2% agarose gel to verify amplification (expected product sizes are 650 bp for Igγ1 vector, 700 bp for Igκ vector, and 590 bp for Igλ vector).
13. Sequence insert PCR products with the expected size using 5′ Ab sense as sequencing primer (see Note 37).
14. Choose colonies with 100% sequence identity to the second PCR products (Subheading 3.2.2) to inoculate 4 ml Terrific Broth with 75 μg/ml ampicillin in round bottom tube. Incubate for 16 h at 37°C, 200 rpm.
15. Isolate plasmid DNA from 2 ml bacteria culture using a plasmid purification kit (see Note 38). Elute DNA in 75 μl elution buffer. Determine plasmid DNA concentrations at A_{260} and A_{260}/A_{280} ratios with a spectrophotometer (see Note 39).

3.4. Expression of Recombinant Monoclonal Antibodies

Using PEI-mediated transfection, antibody concentration in supernatant is on average 40 μg/ml, typically ranging between 10 and 150 μg/ml as determined by ELISA. Depending on the intended antibody binding assays, the antibody may be purified. For some applications, e.g. ELISA, the antibody containing supernatant may be sufficient for reactivity testing. Functional Ig production should be confirmed by sodium-dodecylsulfate polyacrylamide gel electrophoresis (SDS-PAGE) under nonreducing and reducing conditions (30).

3.4.1. Recombinant Antibody Production

1. Grow HEK 293T cells in 150 mm plates and 25 ml D-MEM growth medium under standard conditions (37°C, 5% CO_2; see Note 40). Ensure that cells are evenly spread across the plate.
2. At 80% confluency, remove medium completely and wash with 10 ml D-MEM wash medium. Remove wash medium completely and add 25 ml D-MEM nutridoma medium (see Notes 41 and 42). Put plate back into incubator.
3. Prepare a mix of NaCl, Igγ1 vector and corresponding Igκ/Igλ vector obtained from Subheading 3.3.2, step 15 (ratio 150 mM NaCl:total vector DNA [μl/μg] = 50; see Note 43). Agitate on vortexer for 10 s, add PEI (ratio PEI:total vector DNA [μg/μg] = 3; see Note 43). Immediately agitate on vortexer for 30 s. Incubate transfection mix for 10 min at room temperature.
4. Gently add PEI mixture drop-wise to each plate, aim for even distribution of the mixture. Cultivate cells for 4 days under standard conditions.
5. Collect culture supernatant, centrifuge at 800 × *g* for 10 min and transfer to a new reaction tube. Add sodium azide to a final concentration of 0.05%, store at 4°C (see Notes 44 and 45).

3.4.2. Recombinant Antibody Purification

All steps are conducted on ice while wearing gloves.

1. Equilibrate 25 μl of Protein G beads with PBS. Centrifuge for 10 min at 800 × *g*, 4°C (see Note 46).
2. Incubate 25 μl Protein G beads with 25 ml culture supernatant overnight at 4°C under rotation.
3. Equilibrate chromatography spin column with PBS.
4. Centrifuge Protein G beads/culture supernatant for 10 min at 800 × *g*, 4°C.
5. Carefully remove the supernatant, resuspend beads in residual PBS, transfer to spin column (see Note 47).
6. Wash column twice with 1 ml PBS.
7. Elute recombinant antibody in three fractions (200 μl each) with glycine and collect in tubes, each containing 20 μl Tris/NaN_3 (see Notes 45 and 48).
8. Measure protein concentration in spectrophotometer at A_{280} or in an ELISA (26).

4. Notes

1. Single cell suspensions can be prepared from virtually any tissue or organ by standard procedures (31). The appropriate protocol to obtain mononuclear cells depends on the tissue of interest. Mononuclear cells can be isolated from blood or bone marrow using EDTA or citrate blood collection tubes by Ficoll-Paque® density gradient centrifugation after 1:2 or 1:10 dilution in medium, respectively. Mononuclear cells from tissue can be isolated by collagenase/DNase digest and subsequent Percoll®-based density gradient purification (27). Prior to flow cytometric single cell isolation, mononuclear cell preparations can be frozen. However, freezing is associated with cell loss and may reduce the RT-PCR efficiency. Cell loss after freezing may be particularly high for metabolically active cells, e.g., antibody secreting cells. Therefore, whenever possible, freshly purified mononuclear cells should be used for single cell sorting. Importantly, keep purified mononuclear cells on ice until sorted.
2. For a successful cell sorting, target cells should not be below 0.1–0.5% of total cells. Magnetic cell sorting strategies, e.g., by negative depletion for B cell enrichment are possible prior to isolation by flow cytometry, but result in significant cell loss.
3. Titrate antibodies for use in flow cytometry. If possible, perform the titration on the tissue of interest. Be aware that single cell sorting is associated with considerable cell loss of up to 95%.
4. If unlabeled or biotinylated antibody or antigen is used (32), repeat steps 1–3 accordingly with secondary antibody or streptavidin-conjugate coupled to fluorochrome.
5. 7-AAD is used to exclude dead cells. Alternatively, DAPI may be used.
6. Pay attention to work under RNase-free conditions and on ice.
7. Use skirted 96-well PCR plates with a solid shell for single cell sorting to prevent tilting.
8. The aluminum foil should cover a temperature range from -80°C to 105°C and should be free of DNase, RNase, and DNA.
9. 5′ primer for specific V gene amplification is individually designed based on the V gene germline configuration (http://www.imgt.org/IMGTrepertoire/Proteins/index.php; (26)). The primer contains an *Age*I restriction site and the first 18 nt of V gene FWR1 (Table 1), preferentially ending with C or G. Gene-specific J primer containing *Sal*I restriction site for IgH chain genes and *Bsi*WI restriction site for Igκ chain genes are

listed in Table 1. Igλ chain genes are amplified using the respective 5′ primer listed in Table 1 and 3′ *Xho*I Cλ.

10. IgH γ, IgL κ, or IgL λ expression vectors with human Igγ1, Igκ, or Igλ constant regions, respectively, cloning sites and ampicillin resistance gene are cloned into *E. coli* DH10B bacteria (26). Prepare vectors using a Maxiprep Kit, e.g., HiSpeed Plasmid Maxi (Qiagen) and NucleoBond Xtra Maxi (Macherey-Nagel). Follow the manufacturer's instructions. After plasmid purification, determine DNA concentration and sequentially digest 25 μg of circular vector DNA with 50 U of the appropriate restriction enzymes and an extended incubation for 4 h. Load digested expression vectors onto an 0.8% agarose gel and purify linearized expression vectors by using a gel extraction kit, e.g., QIAquick Gel Extraction (Qiagen) and NucleoSpin Extract II (Macherey-Nagel). To test for successful linearization, perform ligations with and without digested, purified IgV gene PCR product as described in Subheading 3.3.2, step 5.
11. Depending on the frequency of the target population, aliquot the desired number of cells. Do not exceed 10×10^6 cells/100 μl.
12. Simultaneously prepare single stains with the same fluorochrome used in the actual staining procedure. Single stains are important for setting up the instrument and adjusting an appropriate compensation. Use compensation beads instead of cells whenever possible.
13. Incubation with antibody or antigen solution may cross-link cells and thus may cause agglutination. DNase may be used to reduce this effect and therefore use of a cell strainer is essential.
14. This setup ensures that the cell droplet is collected in the 4 μl of lysis buffer. To avoid liquid evaporation, use a cooled stage.
15. Sort at low speed and under low pressure to avoid cell loss.
16. Single cell sort is also possible using 384-well plates. In this case, scale down all required solutions by a factor of 4.
17. Label 96-well plates appropriately ahead of cell sorting. Sort cells using forward versus side scatter with doublet discrimination, apply live cell gate.
18. For parallel processing in a 96-well plate, use a multichannel pipette to speed up the process.
19. Alternatively, plates may be used directly for cDNA synthesis after freezing on dry ice.
20. To avoid contamination, always spin down single cell plates prior to removal of the foil cover. Carefully peel off the PCR film.
21. Allow a surplus of 15% per plate due to the viscosity of Igepal CA-630.

22. Prepare a master mix for all reactions. Allow a surplus of 10%.
23. Alternatively, use 5′ V gene and 3′ J gene primer mixes containing restriction sites (4, 16, 26, 33) and skip Subheading 3.3.1. Using this approach, isotype determination is not possible and amino acid replacement mutations may be introduced due to primer mismatches.
24. Ig gene sequence analysis by IgBLAST comparison with GenBank (http://www.ncbi.nlm.nih.gov/igblast/) will identify germline V(D)J gene segments with the highest identity, the amino acid composition of complementarity determining region 3 (CDR3), the IgH isotype, and the distribution of somatic mutations. The analysis of the Ig gene repertoire and CDR3 characteristics provide information on the clonality of the B cell population and their differentiation stage. Importantly, certain B cell tumors including CLL, mantle cell lymphoma, follicular lymphoma, and primary central nervous system lymphoma frequently express characteristic *IGHV* genes with stereotyped CDR3s or show biased *IGKV* gene usage (14–16, 18, 34–40).
25. Determine IgV and IgJ gene usage for gene-specific PCRs and expression vector cloning.
26. Determine Ig CDR3 length as indicated in the IgBLAST result by counting the amino acid residues following framework region (FWR) 3 up to the conserved tryptophan–glycine motif in all JH gene segments or up to the conserved phenylalanine–glycine motif in Jκ and Jλ gene segments (41).
27. Cells that have undergone affinity maturation in the germinal center show clear signs of antigen-mediated selection, as evidenced by higher ratios of amino acid replacement to silent somatic mutations in V gene CDRs than in FWRs. Unmutated V region genes have been found in mantle cell lymphoma and a subtype of CLL (16, 35, 42). Somatically mutated V region genes are found in multiple myeloma and most types of lymphoma, e.g., follicular lymphoma, MALT lymphoma, hairy cell leukemia, and a subtype of CLL (43).
28. Determine the Ig isotype using the international ImMunoGeneTics database (http://www.imgt.org/).
29. All steps in expression vector cloning can be performed in 96-well plates. If problems occur during cloning, switch to single tubes as this usually results in a better yield.
30. Determine DNA concentration with a spectrophotometer at A_{260}. To ensure for a sufficient yield, at least 300 ng DNA are required for restriction digest. A successful ligation is accomplished using at least 15 ng DNA and aiming for a 3:1 ratio of insert to vector.

31. Confirm that variable regions of somatically mutated antibodies do not contain restrictions sites for the respective restriction enzymes. If so, perform a partial digest with tenfold lower enzyme concentration and 1 min incubation time.
32. Include a restriction control, e.g., linearize 100 ng of the respective expression vector.
33. IgH and Igκ PCR products are double digested in one reaction. Igλ chain PCR products are digested sequentially.
34. Include a ligation control if possible, e.g., purified digested PCR product that has been successfully ligated into the corresponding expression vector.
35. Check three colonies per clone. If none of the inserts has the expected size, it is highly recommended to repeat cloning from the specific PCR.
36. Prepare backup plates for the colonies of interest on fresh LB agar plates containing ampicillin. Inoculate by touching bacterial colony with small pipette tip, touching a spot on the fresh LB agar plate, then dipping pipette tip into PCR plate containing master mix 4. Incubate LB agar plates upside down over night at 37°C.
37. Due to the use of error-prone Taq DNA polymerase, 10% of the inserts will contain mutations as compared to the nucleotide sequences obtained from the respective second PCR product (26). Clones with mutated insert sequences that lead to Ig amino acid exchanges are excluded from further analyses.
38. Use the remaining 2 ml bacterial culture as backup for plasmid DNA isolation. Store as cell pellet at –20°C.
39. Usually, 20–30 μg plasmid DNA are recovered from 2 ml bacteria culture.
40. HEK 293 T cells should be grown and passaged according to the instructions of ATCC. Be aware that HEK 293T cells require at least 24 h to attach firmly to the culture plate. Ensure that passage number for HEK 293T cells is below 30, as cells may otherwise not produce sufficient amounts of antibody.
41. If cells detach while washing, change plate brand.
42. Transfection by calcium-phosphate precipitation is also possible (26). PEI-mediated transfection, however, does not require the change of medium 8–12 h after transfection and results in higher antibody yields.
43. For example, when working with 10 μg HC vector DNA and 10 μg LC vector DNA, 1,000 μl of 150 mM NaCl, and 60 μg of PEI are added.
44. To increase antibody yield, supernatant can be replaced with fresh D-MEM nutridoma medium and harvested again 7 days after transfection.

45. Supernatant and purified antibodies may be stored at 4°C for several months if sodium azide is added. If sodium azide interferes with the intended antibody reactivity assay, collect the supernatant under sterile conditions. Before use in experiments, spin down supernatant or purified antibody in benchtop centrifuge for 5 min at maximum speed and 4°C. Determine antibody concentration after long-term storage.
46. If purifying several antibodies in parallel, equilibrate the required volume of Protein G beads in one tube including 5% surplus.
47. To prevent clogging, cut off pipette tip.
48. Fractions may be pooled after determining protein concentration.

References

1. Tonegawa S (1983) Somatic generation of antibody diversity. Nature 302:575–581
2. Burnet FM (1972) Auto-immunity and autoimmune disease. Medical and Technical Publishing Co Ltd, Lancaster
3. Goodnow CC et al (1995) Self-tolerance checkpoints in B lymphocyte development. Adv Immunol 59:279–368
4. Wardemann H et al (2003) Predominant autoantibody production by early human B cell precursors. Science 301:1374–1377
5. Tiller T et al (2007) Autoreactivity in human IgG+memory B cells. Immunity 26:205–213
6. Scheid JF et al (2011) Differential regulation of self-reactivity discriminates between IgG+human circulating memory B cells and bone marrow plasma cells. Proc Natl Acad Sci USA 108:18044–18048
7. Berek C, Milstein C (1988) The dynamic nature of the antibody repertoire. Immunol Rev 105:5–26
8. Klein U, Dalla-Favera R (2008) Germinal centres: role in B-cell physiology and malignancy. Nat Rev Immunol 8:22–33
9. Muramatsu M et al (2000) Class switch recombination and hypermutation require activation-induced cytidine deaminase (AID), a potential RNA editing enzyme. Cell 102:553–563
10. Rui L et al (2011) Malignant pirates of the immune system. Nat Immunol 12:933–940
11. Küppers R et al (1999) Cellular origin of human B-cell lymphomas. N Engl J Med 341:1520–1529
12. Küppers R (2005) Mechanisms of B-cell lymphoma pathogenesis. Nat Rev Cancer 5:251–262
13. Zenz T et al (2010) From pathogenesis to treatment of chronic lymphocytic leukemia. Nat Rev Cancer 10:37–50
14. Agathangelidis A et al (2011) Unlocking the secrets of immunoglobulin receptors in mantle cell lymphoma: Implications for the origin and selection of the malignant cells. Semin Cancer Biol 21:299–307
15. Messmer BT et al (2004) Multiple distinct sets of stereotyped antigen receptors indicate a role for antigen in promoting chronic lymphocytic leukemia. J Exp Med 200:519–525
16. Hervé M et al (2005) Unmutated and mutated chronic lymphocytic leukemias derive from self-reactive B cell precursors despite expressing different antibody reactivity. J Clin Invest 115:1636–1643
17. Catera R et al (2008) Chronic lymphocytic leukemia cells recognize conserved epitopes associated with apoptosis and oxidation. Mol Med 14:665–674
18. Hadzidimitriou A et al (2009) Evidence for the significant role of immunoglobulin light chains in antigen recognition and selection in chronic lymphocytic leukemia. Blood 113:403–411
19. Bahler DW, Levy R (1992) Clonal evolution of a follicular lymphoma: evidence for antigen selection. Proc Natl Acad Sci USA 89:6770–6774
20. Hussell T et al (1996) Helicobacter pylori-specific tumour-infiltrating T cells provide contact dependent help for the growth of malignant B cells in low-grade gastric lymphoma of mucosa-associated lymphoid tissue. J Pathol 178:122–127
21. Burger JA et al (2009) The microenvironment in mature B-cell malignancies: a target for new treatment strategies. Blood 114:3367–3375
22. Steinitz M et al (1977) EB virus-induced B lymphocyte cell lines producing specific antibody. Nature 269:420–422

23. Lanzavecchia A, Corti D, Sallusto F (2007) Human monoclonal antibodies by immortalization of memory B cells. Curr Opin Biotechnol 18:523–528
24. Köhler G, Milstein C (1975) Continuous cultures of fused cells secreting antibody of predefined specificity. Nature 256:495–497
25. McCafferty J et al (1990) Phage antibodies: filamentous phage displaying antibody variable domains. Nature 348:552–554
26. Tiller T et al (2008) Efficient generation of monoclonal antibodies from single human B cells by single cell RT-PCR and expression vector cloning. J Immunol Methods 329: 112–124
27. Benckert J et al (2011) The majority of intestinal IgA+and IgG+plasmablasts in the human gut are antigen-specific. J Clin Invest 121:1946–1955
28. Binder M et al (2011) B-cell receptor epitope recognition correlates with the clinical course of chronic lymphocytic leukemia. Cancer 117:1891–1900
29. Mouquet H et al (2010) Polyreactivity increases the apparent affinity of anti-HIV antibodies by heteroligation. Nature 467:591–595
30. Tiller T, Busse CE, Wardemann H (2009) Cloning and expression of murine Ig genes from single B cells. J Immunol Methods 350:183–193
31. Gu H, Rajewsky K (2004) B cell protocols. Humana Press Inc, New York
32. Scheid JF et al (2009) A method for identification of HIV gp140 binding memory B cells in human blood. J Immunol Methods 343: 65–67
33. Tsuiji M et al (2006) A checkpoint for autoreactivity in human IgM+memory B cell development. J Exp Med 203:393–400
34. Smilevska T et al (2008) Immunoglobulin kappa gene repertoire and somatic hypermutation patterns in follicular lymphoma. Blood Cells Mol Dis 41:215–218
35. Fais F et al (1998) Chronic lymphocytic leukemia B cells express restricted sets of mutated and unmutated antigen receptors. J Clin Invest 102:1515–1525
36. Thompsett AR et al (1999) V(H) gene sequences from primary central nervous system lymphomas indicate derivation from highly mutated germinal center B cells with ongoing mutational activity. Blood 94: 1738–1746
37. Montesinos-Rongen M et al (1999) Primary central nervous system lymphomas are derived from germinal-center B cells and show a preferential usage of the V4-34 gene segment. Am J Pathol 155:2077–2086
38. Stamatopoulos K et al (1997) Follicular lymphoma immunoglobulin kappa light chains are affected by the antigen selection process, but to a lesser degree than their partner heavy chains. Br J Haematol 96:132–146
39. Noppe SM et al (1999) The genetic variability of the VH genes in follicular lymphoma: the impact of the hypermutation mechanism. Br J Haematol 107:625–640
40. Aarts WM et al (2000) Variable heavy chain gene analysis of follicular lymphomas: correlation between heavy chain isotype expression and somatic mutation load. Blood 95: 2922–2929
41. Kabat EA, Wu TT (1991) Identical V region amino acid sequences and segments of sequences in antibodies of different specificities. Relative contributions of VH and VL genes, minigenes, and complementarity-determining regions to binding of antibody-combining sites. J Immunol 147:1709–1719
42. Hummel M et al (1994) Mantle cell (previously centrocytic) lymphomas express VH genes with no or very little somatic mutations like the physiologic cells of the follicle mantle. Blood 84:403–407
43. Klein U et al (1998) Somatic hypermutation in normal and transformed human B cells. Immunol Rev 162:261–280

Chapter 6

Studying the Replication History of Human B Lymphocytes by Real-Time Quantitative (RQ)-PCR

Menno C. van Zelm, Magdalena A. Berkowska, and Jacques J.M. van Dongen

Abstract

The cells of the adaptive immune system, B and T lymphocytes, each generate a unique antigen receptor through V(D)J recombination of their immunoglobulin (Ig) and T cell receptor (TCR) loci, respectively. Such rearrangements join coding elements to form a coding joint and delete the intervening DNA as circular excision products containing the signal joint. These excision circles are stable structures that cannot replicate and have no function in the cell. Since the coding joint in the genome is replicated with each cell division, the ratio between coding joints and signal joints in a population of B cells can be used as a measure for proliferation. This chapter describes a real-time quantitative (RQ-)PCR-based approach to quantify proliferation through calculating the ratio between coding joints and signal joints of the frequently occurring intronRSS–Kde rearrangements in the *IGK* light chain locus. The approach is useful to study basic B-cell biology as well as abnormal proliferation in human diseases.

Key words: B cell, Immunoglobulin, V(D)J recombination, KREC, Replication history, Proliferation, Real-time quantitative PCR, intronRSS, Kde, *IGK*

Abbreviations

BCR	B-cell antigen receptor
D	Diversity
Ig	Immunoglobulin
IgH	Ig heavy chain
Igκ	Ig kappa light chain
Igλ	Ig lambda light chain
J	Joining
Kde	Kappa-deleting element
KREC	Kappa-deleting recombination excision circle
RAG	Recombination activating gene
RQ-PCR	Real-time quantitative PCR
RSS	Recombination signal sequence
V	Variable

Ralf Küppers (ed.), *Lymphoma: Methods and Protocols*, Methods in Molecular Biology, vol. 971, DOI 10.1007/978-1-62703-269-8_6, © Springer Science+Business Media, LLC 2013

1. Introduction

B cells have a unique role in the vertebrate immune system with the production of antibodies (immunoglobulin; Ig) that are capable of neutralizing invading pathogens. Importantly, antibody responses are highly antigen-specific and generate immunological memory: upon secondary encounter, the same pathogen is eliminated much more efficiently. These adaptive capabilities are the result of two independent stages of development (1). First, each B cell creates a unique antibody during antigen-independent precursor differentiation in bone marrow, which is expressed on the surface membrane: B cell antigen receptor (BCR). Together, all newly produced B cells have a large repertoire of antigen receptors with the potential to specifically recognize many different pathogens. The second phase of development starts when a B cell actually recognizes antigen in peripheral lymphoid organs. These activated B cells undergo enormous clonal proliferation, thereby generating huge numbers of daughter cells with the potential to recognize the same pathogen. This clonal expansion generates effector cells for a strong response and long-term memory in the form of memory B cells and Ig-producing plasma cells. The host requires a highly dynamic immune system, which maintains a tight balance between the production of a large repertoire of B cells with unique receptors and a strong immune response of groups of cells with an antigen-specific and thereby a more limited (selected) repertoire.

While clonal B-cell proliferation is important to generate a strong immune response, it should be tightly controlled to prevent malignant transformation. Until recently, the extent of properly controlled B-cell expansion following antigen recognition was unknown, due to the lack of techniques to assess *in vivo* proliferation. In 2007, we introduced a novel real-time quantitative (RQ-) PCR-based method to assess *ex vivo* the average replication history a population of B cells has undergone *in vivo* (2). Our approach exploits the circular DNA excision products that are generated in precursor B cells that undergo V(D)J recombination.

Ig loci do not contain a functional first exon. Instead, they contain multiple variable (V), diversity (D), and joining (J) genes, which undergo ordered genomic rearrangement during antigen-independent differentiation to form a functional first exon (Fig. 1a) (3, 4). First, D to J gene rearrangements are initiated in the Ig heavy chain locus (*IGH*), followed by V to DJ gene rearrangements. Upon successful completion, Ig light chain rearrangements are induced in the *IGK* locus, leading to direct V to J coupling (Fig. 1a). When *IGK* light chain rearrangements are not successful, V to J gene rearrangements are induced in the *IGL* light chain locus. Thus, mature B cells can exit the bone marrow expressing an antigen receptor with either an Igκ or an Igλ isotype.

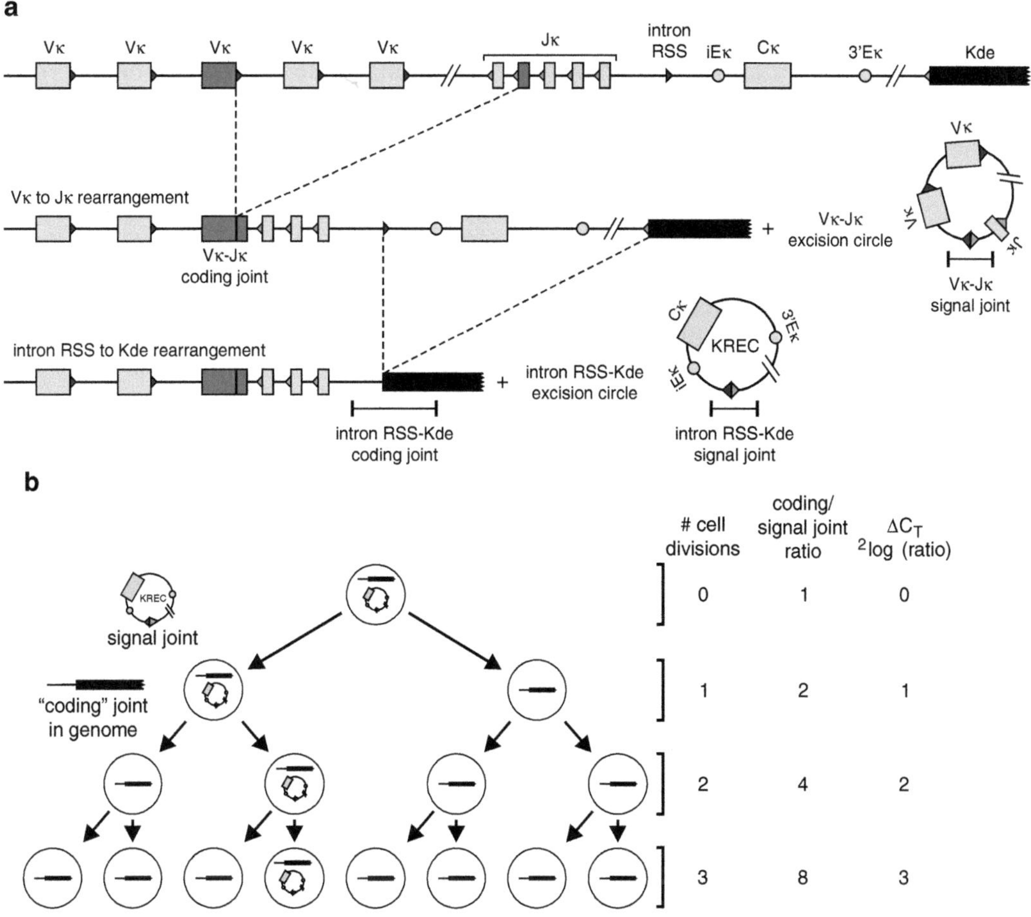

Fig. 1. Detection of IntronRSS–Kde coding joints and signal joints to quantify B-cell replication history. (**a**) V(D)J recombination on the *IGK* locus results in a VκJκ coding joint. Subsequent rearrangement between the intronRSS and the Kde elements can render the *IGK* allele nonfunctional by deleting the Cκ exons and the enhancers. Consequently, the coding joint precludes any further rearrangements in the *IGK* locus and therefore remains present in the genome, whereas an intronRSS–Kde signal joint is formed on an excision circle (KREC). (**b**) When a B lymphocyte with an intronRSS–Kde rearrangement divides, both daughter cells inherit the intronRSS–Kde coding joint in the genome. However, the signal joint, which is on the episomal KREC, will be inherited by only one of the two daughter cells. Crucially, the ΔC_T of the PCRs detecting the coding joint and the signal joint exactly represent the number of cell divisions a B lymphocyte has undergone, because both processes have an exponential increase with base number 2. *Figure panels* are reproduced from Van Zelm et al. (?) with permission from Rockefeller University Press.

V(D)J recombination of Ig loci is initiated by the recombination activating gene products RAG1 and RAG2 that introduce double-stranded DNA breaks on the borders of two genes and their flanking recombination signal sequences (RSS). The resulting coding ends at the side of the gene are ligated together to form a coding joint. The intervening DNA is excised and circularized by ligating the blunt DNA signal ends, thereby forming a signal joint. While multiple of these episomal recombination excision circles are formed in every developing B cell, they seem to have no specific function. They are stable products, but cannot replicate in the cell, do not duplicate during mitosis, and consequently are diluted during cell division, while the corresponding rearrangement on genomic DNA (coding joint) is inherited by all daughter cells (2, 5, 6).

Since a daughter cell either carries a circle or not, it is not possible to study turnover or proliferation of single cells. In fact, measurement of turnover or proliferation via excision circles is performed at the level of a cell population. By comparing the number of detected signal joints on excision circles with the total number of coding joints in chromosomal DNA, the average proliferation is established. The condition to detect signal joints at the population level requires selection of a common rearrangement. However, in each developing B cell, a unique combination of V, D, and J genes is formed to facilitate generation of a large repertoire of Igs. Therefore, most rearrangements are formed to generate a functional B- or T-cell receptor and will consequently not be useful as targets to study proliferation.

Fortunately, most B cells initiate a single step gene rearrangement to render a specific locus nonfunctional. This concerns the rearrangement between the intronRSS and the kappa-deleting element (Kde), which fits all the criteria for optimal measurement of proliferation (2): (a) it is a frequently occurring gene rearrangement that deletes the constant region of 35–40% *IGK* alleles; (b) it is one of the last Ig gene rearrangements to occur in bone marrow (Fig. 1a) (4), ensuring that the corresponding kappa-deletion recombination excision circles (KRECs) are abundantly present in naive B lymphocytes; (c) it is a single-step rearrangement, which allows easy design of RQ-PCR primers and probes for accurate detection of the coding joints and signal joints; and (d) it is an end-stage rearrangement that deletes the enhancers, precluding further rearrangements.

The KREC assay depends on the dilution of signal joints on excision circles: nearly all formed B cells contain a KREC, while only one in many antigen-experienced B cells carries excision circles. Thus, when measuring the replication history of total B cells, the KREC content of newly formed B cells has a dominant effect and will result in a low number of cell divisions. It is therefore recommended to measure the replication history in well-defined, purified B-cell subsets. We and others have used the KREC assay to study the contribution of proliferation to naive B-cell homeostasis and in aberrant B-cell populations (2, 7–9), but also to delineate memory B-cell maturation pathways and defects in immunodeficient patients (2, 10, 11). Furthermore, the KREC assay as described in this chapter can be employed to identify abnormal proliferation in B-cell lymphoproliferations.

2. Materials

2.1. High-Speed Cell Sorting

1. Phosphate-buffered saline (PBS): Dissolve 8 g of NaCl, 0.2 g of KCl, 0.12 g of KH_2PO_4, and 1.67 g of $Na_2HPO_4 \cdot 2H_2O$ in approximately 800 ml of water, adjust pH to 7.8 with HCl, and add water to 1 l.
2. PBS/bovine serum albumin (BSA): PBS supplemented with 0.2% BSA and 0.1% azide.
3. Antibodies: See Table 1.
4. Fluorescence-activated cell sorter (FACS), e.g., FACSAria (BD Biosciences).
5. Fetal calf serum (FCS).
6. CD19 magnetic beads (Miltenyi Biotec, Bergisch Gladbach, Germany).
7. Ficoll.

2.2. DNA Isolation and TaqMan-Based Real-Time Quantitative (RQ)-PCR

1. Column-based miniprep DNA isolation kit, from, e.g., Sigma-Aldrich or Qiagen.
2. Spectrophotometer, e.g., NanoDrop.
3. Universal PCR Master Mix (Applied Biosystems): 5 mM $MgCl_2$, 200 μM dNTP, 0.1 U Taq Gold, 10% glycerol.
4. 2% BSA.
5. Primers, in a 30 pmol/μl working dilution (Table 2).

Table 1
Fluorescently labeled antibodies for high-speed cell sorting of human B cell subsets

Antibody	Clone	Fluorochrome	Supplier
CD19	SJ25C1	PerCP-Cy5.5	BD Biosciences
CD24	gran-B-ly-1	FITC	Sanquin
CD27	L128	APC	BD Biosciences
CD38	HB7	APC-H7	BD Biosciences
IgM	G20-127	Horizon V450	BD Biosciences
IgD	Polyclonal goat	PE	SBA
IgA	IS11-8E10	FITC	Miltenyi Biotec
IgG	G18-145	PE	Miltenyi Biotec

Table 2
Oligonucleotide characteristics for intronRSS–Kde coding joint and signal joint, and albumin control gene quantification

Target (size)	Oligonucleotide	Sequence (5′–3′)
Albumin[a] (118 bp)	F-Alb	CCT.GAT.CCT.CTT.GTC.CCA.CAG
	R-Alb	GGA.TTT.AGA.GTC.TCT.CAG.CTG.GTA.CA
	T-Alb[b]	TGC.TGA.AAC.ATT.CAC.CTT.CCA.TGC.AGA
IntronRSS–Kde coding joint[c] (144 bp)	Intron-F1-TM	CCC.GAT.TAA.TGC.TGC.CGT.AG
	Kde-R1-TM	CCT.AGG.GAG.CAG.GGA.GGC.TT
	T-Kde-cons1[b]	AGC.TGC.ATT.TTT.GCC.ATA.TCC.ACT.ATT.TGG.AGT
IntronRSS–Kde signal joint[c] (148 bp)	Int-Kde KREC F	TCA.GCG.CCC.ATT.ACG.TTT.CT
	Int-Kde KREC R	GTG.AGG.GAC.ACG.CAG.CC
	T-Kde-RSS_2[b]	CCA.GCT.CTT.ACC.CTA.GAG.TTT.CTG.CAC.GG

[a]Refs. (13, 14)
[b]5′-FAM and 3′-TAMRA-labeled probes
[c]Ref. (2)

6. 5′FAM and 3′TAMRA-labeled hydrolysis probes, in a 5 pmol/μl working dilution (Table 2).
7. Genomic DNA of the control cell line U698-DB01 that contains one intronRSS–Kde coding joint and one signal joint per genome (InVivoScribe; see Note 1) (2).
8. Thermocycler that is capable to detect FAM signals from TaqMan-based hydrolysis probes, e.g., ABI Prism 7000 (Applied Biosystems).

3. Methods

3.1. High-Speed Cell Sorting

1. Start with mononuclear cells following Ficoll density separation, or with presorted B cells, e.g., via CD19$^+$ B-cell enrichment with magnetic beads.
2. Resuspend cell pellet after final wash in 500 μl PBS/BSA per 50×10^6 cells.
3. Add appropriate amounts of antibodies of choice (Table 1), vortex shortly and incubate sample for 10 min in the dark at room temperature.
4. Wash sample once with 10–20× labeling volume of PBS/BSA.
5. Resuspend cells to a final concentration of 25×10^6 cells/ml in PBS/BSA.
6. Define B-cell populations with electronic gating on FACS and perform high-speed cell sorting.

7. Collect specified B-cell populations in tubes that contain a small volume (0.1–0.5 ml) of FCS.
8. Upon completion of cell sorting, spin cells down, and start DNA isolation procedure.

3.2. DNA Isolation

1. Perform DNA isolation on 30,000 or more cells using a column-based kit according to manufacturer's instructions (see Note 2).
2. Perform elution of DNA from column with elution buffer pre-heated to 55°C. When starting with 30,000–200,000 cells, elute in 50 μl; 200,000–2×10^6 cells in 100 μl; 2–5×10^6 cells in 200 μl. In case of low cell numbers, the eluate can be reloaded on the column for a consecutive elution to increase the DNA yield.
3. Measure DNA quantity and purity on spectrophotometer.
4. If concentration is >15 ng/μl, prepare working dilution of 10 ng/μl in water.

3.3. TaqMan-Based Real-Time Quantitative (RQ)-PCR

1. Prepare three master mixes for independent RQ-PCR amplification of IntronRSS–Kde coding joints, signal joints and for the albumin control gene. Perform these three reactions in duplicate on each sample, on the control cell line U698-DB01 and on a non-template (water) control (see Note 3).
2. The master mix consists of 12.5 μl Universal PCR Master Mix, 0.75 μl of both the forward and the reverse primer (30 pmol/μl), 0.5 μl hydrolysis probe (5 pmol/μl), 0.5 μl 2% BSA, and 5 μl water per reaction.
3. Add 5 μl template DNA (working dilution 10 ng/μl) or water per reaction well.
4. Adjust the PCR program for FAM detection and TAMRA quenching. The thermocycler program consists of a 50°C step for 2 min and 95°C for 10 s, followed by 50× (95°C for 15 s and 60°C for 1 min).
5. Upon completion of RQ-PCR reaction, set the threshold for signal detection such that all samples are in the exponential amplification phase. For consistency, it is preferred to set the same threshold for all three PCR reactions (IntronRSS–Kde coding joint, signal joint, and albumin).
6. Extract the cycle threshold (C_T) value for each reaction and calculate the average C_T value for duplicate reactions. Duplicates should differ <1 C_T.

3.4. Calculation of Rearranged intronRSS–Kde Alleles and the B-Cell Replication History

1. The ratio between the amount of genomic albumin and the intronRSS–Kde coding joints is indicative for the frequency of *IGK* alleles with an intronRSS–Kde rearrangement. Since the data are derived from independent PCRs, any technical variation is corrected for using the results on the U698-DB01 cell

line which has one intronRSS–Kde coding joint per genome. The frequency of *IGK* alleles containing the intronRSS–Kde coding joint is thus calculated as follows (see Note 4):

$$2^{[(C_{\mathrm{T\,albumin}} - C_{\mathrm{T\,coding\,joint}})_{\mathrm{sample}} - (C_{\mathrm{T\,albumin}} - C_{\mathrm{T\,coding\,joint}})_{\mathrm{cell\,line}}]} \cdot 100\%$$

2. The difference in C_T values for the intronRSS–Kde coding joint and signal joints is directly related to the replication history of the B-cell subset (Fig. 1b). Still, also for this calculation, the results of the U698-DB01 cell line have to be used to correct for differences in PCR efficiencies as follows:

$$(C_{\mathrm{T\,signal\,joint}} - C_{\mathrm{T\,coding\,joint}})_{\mathrm{sample}} - (C_{\mathrm{T\,signal\,joint}} - C_{\mathrm{T\,coding\,joint}})_{\mathrm{cell\,line}}$$

It is important to note that the increased proliferation will lead to reduced signal joints on KRECs (see Note 5).

4. Notes

1. The Igκ$^+$ B-cell line U698-M has two Vκ–Jκ rearranged IGK alleles, one of which is functional and has the intronRSS and Kde elements in germline configuration. The second allele has an out-of-frame VJκ joint and contains an intronRSS–Kde rearrangement. In order to obtain an equal ratio between the intronRSS–Kde coding and signal joints, an intronRSS–Kde signal joint construct was inserted into the genome of the U698-M cell line using retroviral transduction. Individual clones were sorted, and the DB01 clone was found to contain a single insertion per genome, similar to the intronRSS–Kde coding joint (2). The primer-probe sets for both the coding and signal joints are optimized and show nearly equal efficiencies. Still, minimal variation (<1 C_T) can occur. Therefore, it is recommended to correct for this technical variation in each experiment.
2. It is recommended to perform the KREC assay on DNA isolated from 30,000 cells or more for reliable signal joint amplification. When low replication histories are expected, it is still possible to obtain reliable signal joint amplification from fewer cells. However, for DNA isolation from low cell numbers, we recommend to use a direct DNA lysis method (12).
3. For optimal calculation of the frequency of rearranged intronRSS–Kde alleles and B cell replication history it is recommended to perform the coding joint, signal joint and albumin control PCRs in the same run. As with all RQ-PCR experiments, we recommend to perform each reaction at least in duplicate. In addition, each run should contain a non-template control and the U698-DB01 control for correction of technical

variation (see Note 1). When expecting low to undetectable signal joint numbers, it is recommended to include serial dilutions of the U698-DB01 for the coding and signal joint PCRs to determine their quantitative ranges.

4. In addition to the intronRSS–Kde coding to signal joint ratio, we recommend to calculate the frequency of rearranged coding joints with the albumin control gene. While this is not crucial for the actual replication history levels, it provides valuable information on how representative this is for the whole population. In general, 35–50% of *IGK* alleles in mature B cells contain an intronRSS–Kde coding joint. While $Ig\lambda^+$ B cells contain higher levels than $Ig\kappa^+$ B cells (4), these rearrangements are independent of the antigen specificity. Thus, substantially high frequencies of intronRSS–Kde coding joints (>35%) reflect a good representation for the total B cell subset.
5. With each cell division, the V(D)J recombination excision circles are diluted in the total B-cell population. In memory B cell subsets of healthy controls, the replication history can be up to 10–12 cell divisions. This means that one circle is detected for 2^{10}–2^{12} (i.e., ~1,000–4,000) coding joints in the genome. Thus, when high replication histories are expected, it is important to start with high amounts of template DNA to maximize the quantitative range.

Acknowledgments

MAB is supported by a Fellowship from the Ter Meulen Fund–Royal Netherlands Academy of Arts and Sciences. MCvZ is supported by an Erasmus University Rotterdam (EUR)-Fellowship, an Erasmus MC Fellowship, and by Veni grant 916.110.90 from ZonMW/NWO and by grant 689 of the Sophia Children's Hospital Fund.

Conflict of Interest

JJMvD is inventor of the KREC assay, which has been patented (PCT/NL 2005/000761; priority date 25 Oct 2004) and licensed to InVivoScribe Technologies, San Diego, CA; revenues of the patent go to Erasmus MC. The other authors have declared that no conflict of interest exists.

References

1. Berkowska MA, van der Burg M, van Dongen JJM, van Zelm MC (2011) Checkpoints of B cell differentiation: visualizing Ig-centric processes. Ann N Y Acad Sci 1246: 11–25
2. van Zelm MC, Szczepanski T, van der Burg M, van Dongen JJ (2007) Replication history of B lymphocytes reveals homeostatic proliferation and extensive antigen-induced B cell expansion. J Exp Med 204:645–655
3. Ghia P, ten Boekel E, Sanz E, de la Hera A, Rolink A, Melchers F (1996) Ordering of human bone marrow B lymphocyte precursors

by single-cell polymerase chain reaction analyses of the rearrangement status of the immunoglobulin H and L chain gene loci. J Exp Med 184:2217–2229

4. van Zelm MC, van der Burg M, de Ridder D, Barendregt BH, de Haas EF, Reinders MJ, Lankester AC, Revesz T, Staal FJ, van Dongen JJ (2005) Ig gene rearrangement steps are initiated in early human precursor B cell subsets and correlate with specific transcription factor expression. J Immunol 175:5912–5922
5. Livak F, Schatz DG (1996) T-cell receptor alpha locus V(D)J recombination by-products are abundant in thymocytes and mature T cells. Mol Cell Biol 16:609–618
6. Breit TM, Verschuren MC, Wolvers-Tettero IL, Van Gastel-Mol EJ, Hahlen K, van Dongen JJ (1997) Human T cell leukemias with continuous V(D)J recombinase activity for TCR-delta gene deletion. J Immunol 159:4341–4349
7. Moir S, Ho J, Malaspina A, Wang W, DiPoto AC, O'Shea MA, Roby G, Kottilil S, Arthos J, Proschan MA, Chun TW, Fauci AS (2008) Evidence for HIV-associated B cell exhaustion in a dysfunctional memory B cell compartment in HIV-infected viremic individuals. J Exp Med 205:1797–1805
8. Rakhmanov M, Keller B, Gutenberger S, Foerster C, Hoenig M, Driessen G, van der Burg M, van Dongen JJ, Wiech E, Visentini M, Quinti I, Prasse A, Voelxen N, Salzer U, Goldacker S, Fisch P, Eibel H, Schwarz K, Peter HH, Warnatz K (2009) Circulating CD21low B cells in common variable immunodeficiency resemble tissue homing, innate-like B cells. Proc Natl Acad Sci U S A 106:13451–13456
9. van der Burg M, Pac M, Berkowska MA, Goryluk-Kozakiewicz B, Wakulinska A, Dembowska-Baginska B, Gregorek H, Barendregt BH, Krajewska-Walasek M, Bernatowska E, van Dongen JJ, Chrzanowska KH, Langerak AW (2010) Loss of juxtaposition of RAG-induced immunoglobulin DNA ends is implicated in the precursor B-cell differentiation defect in NBS patients. Blood 115: 4770–4777
10. Berkowska MA, Driessen GJ, Bikos V, Grosserichter-Wagener C, Stamatopoulos K, Cerutti A, He B, Biermann K, Lange JF, van der Burg M, van Dongen JJ, van Zelm MC (2011) Human memory B cells originate from three distinct germinal center-dependent and -independent maturation pathways. Blood 118:2150–2158
11. Driessen GJ, van Zelm MC, van Hagen PM, Hartwig NG, Trip M, Warris A, de Vries E, Barendregt BH, Pico I, Hop W, van Dongen JJ, van der Burg M (2011) B-cell replication history and somatic hypermutation status identify distinct pathophysiological backgrounds in common variable immunodeficiency. Blood 118:6814–6823
12. van der Burg M, Kreyenberg H, Willasch A, Barendregt BH, Preuner S, Watzinger F, Lion T, Roosnek E, Harvey J, Alcoceba M, Diaz MG, Bader P, van Dongen JJ (2011) Standardization of DNA isolation from low cell numbers for chimerism analysis by PCR of short tandem repeats. Leukemia 25:1467–1470
13. Pongers-Willemse MJ, Verhagen OJ, Tibbe GJ, Wijkhuijs AJ, de Haas V, Roovers E, van der Schoot CE, van Dongen JJ (1998) Real-time quantitative PCR for the detection of minimal residual disease in acute lymphoblastic leukemia using junctional region specific TaqMan probes. Leukemia 12:2006–2014
14. Van Zelm MC, Van Der Burg M, Langerak AW, Van Dongen JJM (2011) PID comes full circle: applications of V(D)J recombination excision circles in research, diagnostics and newborn screening of primary immunodeficiency disorders. Front Immunol 2:1–9

Chapter 7

The Detection of Chromosomal Translocations Involving the Immunoglobulin Loci in B-Cell Malignancies

Martin J.S. Dyer

Abstract

Chromosomal translocations involving the immunoglobulin (*IG*) loci are frequently seen in most subtypes of B-cell malignancy and have both diagnostic and prognostic utility. These translocations can be detected in clinical samples by several techniques including metaphase cytogenetics, interphase fluorescent *in situ* hybridization, and a variety of PCR methods; interphase FISH is the most commonly used clinical method. Although all the common *IG* translocations have been identified and cloned, new *IG* translocations continue to be identified in both B-cell leukemia and lymphoma. It remains important to identify the involved target genes since they define novel pathogenetic drivers of disease and may represent novel therapeutic targets. This brief chapter outlines methods of detection of chromosomal translocations involving the *IGHJ* segments using long distance inverse (LDI) PCR and their application to clinical lymphoma samples.

Key words: B-cell lymphoma/leukemia, Immunoglobulin translocations, Polymerase chain reaction, Oncogenes

1. Introduction

Acquired genetic abnormalities play a central role in the pathogenesis of malignancy. Structural chromosomal abnormalities can take several different forms including local deletions or amplifications or translocations with interchange of DNA from one chromosomal region to another. More recently, another form of "catastrophic" chromosomal damage, termed chromothripsis has been identified (1). All these abnormalities can now be rapidly detected using modern whole genome sequencing techniques (2). Comprehensive sequence annotation of both genomes and transcriptomes of most major tumor types is now being undertaken. One of the major aims of this huge global enterprise is to define novel therapeutic targets.

Ralf Küppers (ed.), *Lymphoma: Methods and Protocols*, Methods in Molecular Biology, vol. 971,
DOI 10.1007/978-1-62703-269-8_7,

A significant problem remains the identification of genetic events that "drive" the pathogenesis of disease, as opposed to those "passenger" mutations arising nonspecifically from the genetic instability associated with malignancy (3). Interestingly, identical mutations may have different functional consequences in different cell types according to the nature of concurrent genetic abnormalities (4); genetic identification of key target mutations alone may therefore not be sufficient to indicate sensitivity to targeted therapies.

In hematological malignancies, analysis of simple chromosomal translocations, initially identified by metaphase cytogenetics, has been successful in defining key central pathogenetic events; this in turn has lead to the development of new targeted therapies. There are currently two examples of this paradigm. Firstly, the identification of the t(9;22)(q34;q11) chromosomal translocation in chronic myeloid leukemia (CML) in Philadelphia and Edinburgh in 1960, eventually lead to the development of kinase inhibitors that successfully target the resulting chimeric BCR-ABL fusion protein; these drugs have transformed the management of CML (5). Similarly, molecular cloning of t(14;18)(q32;q21) involving *BCL2* in cases of follicular and diffuse large B-cell lymphoma (DLBCL) has lead to the development of rationally designed BCL2 inhibitors such as navitoclax (ABT-263) that can, under some circumstances, induce apoptosis at nanomolar concentrations with minimal toxicity (6). It is noteworthy that despite the wide expression of BCL2 in normal and malignant tissues, some B-cell malignancies are specifically "addicted" to maintained BCL2 functions, and inhibition results in rapid loss of tumor cell viability without obvious effect on normal cells (7). Whether the success of these two paradigms can be repeated with other genetically defined targets remains to be seen, but there are now many molecules undergoing evaluation in different clinical settings (8). A significant problem is the large number of different mutations identified in individual tumor types, as well as the huge genetic complexity within the one neoplastic clone. In B-cell malignancies there are many chromosomal translocations targeting the *IG* loci, involving potential therapeutic targets as various as transcription factors, signaling molecules, miRNAs, and cytokine receptors (9). Somewhat surprisingly, many previously unknown *IG* chromosomal translocations continue to be described and in novel diseases subtypes such as B-cell precursor acute lymphoblastic leukemia (BCP-ALL) (10). Table 1 shows a compendium of *IG* translocations; recently identified/cloned translocations are shaded. Each of these translocations may occur in only a small number of cases, although involvement of the same target gene may also occur *via* other genetic and epigenetic mechanisms, including gene amplification and promoter substitution, as well as gene mutation. An excellent example of several different pathogenetic mechanisms targeting the one gene is found in the *CRLF2* (TSLP receptor) gene in cases

Table 1
IG chromosomal translocations in B-cell malignancies and their target genes

BCP-ALL	BL	MCL	FL	DLBCL	CLL	MALT/MZL	Myeloma
t(X/Y;14)(p11;q32) *CRLF2*	t(8;14) (q24;q32) *MYC*	t(11;14) (q13;q32) *CCND1*	t(14;18)(q32;q21) *BCL2*	t(14;18)(q32;q21) *BCL2*	t(14;18)(q32;q21)[a] *BCL2*	t(1;14)(p22;q32) *BCL10*	t(11;14) (q13;q32) *CCND1*
Various *CEBPA/B/D/E*		t(8;14) (q24;q32) *MYC*	t(3;14)(q27;q32) *BCL6*	t(3;14)(q27;q32) *BCL6*	t(14;19)(q32;q13) *BCL3*	t(14;18)(q32;q21) *MALT1*	t(4;14)(p16;q32) *FGFR3* *MMSET*
t(6;14)(p21;q32) *ID4*			t(8;14)(q24;q32) *MYC*	t(8;14)(q24;q32) *MYC*	t(2;14)(p13;q32) *BCL11A*	t(3;14)(p21;q32) *FOXP1*	t(6;14)(p25;q32) *IRF4*
t(14;19)(q32;p13) *EPOR*			t(1;22)(q21;q12) *FCRG2B*	t(9;14)(p13;q32) *PAX5*	interstitial del(14) (q24.1q32.33) *TRAF3?*	t(7;14)(q21;q32) *CDK6*	t(14;16) (q32;q23) *CMAF*
t(5;14)(q31;q32) *IL3*				t(11;14)(q23;q32) *RCK*		t(X;14)(p11;q32) *GPR34*	t(1;14)(q21;q32) *FCRL4/5*
t(14;17)(q32;q21) *IGF2BP1*				t(12;14)(q24;q32) *BCL7A*		t(10;14)(q24;q32) *NKX2-3*	
t(1;14)(q21;q32) *BCL9*				t(12;22)(p13;q11) *CCND2*			
t(11;14)(q23;q32) *MIR125B*				t(6;14)(p21;q32) *CCND3*			
t(1;14)(q25;q32) *LHX4*				t(14;19)(q32;q13) *SPIB*			
t(10;14)(q26;q32) *DUX4*				t(14;19)(q32;q12) *CCNE*			

(continued)

Table 1 (continued)

BCP-ALL	BL	MCL	FL	DLBCL	CLL	MALT/MZL	Myeloma
t(5;14)(p13;q32) *hTERT*				t(5;14)(q23;q32) *PRDM6*			
				t(14;16)(q32;q24) *CBFA2T3/ACSF3*			

Some B-cell malignancies are associated with a specific *IG* translocation found in the majority of cases, whereas others are associated with multiple different translocations as noted above. Shaded translocations represent those that have been identified and cloned within the past 3 years. (1) *BCP-ALL* B-cell precursor acute lymphoblastic leukemia. These translocations have been reviewed (10). *IG* translocations involving *hTERT* are not specific seen in several subtypes of B-cell malignancy (15); *DUX4* translocations, MJSD, Akasaka and Siebert, unpublished observations. (2) *BL* Burkitt lymphoma. All BL exhibit *MYC* translocations to either *IGH* or *IGL* loci. (3) *MCL* mantle cell lymphoma. (4) *DLBCL* diffuse large B-cell lymphoma. Like BCP-ALL, DLBCL is characterized by a plethora of *IG* translocations many of which remain to be characterized. *SPIB* translocations (16), *CCNE1* translocations (17), and CBFA2T3 translocations (18). *PRDM6* translocations, MJSD et al., unpublished observations. (5) *CLL* chronic lymphocytic leukemia. (6) *MALT lymphoma*; *MZL* marginal zone lymphoma. The t(X;14)(p11;q32) chromosomal translocation involving upregulation of *GPR34* has recently been identified by two separate groups (20, 21). (7) *Myeloma IG* chromosomal translocations reviewed (22)

[a]Note that *BCL2/IGH* translocations can occur in CLL as a secondary event, i.e., present only in a subclone of cells. Breakpoints may occur at either the 5′ end (telomeric) of the gene or within the 3′ region as typically found in FL and other malignancies. BCL2 translocations in CLL can involve either *IGH* or *IGL* sequences. The interstitial deletion del(14)(q24.1q32.33) involves the *IGH* locus with more centromeric sequences (19)

of BCP-ALL (10, 11). Some of these target genes may represent novel therapeutic targets; CRLF2 is a prime example.

In this chapter I describe an LDI-PCR method developed in my laboratory for the molecular cloning of chromosomal translocations directly involving the *IGHJ* segments (12, 13). Whilst this and related methods have proved to be invaluable over the past 15 years, it is likely they will shortly be superseded by genome sequencing methods.

LDI-PCR for translocations involving *IGHJ* segments is a relatively robust method and usually free from artifacts. However it is not a "routine" laboratory technique and should therefore only be performed in attempt to address specific scientific questions. We usually perform this technique following demonstration of *IGH* translocations involving unknown partner regions by a combination of metaphase cytogenetics and interphase FISH. In addition, fresh, frozen material is essential not only to allow extraction of high molecular weight DNA but also to demonstrate over-expression of the targeted oncogene(s) by RT-PCR. My laboratory has focused solely on the use of LDI-PCR for cloning *IGH* translocations. However, it is possible to adapt the method for any translocation where breaks are clustered within a specific region immediately adjacent to the translocation breakpoint. A recent review of the application of LDI-PCR to a variety of other translocations may be found online: http://atlasgeneticsoncology.org/Deep/LDI-PCRinCancerID20087.html.

The principles behind LDI-PCR to *IGH* translocations have been described previously (12, 13). Inverse PCR is a method involving the amplification of restriction endonuclease-digested DNA that has been self-ligated at low concentration to form monomeric circles. It is a simple strategy for amplifying DNA sequences flanked on the one side by a region of known sequence. High molecular weight DNA is digested with restriction enzymes to produce fragments <20 kb; digested DNA is then allowed to re-anneal at very low concentrations that promote self-annealing to produce circles of DNA rather than annealing to other DNA fragments. This is probably the most crucial step in the procedure; poor DNA quality is a major limitation. DNA primers designed in opposite orientations within the known DNA sequence allow PCR amplification around the DNA circle. The unknown sequence is flanked on both sides by known sequences following re-ligation in the resultant amplicon. Using this technique it has been possible to amplify virtually all translocations to the *IGHJ* segments (please see Table 1). Both alleles (translocated allele and productively rearranged *VDJ* allele) are usually obtained in the same PCR mixture despite very different molecular weights (see example Fig. 1 in (14)). With the sequence of the human genome, DNA sequence analysis of even very short regions from the translocation breakpoint is nowadays usually sufficient to identify the breakpoint unequivocally. We have subsequently developed LDI-PCR methods for the *IGHS* regions; this is technically more difficult and more prone to artifacts perhaps due to the repetitive nature of the *IGHS* sequences (13). We have not developed LDI-PCR methods for the *IGL* translocations, although these can be detected by FISH techniques. Similarly, we have not attempted to develop methods for DNA purified from paraffin sections.

2. Materials

1. *High molecular weight DNA.* Ideally, at least 50 μg of high molecular weight DNA should be prepared (see Note 1). Each LDI-PCR reaction usually requires approximately 500 ng of DNA, but it is possible to use much lower input DNA amounts (down to 50 ng). Genomic DNA in TE (10 mM Tris, pH 7.5; 1 mM ethylenediamine tetraacetic acid pH 8.0) is stable at +4°C for at least 2 years.
2. *Restriction enzymes. Bgl*II, *Pst*1, *Hin*dIII, and *Xba*I can all be used; LDI-PCR products should ideally be obtained from multiple different restriction digests. All these enzymes cleave within the intronic *IGH* enhancer and thus produce suitable fragments for LDI-PCR in most instances. Digests of genomic DNA are performed according to the manufacturer's instructions, using buffers provided. Digested DNA can be stored at –20°C if necessary but is better used immediately.
3. *T4 DNA ligase.* Several DNA ligases and buffers are available commercially.
4. *Silica-based DNA purification.* Self-ligated DNA must be purified to remove excess salt from the ligation mixture. Commercially available silica-based DNA purification kits are available for this purpose (Qiagen). Purified circularized DNA is eluted in distilled, sterile water and may be stored at –20°C.
5. *Thermostable DNA polymerases and primers.* Many thermostable DNA polymerases suitable for long-distance amplification are suitable for LDI-PCR. First round PCR amplification is not sufficient to generate a band on agarose gel electrophoresis and nested PCR is necessary to generate a visible product. The primer sequences for LDI-PCR from the human *IGHJ* sequences are shown in Table 2.

Table 2
Sequences of DNA oligonucleotide primers for LDI-PCR from IGHJ sequences

Primer	Nucleotide sequence (5′–3′)	Application
J6E	CCCACAGGCAGTAGCACAAAACAA	External JH6 primer
J6I	GCAGAAAACAAAGGCCCTAGAGTG	Internal JH6 primer
JBE	GAAGCAGGTCACCGCGAGAGT	External primer following *Bgl*II digestion
JBI	CTTCTGGTTGTGAAGAGGGGTTTTG	Internal primer following *Bgl*II digestion
JHE	TGGGATGCGTGGCTTCTGCT	External primer following *Hin*dIII digestion
JHI	GCCCTTGTTAATGGACTTGGAGGA	Internal primer following *Hin*dIII digestion
JXE	CACTGGCATCGCCCTTTGTCTAA	External primer following *Xba*I or *Pst*I digestion
JXI	CCCTGCCTTCCAAAGCGATT	Internal primer following *Xba*I or *Pst*I digestion

3. Methods

Detailed methods for the preparation of high molecular weight DNA, Southern blot analysis, and DNA sequencing may be obtained elsewhere.

3.1. Digestion of High Molecular Weight DNA with Restriction Endonucleases

Use 20 units of restriction endonuclease for 1 μg of DNA. Excess enzymatic activity may cause nonspecific cleavage (star activity) of DNA and therefore high concentrations of enzyme should be avoided. Several enzymes may require the addition of a small amount of BSA for maximum activity. Prepare the following reaction mix in a 1.5 ml microcentrifuge tube on ice.

Reaction mix

10× buffer	2 μl
Restriction endonuclease (20 unit/ml)	1 μl
DNA (500 ng/ml)	2 μl
Distilled sterile H_2O	15 μl

Incubate the reaction mix at the appropriate temperature for 18–24 h. It is important to check that the DNA digestion has gone to completion by agarose gel electrophoresis of a small aliquot on a minigel; completely digested DNA should show a typical satellite bands at about 1 kb. Heat inactivation of the restriction enzyme should be performed.

3.2. Self-Ligation of DNA

Several T4 DNA ligase kits are commercially available. We have used a T4 DNA ligation kit from New England Biolabs but all are suitable. Prepare the following reaction mix in a 1.5 ml microcentrifuge on ice.

	Ligation mix
10× buffer	12.5 μl
Digested DNA	2 μl (100 ng)
T4 ligase (3 U/ml)	0.5 μl
Distilled sterile H_2O	110 μl

Incubate the reaction at 15°C for 16–24 h. The final concentration should be 0.8 ng/μl (see Note 2).

3.3. Purification of the Ligation Mix Using a Silica-Based Column

Before using the self-ligated circular DNA as a template for LDI-PCR, purification of the ligated DNA must be conducted to remove excess salts. Various silica-based column kits are available. We have used a PCR purification kit from Qiagen for this purpose; detailed procedure is as per the manufacturer's instructions. Purified DNA is eluted with 50 μl of sterile distilled H_2O and can be stored at −20°C.

3.4. LDI-PCR

The components of the LDI-PCR reaction should be made according to the thermostable DNA polymerase manufacturer's instructions. In particular, the manufacturers have optimized the Mg concentration as well as annealing/extension times according to the size of the DNA target. The annealing and extension temperature should be fixed at 68°C to amplify specific and long DNA targets. One procedure for LDI-PCR of purified self-ligated circular DNA using primers to amplify from *IGHJ* is as follows:

	PCR reaction mix	Premix (3.2×)
3.3× GeneAmp XL Buffer II	15 μl	48 μl
10 μM dNTPs	1 μl	3.2 μl
10 μM forward primer	1 μl	3.2 μl
10 μM reverse primer	1 μl	3.2 μl
Purified circularized DNA	5 μl (10 ng)	NIL
25 mM $Mg(OAc)_2$	5 μl	NIL
rTth thermostable DNA polymerase	1 μl	3.2 μl
Distilled sterile H_2O	21 μl	67.2 μl

Prepare the reaction premix in a 1.5 ml microcentrifuge tube on ice. To achieve a "hot start" PCR method omit the $Mg(OAc)_2$ at this stage. Mix by gentle inversion and collect the components by brief centrifugation. Transfer 40 μl of the premix into a thin-walled PCR tube and then add 5 μl (about 10 ng) of the purified circular DNA. Heat the reaction mix at 85°C for 3 min and then add 5 μl of 10× $Mg(OAc)_2$. The PCR program is 94°C for 5 min followed by 30 cycles of 94°C for 1 min and 68°C for 6 min followed by terminal extension at 68°C for 7 min. One microliter of the PCR reaction is then re-amplified for a further 30 cycles using identical conditions with nested primers. Amplification may be performed on a variety of programmable heating blocks.

3.5. PCR Product Analysis and DNA Sequencing

5 μl of the 50 μl PCR product is analyzed on a 0.8 % agarose gel in 1× TAE buffer. When a PCR product of potential interest is obtained, run the whole nested product on an agarose gel and isolate DNA from a gel slice using a silica-based DNA purification kit (Qiagen). It is usually possible to obtain the DNA sequence directly using the nested primers. However, when the PCR product is of low level or of a large size, cloning into a plasmid is recommended. The TA cloning kit from Invitrogen is ideal for this application. On occasions, direct sequencing using nested primers fails to reach the *IGHJ* breakpoint. In such cases further primers will need to be designed. After sequencing standard NBCI Blast programs (http://www.ncbi.nlm.nih.gov/BLAST/) are used to identify the *IGHJ* sequences and any additional "unknown" sequences in the LDI-PCR product (see Note 3). A full description of the methods necessary to analyze the sequences obtained is beyond the scope of this chapter. However, the human genome sequence has greatly facilitated analysis and now only a few base pairs of sequence beyond the translocation breakpoint are necessary. Once the translocation breakpoint has been defined the involved gene is usually readily identifiable as the breaks usually fall either immediately 5′ or 3′ to the gene of interest. Breakpoints may fall with either 5′ or 3′ untranslated regions, but there has been no example to my knowledge of the protein coding region of the gene being disrupted. It should be remembered however that on occasion, breakpoints may fall many hundreds of kilobases distant of the target gene (for example in *MYC/IGH* translocations); this will require detailed analysis of the transcriptome. Finally, the possibility of two target genes being involved should be considered, especially in the context of *IGHS* translocations, since the intronic enhancer is translocated to the derivative partner chromosome, whilst the 3′ enhancers are retained on the derivative chromosome 14, thus causing two target genes to be deregulated simultaneously.

4. Notes

1. Quality of the extracted DNA is paramount for successful LDI-PCR, particularly when long distance targets have to be amplified.
2. The concentration of digested DNA used in the circularization reaction is crucial as excess digested DNA may inhibit circle formation. This is difficult to control; performing LDI-PCR from a cell line with known rearrangements concurrently may help.
3. We used to advise Southern blotting of genomic DNA to ensure that products obtained by LDI-PCR were not artifactual, but with the availability of the human genome sequence this should not now be necessary particularly if the chromosomal breakpoint has been localized to specific region by interphase FISH.

Dedication and Acknowledgments

This chapter is dedicated to all those who have worked in my laboratory over the past 20 years and have contributed to the development of LDI-PCR, including Drs Tony Willis, Loraine Karran, Takashi Sonoki, and last but certainly not least, Takashi Akasaka; it has been a privilege to work with you. I gratefully acknowledge generous financial support from the Medical Research Council, Leukemia Lymphoma Research, the Hope Against Cancer. Additional references and offprints of selected articles may be obtained on request.

References

1. Maher CA, Wilson RK (2012) Chromothripsis and human disease: piecing together the shattering process. Cell 148:29–32
2. Stratton MR (2011) Exploring the genomes of cancer cells: progress and promise. Science 331:1553–1558
3. Bozic I, Antal T, Ohtsuki H, Carter H, Kim D, Chen S, Karchin R, Kinzler KW, Vogelstein B, Nowak MA (2010) Accumulation of driver and passenger mutations during tumor progression. Proc Natl Acad Sci U S A 107:18545–18550
4. Prahallad A, Sun C, Huang S, Di Nicolantonio F, Salazar R, Zecchin D, Beijersbergen RL, Bardelli A, Bernards R (2012) Unresponsiveness of colon cancer to BRAF(V600E) inhibition through feedback activation of EGFR. Nature 483:100–103
5. Druker BJ (2008) Translation of the Philadelphia chromosome into therapy for CML. Blood 112:4808–4817
6. Oltersdorf T, Elmore SW, Shoemaker AR, Armstrong RC, Augeri DJ, Belli BA, Bruncko M, Deckwerth TL, Dinges J, Hajduk PJ, Joseph MK, Kitada S, Korsmeyer SJ, Kunzer AR, Letai A, Li C, Mitten MJ, Nettesheim DG, Ng S, Nimmer PM, O'Connor JM, Oleksijew A, Petros AM, Reed JC, Shen W, Tahir SK, Thompson CB, Tomaselli KJ, Wang B, Wendt MD, Zhang H, Fesik SW, Rosenberg SH (2005) An inhibitor of Bcl-2 family proteins induces regression of solid tumours. Nature 435:677–681
7. Letai AG (2008) Diagnosing and exploiting cancer's addiction to blocks in apoptosis. Nat Rev Cancer 8:121–132

8. Dawson MA, Prinjha RK, Dittmann A, Giotopoulos G, Bantscheff M, Chan WI, Robson SC, Chung CW, Hopf C, Savitski MM, Huthmacher C, Gudgin E, Lugo D, Beinke S, Chapman TD, Roberts EJ, Soden PE, Auger KR, Mirguet O, Doehner K, Delwel R, Burnett AK, Jeffrey P, Drewes G, Lee K, Huntly BJ, Kouzarides T (2011) Inhibition of BET recruitment to chromatin as an effective treatment for MLL-fusion leukaemia. Nature 478:529–533
9. Willis TG, Dyer MJ (2000) The role of immunoglobulin translocations in the pathogenesis of B-cell malignancies. Blood 96:808–822
10. Dyer MJ, Akasaka T, Capasso M, Dusanjh P, Lee YF, Karran EL, Nagel I, Vater I, Cario G, Siebert R (2010) Immunoglobulin heavy chain locus chromosomal translocations in B-cell precursor acute lymphoblastic leukemia: rare clinical curios or potent genetic drivers? Blood 115:1490–1499
11. Russell LJ, Capasso M, Vater I, Akasaka T, Bernard OA, Calasanz MJ, Chandrasekaran T, Chapiro E, Gesk S, Griffiths M, Guttery DS, Haferlach C, Harder L, Heidenreich O, Irving J, Kearney L, Nguyen-Khac F, Machado L, Minto L, Majid A, Moorman AV, Morrison H, Rand V, Strefford JC, Schwab C, Tonnies H, Dyer MJ, Siebert R, Harrison CJ (2009) Deregulated expression of cytokine receptor gene, CRLF2, is involved in lymphoid transformation in B-cell precursor acute lymphoblastic leukemia. Blood 114:2688–2698
12. Willis TG, Jadayel DM, Coignet LJ, Abdul-Rauf M, Treleaven JG, Catovsky D, Dyer MJ (1997) Rapid molecular cloning of rearrangements of the IGHJ locus using long-distance inverse polymerase chain reaction. Blood 90:2456–2464
13. Sonoki T, Willis TG, Oscier DG, Karran EL, Siebert R, Dyer MJ (2004) Rapid amplification of immunoglobulin heavy chain switch (IGHS) translocation breakpoints using long-distance inverse PCR. Leukemia 18:2026–2031
14. Willis TG, Jadayel DM, Du MQ, Peng H, Perry AR, Abdul-Rauf M, Price H, Karran L, Majekodunmi O, Wlodarska I, Pan L, Crook T, Hamoudi R, Isaacson PG, Dyer MJ (1999) Bcl10 is involved in t(1;14)(p22;q32) of MALT B cell lymphoma and mutated in multiple tumor types. Cell 96:35–45
15. Nagel I, Szczepanowski M, Martin-Subero JI, Harder L, Akasaka T, Ammerpohl O, Callet-Bauchu E, Gascoyne RD, Gesk S, Horsman D, Klapper W, Majid A, Martinez-Climent JA, Stilgenbauer S, Tonnies H, Dyer MJ, Siebert R (2010) Deregulation of the telomerase reverse transcriptase (TERT) gene by chromosomal translocations in B-cell malignancies. Blood 116:1317–1320
16. Lenz G, Nagel I, Siebert R, Roschke AV, Sanger W, Wright GW, Dave SS, Tan B, Zhao H, Rosenwald A, Muller-Hermelink HK, Gascoyne RD, Campo E, Jaffe ES, Smeland EB, Fisher RI, Kuehl WM, Chan WC, Staudt LM (2007) Aberrant immunoglobulin class switch recombination and switch translocations in activated B cell-like diffuse large B cell lymphoma. J Exp Med 204:633–643
17. Nagel I, Akasaka T, Klapper W, Gesk S, Bottcher S, Ritgen M, Harder L, Kneba M, Dyer MJ, Siebert R (2009) Identification of the gene encoding cyclin E1 (CCNE1) as a novel IGH translocation partner in t(14;19)(q32;q12) in diffuse large B-cell lymphoma. Haematologica 94:1020–1023
18. Salaverria I, Akasaka T, Gesk S, Szczepanowski M, Burkhardt B, Harder L, Damm-Welk C, Oschlies I, Klapper W, Dyer MJ, Siebert R (2012) The CBFA2T3/ACSF3 locus is recurrently involved in IGH chromosomal translocation t(14;16)(q32;q24) in pediatric B-cell lymphoma with germinal center phenotype. Genes Chromosomes Cancer 51:338–343
19. Nagel I, Bug S, Tonnies H, Ammerpohl O, Richter J, Vater I, Callet-Bauchu E, Calasanz MJ, Martinez-Climent JA, Bastard C, Salido M, Schroers E, Martin-Subero JI, Gesk S, Harder L, Majid A, Dyer MJ, Siebert R (2009) Biallelic inactivation of TRAF3 in a subset of B-cell lymphomas with interstitial del(14)(q24.1q32.33). Leukemia 23:2153–2155
20. Remstein ED, Novak AJ, Akaska T, Manske M, Gupta M, Witzig TE, Dyer MJS, Ansell SM, Dogan A (2009) A novel translocation in MALT lymphoma involving a G-protein coupled receptor (GPR34) and immunoglobulin heavy chain gene activates NFKB pathway. Modern Pathol 22:282a
21. Baens M, Finalet Ferreiro J, Tousseyn T, Urbankova H, Michaux L, de Leval L, Dierickx D, Wolter P, Sagaert X, Vandenberghe P, De Wolf-Peeters C, Wlodarska I (2012) t(X;14)(p11.4;q32.33) is recurrent in marginal zone lymphoma and up-regulates GPR34. Haematologica 97:184–188
22. Zhan F, Huang Y, Colla S, Stewart JP, Hanamura L, Gupta S, Epstein J, Yaccoby S, Sawyer J, Burington B, Anaissie E, Hollmig K, Pineda-Roman M, Tricot G, van Rhee F, Walker R, Zangari M, Crowley J, Barlogie B, Shaughnessy JD (2006) The molecular classification of multiple myeloma. Blood 108:2020–2028

Chapter 8

Stereotyped B Cell Receptors in B Cell Leukemias and Lymphomas

Nikos Darzentas and Kostas Stamatopoulos

Abstract

Recent research has revealed the existence of subsets (clusters) of patients with different types of B-cell lymphomas and leukemias with restricted, "stereotyped" immunoglobulin (IG) variable heavy complementarity-determining region 3 (VH CDR3) sequences within their B cell receptors (BcR), suggesting selection by common epitopes or classes of structurally similar epitopes. BcR stereotypy was initially described in chronic lymphocytic leukemia (CLL), where it constitutes a remarkably frequent feature of the IG repertoire, and subsequently identified in other malignancies, including mantle cell lymphoma and splenic marginal-zone lymphoma. Of note, at least in CLL, emerging evidence indicates that the grouping of cases into distinct clusters with stereotyped BcR is functionally and prognostically relevant. Hence, the reliable identification of BcR stereotypy may assist in the investigation of the nature of the selecting antigens and immune pathways leading to lymphoma development, and also potentially pave the way for tailored treatment strategies applicable to each major stereotyped subset. In this chapter, we provide an overview of BcR stereotypy in human B-cell malignancies, and outline previous and current methodological approaches used for its identification.

Key words: B cell receptor, Immunoglobulin gene, CDR3, Antigen, Pattern, Stereotypy, Bioinformatics

1. Introduction

1.1. Immunogenetic Evidence for Antigen Selection in Malignancies of Mature B Cells

Research over the last two decades indicates the important role of antigen recognition and/or selection through the clonotypic BcR in the immune pathogenesis of malignancies of mature B cells. Immunogenetic studies have been instrumental in shaping this notion by reporting restricted repertoires of immunoglobulin (IG) heavy and light variable genes (*IGHV* and *IGKV/IGLV*, respectively) for different B-cell malignancies (1, 2). This finding is widely taken as evidence that antigens and/or superantigens may be involved in lymphoma development by stimulating proliferation and expansion of B cells expressing distinctive BcR. Furthermore, many lymphomas exhibit imprints of SHM on IG genes typical of selection by antigen,

Ralf Küppers (ed.), *Lymphoma: Methods and Protocols*, Methods in Molecular Biology, vol. 971,
DOI 10.1007/978-1-62703-269-8_8, © Springer Science+Business Media, LLC 2013

while in some cases there is evidence for intraclonal diversification through ongoing mutational activity post-transformation (3). Finally, most lymphoma malignant cells do not survive or proliferate autonomously in vitro, indicating that they are still engaging on a crosstalk with their microenvironment and are largely dependent on external signals for their expansion (4). Although the precise nature of these signals is still mostly unknown, in at least some instances it might involve antigenic stimulation.

From a clinical standpoint, the importance of BcR-mediated stimulation in promoting the development and evolution and also dictating the eventual clinical outcome is underscored by the therapeutic efficacy of small-molecule antagonists of BcR-associated kinases (spleen tyrosine kinase (Syk), Bruton's tyrosine kinase (Btk), and PI3Kδ) for targeting specific interactions between lymphoma cells and the microenvironment, at least in chronic lymphocytic leukemia (CLL) (5) and mantle cell lymphoma (MCL) (6).

1.2. B Cell Receptor Stereotypy: CLL Paves the Way

In recent years we have witnessed a transformation in the concepts concerning the immune development of CLL. What was previously thought a disease of naïve, immune incompetent B cells is now considered a prototype for cancers where both intrinsic (genetic) and extrinsic (microenvironmental) factors are known to play a key role in the onset, expansion and progression of the disease (7).

The first hints that CLL may express a restricted *IGHV* gene repertoire originated from studies in the early 1990s (8). However, a milestone in this line of research came in 1998 when Chiorazzi's group showed that the *IGHV* gene repertoire of CLL is nonrandom and distinct from the normal B-cell repertoire and also provided evidence for *IGHV* subgroup and gene-related differences in SHM status and heavy chain CDR3 (VH CDR3) composition and features (9). Since the original publication, all relevant studies have reached concordant results, while, some years later, similar asymmetries were shown to exist for the light chain genes as well (10), overall, strongly supporting the hypothesis that CLL cells may be antigen experienced and driven.

A role for antigen in CLL was also inferred by the fact that SHM present in rearranged *IGHV* genes in CLL defines two disease subtypes associated with a different clinical course and outcome. In particular, patients carrying mutated *IGHV* genes have consistently been found to follow a more indolent course than those with unmutated *IGHV* genes, who tend to show evidence of advanced, progressive disease, adverse cytogenetic features, clonal evolution, and resistance to therapy (11, 12).

IG gene sequence analysis is now widely utilized for risk stratification in CLL, paving the way for a biologically oriented assessment of prognosis. In fact, the definition of standardized procedures has allowed for reliable and reproducible results and also transformed what was previously thought a research tool into a routine diagnostic procedure (13, 14). Through this paradigm shift, a large body of IG sequence data became available, hence enabling to perform meaningful immunogenetic studies which have helped reveal some of the secrets of the IG genes.

One of the best-kept such secrets concerned the remarkable sequence restrictions of the antigen-binding sites in CLL. The first hints pointing to this direction originated in the mid-1990s when it was reported that unrelated CLL cases may carry highly similar, distinctive VH CDR3, prompting speculation that a similar type of antigenic drive might have been involved in their diversification and selection processes (15–17). A little later, the restricted combined usage of *IGVH/IGHD/IGHJ* genes in CLL was corroborated by the identification of a few sets of "prototypic" BcRs with highly similar VH CDR3s. For instance, a "prototypic" *IGHV1-69* BcR was predominantly unmutated and used an *IGHD3-3* gene and an *IGHJ6* gene, encoding a long, tyrosine-rich highly acidic VH CDR3 (18). However, the turning point in this line of research came in 2003, when it was demonstrated that approximately half of CLL cases utilizing the *IGHV3-21* gene display distinctive IGs with a short and highly similar, if not identical, VH CDR3 and remarkably biased pairing with lambda light chains utilizing the *IGLV3-21* gene (19) (Fig. 1). This observation was justifiably interpreted as evidence for common antigenic drive, perhaps of pathogenic significance.

Soon thereafter, several groups from both Europe and the US reported subsets of CLL cases carrying quasi-identical antigen-binding sites among both mutated and unmutated cases (20–23). Hence, the fact that distinct prototypic BcRs could be repeated with minimum or no variation in different subsets of patients

Fig. 1. Restricted VH CDR3s among unrelated CLL cases utilizing the IGHV3-21 gene. In 2003, Tobin et al. (19) reported that almost half of all IGHV3-21 CLL cases display distinctive IGs with a short and highly similar, if not identical, VH CDR3 leading to their clustering in a subset, nowadays referred to as subset #2. The sequence logo presented here depicts the VH CDR3s of all cases assigned to subset #2 in our recent study of 7,424 cases with CLL (39). In this logo, each VH CDR3 position is displayed as a stack of upright amino acid symbols. The height of each one-letter amino acid symbol is directly proportional to the relative frequency of that amino acid at a given position in the alignment among all sequences of the subset. Rearrangements belonging to this subset can be simply identified by a nine amino-acid (aa) long VH CDR3 with an acidic residue (aspartic acid D) at position 107.

with CLL was irrefutable. Considering the fact that the possibility of two independent B-cell clones carrying exactly the same BcR is virtually negligible (1:10^{-9} to 1:10^{-12}), this finding is at variance with the "logistics" of IG synthesis. From a biologic perspective, it clearly argues against serendipity and, instead, favors an antigen-driven pathway to CLL development. To describe this phenomenon, Chiorazzi's group applied the term "stereotyped" (22): indeed, the quasi-identical CLL BcRs in unrelated and geographically distant cases fulfill the definition of stereotype as "something conforming to a fixed or general pattern."

The study of stereotypy in large cohorts of patients has proven extremely rewarding for understanding the pathogenesis of CLL. Recent studies by us and others have conclusively proved the existence of different subsets of cases with distinct stereotyped VH CDR3 sequences within their BcRs collectively accounting for almost one-third of the CLL repertoire (24–27). Of note, the *IGHV* gene repertoire restrictions typical of CLL were in essence shown to be a property confined to stereotyped cases, clearly segregating them from the non-stereotyped ones, leading to speculations about differential ontogeny and/or selective processes underlying the development of the stereotyped versus non-stereotyped CLL categories (27). From another standpoint, these studies have also shown that stereotypy may extend from VH CDR3 sequence patterns to shared biological and clinical characteristics and, perhaps, outcome, suggesting that a particular antigen-binding site can be critical in determining the clinical features and prognosis of subsets of CLL patients (24, 25). Therefore, the study of BcR stereotypy has important implications not only for elucidating the pathogenesis of CLL but also for improving patient group stratification towards personalized therapeutic applications.

1.3. B Cell Receptor Stereotypes in Mature B Cell Malignancies Other Than CLL

For some time after the initial publications, BcR stereotypy was thought to be exclusive to CLL since limited or no VH CDR3 similarity among unrelated cases was found in other lymphomas. A caveat of the early studies was the relatively small number of sequences available for examination and cross-entity comparisons. However, more recently, the situation has changed, enabling to conduct this type of analysis in sizeable cohorts of cases with lymphomas other than CLL, most notably MCL and splenic marginal-zone lymphoma (SMZL).

The study of the IG gene repertoire in MCL has focused mainly on *IGH* gene rearrangements. A consistent finding in all series concerns the strong bias to the usage of the certain *IGHV* genes, in particular, *IGHV4-34*, *IGHV3-21*, and *IGHV3-23*. We recently confirmed and significantly extended these observations in a cohort of 807 patients with MCL, showing that only four genes (*IGHV3-21*, *IGHV4-34*, *IGHV1-8*, and *IGHV3-23*) collectively accounted for almost half of the cases (28). On these grounds, we suggested that selection by antigens and/or superantigens may be specifically implicated in MCL development (29).

Biases in the *IGHD* and *IGHJ* gene repertoires and other distinctive VH CDR3 features were already apparent in the early immunogenetic studies of MCL. In our recent study, we found restricted associations of specific *IGHV* genes with certain *IGHD* or *IGHJ* genes (28). For instance, more than 30% of *IGHV3-21* rearrangements utilized the *IGHD3-3* gene. Similar, though less pronounced biases were seen for *IGHV1-8* and *IGHV4-34* rearrangements, which were frequently recombined to the *IGHD2-2* or the *IGHD2-15* gene. Furthermore, differential patterns of associations with *IGHJ* genes were evident for certain *IGHV* genes: in fact, almost two-thirds of *IGHV3-21* rearrangements utilized the IGHJ6 gene, compared to <10% of *IGHV3-23* rearrangements ($p<0.001$). These restricted associations of *IGHV*, *IGHD*, and *IGHJ* genes underlie the formation of distinctive antigen-binding sites with highly similar (stereotyped) VH CDR3 motifs, especially among subgroups of *IGHV3-21*- and *IGHV4-34*-expressing cases (28).

Given how much attraction stereotyped BcRs have held not only among scientists working on CLL but also for the scientific community at large, it was only natural to ask if and how MCL stereotypes might relate to those identified in CLL. In order to address this question, we performed detailed comparison of the amino acid motifs defining VH CDR3 stereotypes in MCL versus CLL cases utilizing the *IGHV3-21* and *IGHV4-34* genes and found that the observed sequence restrictions were indeed "disease-biased." Hence, the stereotyped BcRs in MCL differed from those in CLL to such an extent that distinct immune-mediated mechanisms of lymphomagenesis could be invoked (28, 29).

Studies by several groups, including ours, have shown that SMZL exhibits a remarkably biased IG gene repertoire. In fact, very recently we demonstrated that over 30% of cases with a diagnosis of SMZL based on splenic histopathology express distinctive IG receptors that utilize a single polymorphic variant of the *IGHV1-2* gene (*IGHV1-2*04*) and also exhibit a generally low impact of SHM (with the great majority showing an identity to the germline ranging from 97 to 99.7%) (30). The *IGHV1-2*04* rearrangements in SMZL were found to carry long, electropositive VH CDR3s and showed biased usage of the *IGHD3-3* and *IGHD3-10* genes, mostly in reading frame-3, resulting in VH CDR3s with common "IGHD-derived" motifs (30). Although a few such rearrangements were indeed stereotyped (a finding independently reported in another recent study (31)), the vast majority of *IGHV1-2*04* rearrangements did not meet the established criteria for VH CDR3 stereotypy, as previously defined in CLL, mainly because of variable IGHJ gene usage and heterogeneous junctional residues. Hence, altogether, these results might be taken as evidence for an "SMZL-biased" type of IG stereotypy, defined by restrictions in

VH CDR3 length along only a few conserved amino acids in the apex of the VH CDR3.

With the identification of stereotyped BcRs not only in CLL but also in other lymphomas as well, namely MCL and SMZL, albeit at significantly lower frequencies than CLL, it was only reasonable to ask the question whether the various stereotypes are "disease-biased" or else "common." In order to answer this question, we performed cross-entity comparisons of the identified VH CDR3 sequence patterns to search for potential commonalities between their antigen-binding sites. Focusing on stereotypes utilizing *IGHV* genes frequent in each lymphoma subtype (CLL versus MCL versus SMZL), we found differences so fundamental that the current answer can only be "disease-biased."

With all this in mind, we must consider a fundamental issue. How can BcR stereotypes be identified?

2. Criteria and Methods for the Identification of BcR Stereotypy

In all studies dealing with BcR stereotypy, great emphasis is given on the IG heavy chains, especially the VH CDR3, the region of the heavy chain that plays a very critical role in the makeup of the antigen-binding site (32) to such an extent that the more similar the primary VH CDR3 sequences of two IGs, the more similar their folding and, likely, their specificities (33).

The detection of stereotypy in BcRs had an obvious starting point, the initial analysis of and comparison between rearranged primary sequences of the receptors themselves. Bioinformatics resources such as the IMGT, the international ImMunoGeneTics information system, provide consistent germline gene and reading frame assignment, robust alignment, and codon numbering, translation to amino acid sequence, as well as a detailed decoding of the junctional sequence.

2.1. Messmer 2004

Based on such basic information, the first set of criteria for the definition of subsets was proposed in 2004 by Messmer et al. (22). It required potentially stereotyped receptors to:

1. Use the same *IGHV*, *IGHD*, *IGHJ* germline genes.
2. Use the same *IGHD* gene reading frame.
3. Feature VH CDR3 amino acid identity between them equal to or >60%, a threshold in line with established bioinformatics concepts such as amino acid substitution matrices and in particular BLOSUM62 (34).

In an alignment of sequences, studying only identical amino acids may not be adequate to understand their evolutionary and functional relationship. When there are differences in amino acid composition, positions that are important for the structure and function of the sequence seem to only tolerate specific substitutions, termed conservative, where an amino acid residue is replaced by another having similar physicochemical properties. This phenomenon has been quantified through the study of nonidentical residues in short, highly conserved sequence blocks, resulting in now ubiquitous amino acid substitution scoring systems such as BLOSUM (blocks substitution matrix). In particular, BLOSUM62 was generated from sequences that were at least 62% identical between them, and while similar tables have been generated at the 50% or 80% thresholds, BLOSUM62 has been shown to be the most reliable scoring system over a wide range of protein families—thus also initially applicable to the study of IG receptors.

2.2. Stamatopoulos 2007

Despite this representing a major step forward, it soon became apparent that the first criterion could not be met in all cases. In fact, even in the original publication by Messmer et al., stereotyped cases were identified using different *IGHV* genes. Subsequently, in a study from our group including 916 CLL cases of Mediterranean origin, an extended methodology was devised (24), where criteria comprised of:

1. Minimum VH CDR3 amino acid sequence identity of 60%.
2. Functional conservation of amino acids (based on the BLOSUM theory outlined above).
3. Maximum VH CDR3 length difference of three amino acids.
4. Usage of different *IGHV* genes.

An illustrative example of the validity of this approach was provided by a cluster now defined as subset #1 (24, 25) which was characterized by usage of the *IGHD6-19* and *IGHJ4* genes in association with different *IGHV* genes (namely *IGHV1-2*, *IGHV1-3*, *IGHV1-18*, *IGHV1-8*, *IGHV5-a*, *IGHV7-4-1*) (Fig. 2). Of note, in the case of subset #1, several lines of evidence support the clustering approach based on primary IG sequences documenting

Fig. 2. Restricted VH CDR3s in stereotyped BcRs of subset #1 CLL cases. This subset is characterized by usage of the *IGHD6-19* and *IGHJ4* genes in association with different *IGHV* genes (namely *IGHV1-2*, *IGHV1-3*, *IGHV1-18*, *IGHV1-8*, *IGHV5-a*, *IGHV7-4-1*), all however members of phylogenetic clan I. The sequence logo presented here depicts the VH CDR3s of all cases assigned to subset #1 in our recent study of 7,424 cases with CLL (39).

that it extends from sequence similarity to shared functional (35) and clinical (24, 26) features. Interestingly, all the *IGHV* genes utilized in *IGHV/IGHD/IGHJ* rearrangements assigned to subset #1 are members of the phylogenetic clan I (36), which means that their germline sequences are closely related (also see below).

2.3. Darzentas 2010

Soon however, the accumulation of sequence data for thousands of B cell leukemia and lymphoma patients rendered such, mainly manual, approaches unrealistic with regards to efficiency, robustness, and, importantly, sensitivity. To meet the challenge, our group developed purpose-built bioinformatics methods based on de novo sequence pattern (motif) discovery with the TEIRESIAS algorithm from IBM Research (37), enabling sophisticated identification of VH CDR3 sequence similarities regardless of the *IGHV/IGHD/IGHJ* genes used.

A highly significant development introduced with this approach was the establishment of more distant relationships between sequences in a tree-like fashion, with first (ground) level clusters merging, under certain conditions, into higher level clusters with less stringent membership criteria and more members.

In summary, the algorithm discovers common sequence patterns across curated VH CDR3 sequences, takes the highest scoring pattern-based connections between sequences and starts building, in a highly controlled manner, clusters on the ground level (before potentially merging them as mentioned above and explained below), in which sequences are guaranteed to:

1. Exhibit at least 50% amino acid identity and 70% amino acid similarity against all other members of that cluster.
2. Have sequence lengths no more than two amino acids different.
3. Display the identified sequence patterns in positions no more than two amino acids apart.

In more detail, the steps up to the clustering (explained extensively with Fig. 3) are as follows.

1. Sequences are curated to be free of ambiguities, with proper length and characterized by a unique identifier.
2. The input to the pattern discovery algorithm, TEIRESIAS, is a FASTA format file, containing VH CDR3 sequences, e.g.,
```
> sequence A
CAHSGITGTTGFDYW
> sequence B
CARGGWDLNYYFDYW
```
3. For reasons of clarity and computational efficiency, the C and W residues, the bookends of the VH CDR3, are removed prior to the pattern discovery process, e.g.,
```
> sequence A
C AHSGITGTTGFDY W
```

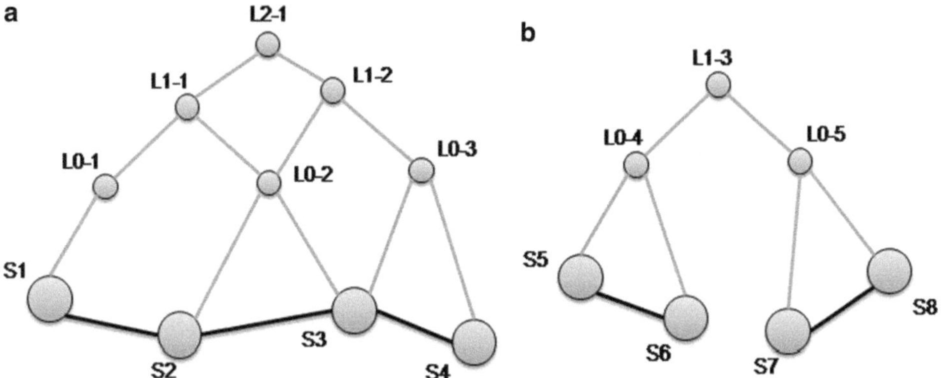

Fig. 3. A network representation of two clustering examples, namely the formation of clusters L2-1 and L1-3 with four samples each. Black nodes are either clusters (smaller nodes) or samples (larger nodes). *Gray lines* connect clusters to samples and clusters to clusters, while *black lines* connect samples between themselves. (**a**) Level 2 cluster L2-1 with samples S1, S2, S3, and S4. (**b**) Level 1 cluster L1-3 with samples S5, S6, S7, and S8. The clustering procedure for these samples is as follows: sample S1 (VH CDR3 sequence CARGPDESGWCGFRYW) is connected to S2 (VH CDR3 sequence CARGPDISGWNGFEYW) by pattern ARGPD.SGW.GF.Y providing 78.57% identity, forming the Level 0 cluster L0-1. S2 is then found to be connected to S3 (CARGPDTSGWNSLDYW) by ARGPD.SGWN..[DE]Y providing 71.43% identity and 78.57% similarity, making S3 a candidate for membership in cluster L0-1. However, S3 is not found to be connected to the other member of the cluster, S1, and therefore is not allowed to join the cluster. Consequently, S3 forms a new Level 0 cluster L0-2 by "borrowing" S2. S3 (CARGPDTSGWNSLDYW) is finally found to be connected to S4 (CARGPDESGWLALAYW) by ARGPD.SGW..L.Y with 71.43% identity, making S4 a candidate for membership in cluster L0-2. As in the previous case though, S4 is not found to be connected to the other member of the cluster, S2, therefore it has to "borrow" S3 and form yet another new Level 0 cluster L0-3. Since clusters L0-1 and L0-2 share S2 they are connected on Level 1 to form cluster L1-1; similarly, clusters L0-2 and L0-3 share S3 and are thus connected on Level 1 to form cluster L1-2. Finally, since clusters L1-1 and L1-2 share cluster L0-2, they form the Level 2 cluster L2-1 (**a**). Samples S5 (CARAGEMATLMGLGAFDIW) and S6 (CARAGEMATVFGRGAFDIW) are connected by pattern ARAGEMAT[AVLI].G.GAFDI providing 82.35% identity and 88.24% similarity to form Level 0 cluster L0-4. With the same identity and similarity values, the pattern REGEMAT.[KRH] GFGAFDI connects samples S7 (CGREGEMATQRGFGAFDIW) and S8 (CAREGEMATMKGFGAFDIW) to form cluster L0-5. Pattern AR.GEMAT..G.GAFDI connects S8, S6, and S5 (76.47% identity), and all four samples share pattern R.GEMAT..G.GAFDI (70.59% identity). Therefore the two Level 0 clusters reconnected into the L1-3 cluster (**b**). The second example (**b**) provides the opportunity to make some important remarks regarding the clustering procedure. One could argue that S5, S6, S7, and S8 should be in the same L0 cluster since they are all connected between themselves by the same pattern albeit with a lower score. However, this would lead to a loss of information, in this case the fact that S7 and S8 are more similar between themselves than across to S5 and S6, and vice versa. The issue is addressed by higher-level clusters, with the end result that if two sequences are above-threshold identical and similar they are guaranteed to reside within a cluster of some level, i.e., the two sequences could be in different L0 and L1 clusters but in the same L2 cluster.

```
> sequence B
C ARGGWDLNYYFDY W
```

4. The parameters TEIRESIAS uses are amino acids in the pattern (-*l*), number of overlapping characters in the convolved pattern (-*c*), maximum length of an elementary pattern (-*w*), minimum number of appearances of the pattern (-*k*), maximum number of brackets (indicating equivalent amino acids) allowed in the pattern (-*n*), flag for the support *k* to be the minimum number of sequences in which a pattern should appear (compared to the minimum number of instances of the pattern) (-*v*), flag for the algorithm to use amino acid equivalences (-*b*<equivalences

file>), and flag for the algorithm to consider only the uppercase characters during pattern discovery (-*u*). Default parameters are [$l=3$] [$c=2$] [$w=6$] [$k=2$] [$n=2$] [-*v*] [-*b*<equivalences file>] [-*u*]. These values ensure that the procedure will be extremely sensitive, trying to connect all pairs ($k=2$) of sequences that share at least 50% identity, defined by the l/w ($3/6=0.5$) ratio—a higher similarity threshold is set by the use of amino acid equivalences ($n=2$), i.e., by allowing some ambiguity (37). The equivalences used are AVLI, CM, ST, KRH, DE, NQ, F, W, P, G, Y, based on previous IMGT research (38). To better explain the format, [DE] means that D and E are considered equivalent amino acids, and can occupy the same position in different patterns, and the patterns would be considered the same, i.e., ACD and ACE could both be described by pattern AC[DE].

5. The output is a list of patterns, which are pre-filtered in order to pass the 50% identity and 70% similarity thresholds, e.g.,
   ```
   ARD.NGMDV
   A[KRH]D..GMDV
   DR[ST]GMDV
   E.[AVLI]PFDAFDI
   ```
6. Since sequences with (very) different lengths may not fold the same way, we do not allow a pattern to describe sequences that differ by more than two amino acids in length (i.e., one sequence in a cluster can be 15 amino acids in length and the other 17 but not 18), e.g., schematically ("#" is a position covered by a pattern)
   ```
   - # # # # - - -               ✓
   - # # # # - - - - -       ✓
   - # # # # - - - -         ✓
   - # # # # -               ✗
   ```
7. A pattern in different parts of different sequences may not represent a meaningful and consistent participation in folding, so we also do not allow a pattern to appear in offsets on its sequences that differ by more than two amino acids (i.e., a pattern can appear in position 5 of one sequence and position 7, but not 8, of another), e.g., schematically
   ```
   - - # # # # - - - - -           ✓
   - - # # # # - - - - - - -   ✓
   - - - # # # # - - -         ✓
   - - - - - - # # # # - -     ✗
   ```

Critically, sequences can be members of more than ground level cluster, or share patterns across ground level clusters, which can enable their aforementioned grouping in clusters at progressively higher levels of hierarchy—this has been shown to lead to the conservation of just a few critically positioned residues in the VH

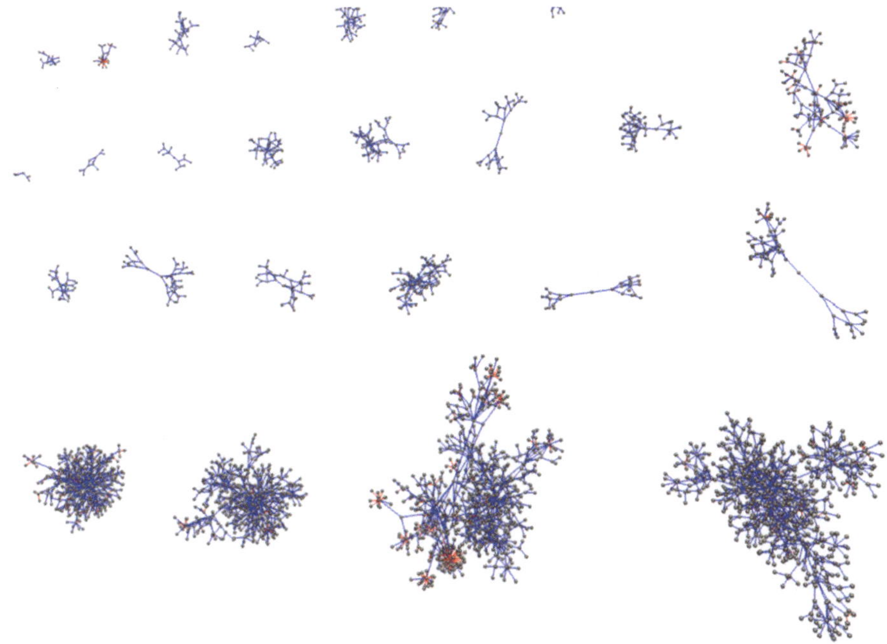

Fig. 4. Partial screenshot taken with Biolayout3D of the clustering of VH CDR3 sequences of 7,423 CLL patients. The *gray spherical nodes* can either represent sequences or clusters at any level; the connecting lines are *colored* according to the score of the connection ranging from *blue* (low score) to *red* (high score).

CDR3 consensus, still readily and specifically characterizing a subset (Fig. 3). Following this approach, we demonstrated that, while ground-level clusters provide a high-resolution picture of VH CDR3 stereotypy, high-level clusters can be considerably larger in size and able to capture and describe more distant sequence relationships in the form of more widely shared, though more "degenerate," sequence patterns (27) (Fig. 4). Intriguingly, high-level clusters were found to be still characterized by striking *IGHV* repertoire restrictions, which, as was revealed by phylogenetic analysis of functional human *IGHV* genes, reflected the branching of sequence distance trees (27).

2.4. Agathangelidis 2012

The knowledge from the first large-scale application of the methodology on 2,662 CLL cases, and other subsequent studies in different contexts, provided the basis for an update in criteria (Table 1). The major change in our approach (39) stepped on the aforementioned realization that phylogenetically related *IGHV* genes are reflected in the gene repertoire of subsets. Consequently, we now require that only sequences carrying *IGHV* genes of the same clan can be assigned to the same cluster, thus taking into account the role of the *IGHV* region in antigen recognition. In particular, three ancestral phylogenetic clans were imported in the pipeline: *IGHV1/5/7* subgroup genes from clan I, *IGHV2/4/6* from clan II, and finally, *IGHV3* genes from clan III (36).

Table 1
Evolution of the criteria for the identification of BcR stereotypy in human B-cell malignancies

	Messmer 2004	Stamatopoulos 2007	Darzentas 2010	Agathangelidis 2012
VH CDR3 amino acid identity (%)	≥60	≥60	≥50	≥50
VH CDR3 amino acid similarity (%)	n/a	n/a	≥70	≥70
IGHV genes	Same	Any	Any	Clan
VH CDR3 length difference	?	≤3	≤2	=0
Offset of the pattern	n/a	n/a	≤2	=0
IGHD gene reading frame	Same	n/a	n/a	n/a

n/a: non-applicable
Clan: phylogenetic clan of IGHV genes

Additionally, more stringent criteria related, indirectly, to the three-dimensional structure of the BcR included the requirement for identical VH CDR3 lengths and identical pattern offsets (i.e., exact locations within the VH CDR3 region) between connected sequences (Table 1).

2.5. Subset Inference

Such de novo processes can subsequently become the basis for a rule-based inference system to assign new sequences to existing, predefined subsets. The rules are made up of sequence features of subset members and can be accompanied by a confidence value—e.g., if 8 out of 10 members of subset X have an arginine in IMGT position 4 of their VH CDR3, then if a new sequence also has arginine there, then one can claim with 80% confidence that the new sequence is a member of subset X.

3. Future Developments and Perspective

In the future, we plan to support and extend the methodology with more in-depth knowledge about what each nucleotide and amino acid may, or may not, do at each position of the receptor sequence, including how the primary sequence arranges itself in 3D space, and if and how any 1D or 3D variability actually affects the in vivo functionality of the receptor. Indeed, at least for a number of subsets, there seems to be considerably redundancy in the sequence make up, so it is arguably almost certain that we are looking at an illusion of diversity which begs for dissolution.

In the wake of ever-evolving high-throughput and powerful technologies in the hands of dedicated international consortiums, a prerequisite for sustaining such research on stereotyped BcRs and translating it into patient-side know-how is to implement, as a matter of urgency, a bioinformatics-supported systematic collection, curation and analysis of patient, sequence, laboratory, and clinical information. In fact, breakthroughs described in this chapter would not have been possible without efforts to this direction, including the IMGT-based tools and databases (such as the IMGT/CLL-DB initiative behind the latest published work on BcR stereotypy in CLL) (http://www.imgt.org/CLLDBInterface/Welcome.do), and all the people and expertise of the IgCLL group (www.igcll.org) and its workshops.

References

1. Stevenson FK, Sahota SS, Ottensmeier CH, Zhu D, Forconi F, Hamblin TJ (2001) The occurrence and significance of V gene mutations in B cell-derived human malignancy. Adv Cancer Res 83:81–116
2. Chiorazzi N, Ferrarini M (2003) B cell chronic lymphocytic leukemia: lessons learned from studies of the B cell antigen receptor. Annu Rev Immunol 21:841–894
3. Dunn-Walters D, Thiede C, Alpen B, Spencer J (2001) Somatic hypermutation and B-cell lymphoma. Philos Trans R Soc Lond B Biol Sci 356:73–82
4. Ghia P, Granziero L, Chilosi M, Caligaris-Cappio F (2002) Chronic B cell malignancies and bone marrow microenvironment. Semin Cancer Biol 12:149–155
5. O'Brien S, Burger J, Blum KA et al (2011) The Bruton's Tyrosine Kinase (BTK) inhibitor PCI-32765 induces durable responses in relapsed or refractory (R/R) chronic lymphocytic leukemia/small lymphocytic lymphoma (CLL/SLL): follow-up of a phase Ib/II study. Blood 118 (Abstract 983)
6. Wang L, Martin P, Blum KA et al (2011) The Bruton's Tyrosine Kinase inhibitor PCI-32765 is highly active as single-agent therapy in previously-treated mantle cell lymphoma (MCL): preliminary results of a phase II trial. Blood 118 (Abstract 442)
7. Chiorazzi N, Ferrarini M (2011) Cellular origin(s) of chronic lymphocytic leukemia: cautionary notes and additional considerations and possibilities. Blood 117:1781–1791
8. Schroeder HW Jr, Dighiero G (1994) The pathogenesis of chronic lymphocytic leukemia: analysis of the antibody repertoire. Immunol Today 15:288–294
9. Fais F, Ghiotto F, Hashimoto S et al (1998) Chronic lymphocytic leukemia B cells express restricted sets of mutated and unmutated antigen receptors. J Clin Invest 102:1515–1525
10. Stamatopoulos K, Belessi C, Hadzidimitriou A et al (2005) Immunoglobulin light chain repertoire in chronic lymphocytic leukemia. Blood 106:3575–3583
11. Damle RN, Wasil T, Fais F et al (1999) Ig V gene mutation status and CD38 expression as novel prognostic indicators in chronic lymphocytic leukemia. Blood 94:1840–1847
12. Hamblin TJ, Davis Z, Gardiner A, Oscier DG, Stevenson FK (1999) Unmutated Ig V(H) genes are associated with a more aggressive form of chronic lymphocytic leukemia. Blood 94:1848–1854
13. Ghia P, Stamatopoulos K, Belessi C et al (2007) European Research Initiative on CLL. ERIC recommendations on IGHV gene mutational status analysis in chronic lymphocytic leukemia. Leukemia 21:1–3
14. Langerak AW, Davi F, Ghia P et al (2011) European Research Initiative on CLL (ERIC). Immunoglobulin sequence analysis and prognostication in CLL: guidelines from the ERIC review board for reliable interpretation of problematic cases. Leukemia 25:979–984
15. Hashimoto S, Dono M, Wakai M et al (1995) Somatic diversification and selection of immunoglobulin heavy and light chain variable region genes in IgG+ CD5+ chronic lymphocytic leukemia B cells. J Exp Med 181:1507–1517
16. Efremov DG, Ivanovski M, Siljanovski N et al (1996) Restricted immunoglobulin VH region repertoire in chronic lymphocytic leukemia patients with autoimmune hemolytic anemia. Blood 87:3869–3876

17. Johnson TA, Rassenti LZ, Kipps TJ (1997) Ig VH1 genes expressed in B cell chronic lymphocytic leukemia exhibit distinctive molecular features. J Immunol 158:235–246
18. Widhopf GF 2nd, Kipps TJ (2001) Normal B cells express 51p1-encoded Ig heavy chains that are distinct from those expressed by chronic lymphocytic leukemia B cells. J Immunol 166:95–102
19. Tobin G, Thunberg U, Johnson A et al (2003) Chronic lymphocytic leukemias utilizing the VH3-21 gene display highly restricted Vlambda2-14 gene use and homologous CDR3s: implicating recognition of a common antigen epitope. Blood 101:4952–4957
20. Widhopf GF 2nd, Rassenti LZ, Toy TL, Gribben JG, Wierda WG, Kipps TJ (2004) Chronic lymphocytic leukemia B cells of more than 1% of patients express virtually identical immunoglobulins. Blood 104:2499–2504
21. Ghiotto F, Fais F, Valetto A et al (2004) Remarkably similar antigen receptors among a subset of patients with chronic lymphocytic leukemia. J Clin Invest 113:1008–1016
22. Messmer BT, Albesiano E, Efremov DG et al (2004) Multiple distinct sets of stereotyped antigen receptors indicate a role for antigen in promoting chronic lymphocytic leukemia. J Exp Med 200:519–525
23. Tobin G, Thunberg U, Karlsson K et al (2004) Subsets with restricted immunoglobulin gene rearrangement features indicate a role for antigen selection in the development of chronic lymphocytic leukemia. Blood 104:2879–2885
24. Stamatopoulos K, Belessi C, Moreno C et al (2007) Over 20% of patients with chronic lymphocytic leukemia carry stereotyped receptors: pathogenetic implications and clinical correlations. Blood 109:259–270
25. Murray F, Darzentas N, Hadzidimitriou A et al (2008) Stereotyped patterns of somatic hypermutation in subsets of patients with chronic lymphocytic leukemia: implications for the role of antigen selection in leukemogenesis. Blood 111:1524–1533
26. Bomben R, Dal Bo M, Capello D et al (2009) Molecular and clinical features of chronic lymphocytic leukaemia with stereotyped B cell receptors: results from an Italian multicentre study. Br J Haematol 144:492–506
27. Darzentas N, Hadzidimitriou A, Murray F et al (2010) A different ontogenesis for chronic lymphocytic leukemia cases carrying stereotyped antigen receptors: molecular and computational evidence. Leukemia 24:125–132
28. Hadzidimitriou A, Agathangelidis A, Darzentas N et al (2011) Is there a role for antigen selection in mantle cell lymphoma? Immunogenetic support from a series of 807 cases. Blood 118:3088–3095
29. Agathangelidis A, Hadzidimitriou A, Rosenquist R, Stamatopoulos K (2011) Unlocking the secrets of immunoglobulin receptors in mantle cell lymphoma: implications for the origin and selection of the malignant cells. Semin Cancer Biol 21:299–307
30. Bikos V, Darzentas N, Hadzidimitriou A et al (2012) Over 30% of patients with splenic marginal zone lymphoma express the same immunoglobulin heavy variable gene: ontogenetic implications. Leukemia. 26:1638–1646]
31. Zibellini S, Capello D, Forconi F et al (2010) Stereotyped patterns of B-cell receptor in splenic marginal zone lymphoma. Haematologica 95:1792–1796
32. Xu JL, Davis MM (2000) Diversity in the CDR3 region of V(H) is sufficient for most antibody specificities. Immunity 13:37–45
33. Barrios Y, Jirholt P, Ohlin M (2004) Length of the antibody heavy chain complementarity determining region 3 as a specificity-determining factor. J Mol Recognit 17:332–338
34. Henikoff S, Henikoff JG (1993) Performance evaluation of amino acid substitution matrices. Proteins 17:49–61
35. Arvaniti E, Ntoufa S, Papakonstantinou N et al (2011) Toll-like receptor signaling pathway in chronic lymphocytic leukemia: distinct gene expression profiles of potential pathogenetic significance in specific subsets of patients. Haematologica 96:1644–1652
36. Kirkham PM, Mortari F, Newton JA, Schroeder HW Jr (1992) Immunoglobulin VH clan and family identity predicts variable domain structure and may influence antigen binding. EMBO J 11:603–609
37. Rigoutsos I, Floratos A (1998) Combinatorial pattern discovery in biological sequences: the TEIRESIAS algorithm. Bioinformatics 14: 55–67
38. Pommie C, Levadoux S, Sabatier R, Lefranc G, Lefranc MP (2004) IMGT standardized criteria for statistical analysis of immunoglobulin V-REGION amino acid properties. J Mol Recognit 17:17–32
39. Agathangelidis A, Darzentas N, Hadzidimitriou A et al (2012) Stereotyped B-cell receptors in one-third of chronic lymphocytic leukemia: a molecular classification with implications for targeted therapies. Blood 119:4467–4475

Chapter 9

Flow Cytometric MRD Detection in Selected Mature B-Cell Malignancies

Sebastian Böttcher, Matthias Ritgen, and Michael Kneba

Abstract

The quantification of submicroscopic minimal residual disease (MRD) after therapy proved to have independent prognostic significance in many mature B-cell malignancies. With the advent of routine bench-top cytometers capable of simultaneously analyzing ≥4 colors and with improved standardization, flow cytometry has become the method of choice for MRD assessments in some lymphoma entities. Herein we describe general aspects of flow cytometric standardization. Using chronic lymphocytic leukemia (CLL) and mantle cell lymphoma (MCL) as examples we explain in detail the application of flow cytometry for MRD detection.

Key words: Minimal residual disease, Chronic lymphocytic leukemia, Mantle cell lymphoma, Flow cytometry, Standardization

1. Introduction

Due to improved treatment efficacy complete remissions without morphological evidence of disease can now be achieved in many patients with mature B-cell malignancies. Nevertheless, relapses occur even in patients with complete responses suggesting the persistence of lymphoma cells at submicroscopic levels. Patients with detectable residual lymphoma cells at very low levels ("minimal residual disease," MRD) carry an increased relapse risk in mantle cell lymphoma (MCL) (1, 2), follicular lymphoma (3–6), chronic lymphocytic leukemia (CLL) (7–14), and multiple myeloma (15–19). MRD is therefore increasingly being used to tailor treatment intensity according to individual patients' relapse probabilities (20, 21). Moreover, characterizing MRD is also an attractive tool

Ralf Küppers (ed.), *Lymphoma: Methods and Protocols*, Methods in Molecular Biology, vol. 971,
DOI 10.1007/978-1-62703-269-8_9, © Springer Science+Business Media, LLC 2013

to better understand mechanisms of resistance to therapy and to develop new treatment protocols.

MRD can be assessed in mature B-cell non-Hodgkin's lymphomas (NHL) using either flow cytometry (MRD flow) or polymerase chain reaction (PCR). The first methods for MRD measurements in lymphomas were PCR-based, targeting either recurrent chromosomal translocations or the rearranged immunoglobulin genes of individual patients. The sensitivity of more simple PCR methods (e.g., consensus primer immunoglobulin heavy chain PCR) is dependent on the number of polyclonal benign B-cells in the sample as well as on the structure of the rearranged immunoglobulin gene of the individual patient (22, 23). This disadvantage was overcome by the introduction of real-time quantitative (RQ-) PCR technology that allowed the assessment of quantitative disease levels. RQ-PCR is still considered the gold standard for sensitive, quantitative, and reproducible MRD measurements in many lymphoma entities (24).

Comparative analyses between MRD flow and RQ-PCR targeting the rearranged immunoglobulin genes of the individual patients were performed in multiple myeloma and in CLL (22, 25, 26). While flow cytometric MRD assessments in CLL (sensitivity at least 10^{-4}) are often qualitatively less sensitive than PCR (sensitivity up to 10^{-5}), both methods yield quantitative results in comparable sensitivity ranges (typically 10^{-4}). Within the quantifiable ranges of the two methods disease levels assessed by flow cytometry and by RQ-PCR correlate very well (9, 25). Advantages of flow cytometric MRD assessments are very short turn-around time (less than 6 h), relatively low costs and broad availability of the equipment.

MRD flow relies on the identification of expression patterns of antigens that are not present in benign tissues (aberrant immunophenotypes). MRD flow approaches were pioneered in multiple myeloma and CLL (27, 28), as in those B-cell neoplasms well-defined normal counterparts exist and many markers to distinguish malignant from benign cells (27, 29, 30) have been identified. Flow cytometric MRD assessments have to incorporate markers to define the population of interest (e.g., $CD19^{+}CD5^{+}$ B-cells) as well as markers to detect aberrations within that population (e.g., $CD20^{low}$). Both objectives are best achieved using multiparameter measurements, so that MRD flow requires at least four-color equipment. Most immunophenotypic aberrations in B-NHL are quantitative, i.e., malignant cells are identified by over- or under-expression of antigens relative to the most similar normal B-cell population. Sensitive MRD assessments require markers with a large difference in staining intensity between malignant and normal cells. If such clear immunophenotypic aberrations exist in a given entity, the sensitivity of MRD flow primarily depends on the total number of acquired leukocytes for analysis (22, 25).

The determination of over- or underexpression to detect aberrations is the mainstay of MRD flow in B-NHL. Nevertheless, the detection of quantitative differences in antigen expression has so far relied on operator experience and was not unequivocally defined (27). To allow the reproducible quantification of fluorescence intensities the international EuroFlow consortium developed and tested optimized flow cytometer settings and immunophenotyping protocols, covering all technical aspects of flow cytometry in hematological malignancies (www.euroflow.org, see "protocols"). The application of these standard operating procedures allows now to directly compare expression levels measured in different laboratories and within one laboratory over time (31).

We describe herein standardized staining, instrument setup and analysis algorithms for MRD flow in CLL and MCL. CLL serves as an example for a disease in which MRD flow data are mature and have already been validated in large clinical trials (14), whereas all available clinical data on MRD in MCL come from PCR-based approaches. Staining principles and instrument setup are based on EuroFlow standardization. The 4-color MRD antibody combination in CLL and the corresponding gating strategy represents our implementation of the current standard defined by an international consortium (27). This version of MRD flow has been validated against RQ-PCR in 530 samples by our group (25) and is the current routine method of the central MRD laboratory of the German CLL study group. The 6-color MRD flow in MCL described herein was developed based on our experiences with 4-color panels (23) and on an unpublished systematic analysis of aberrant markers in MCL. At present we evaluate pretreatment samples for the presence of minimal infiltrations in peripheral blood (PB) or bone marrow (BM) using this 6-color MRD approach in MCL. Its utility for MRD detection is currently compared to RQ-PCR in posttreatment samples by our group.

2. Materials

2.1. Staining for MRD Flow in CLL

1. PBS/BSA (bovine serum albumin)/NaN_3 (sodium azide): Dissolve 4 PBS tablets (Life technologies, Darmstadt, Germany), 10 g BSA and 1.8 g NaN_3 at room temperature (RT) in 2 L of distilled water. Store at 4°C for a maximum of 14 days.
2. PBS/NaN_3: Dissolve 1 PBS tablet (Life technologies) and 0.45 g NaN_3 at RT in 0.5 L of distilled water. Store at RT for a maximum of 28 days.
3. PBS: Dissolve 4 PBS tablets (Life technologies) in 2 L of distilled water. Store at RT for a maximum of 14 days.

4. FACS Lysing Solution (Becton Dickinson, Heidelberg, Germany, BD): Prepare working solution by diluting 50 mL reagent with 450 mL distilled water.
5. Calculate reagents necessary for monoclonal antibody (moab) premixes for tubes 2, 3, 5, 7, and 9 by multiplying volumes of individual moabs and PBS/NaN_3 from Table 1 by the number of tests. Combine moabs (clones, labels, suppliers outlined in Table 2) and PBS/NaN_3 for each premix in a capped Eppendorf tube. Store at 4°C protected from light for a maximum of 14 days.

2.2. Staining for MRD Flow in MCL

Reagents 1–4 are prepared as described in Subheading 2.1.

5. Moab premixes contain FITC, PE, PerCP Cy5.5, and APC labeled moabs only. Calculate reagents necessary for moab premixes for tubes 3, 5, and 7 by multiplying volumes of individual moabs and PBS/NaN_3 from Table 3 for a single test by the number of tests. Combine moabs (Table 4) and PBS/NaN_3 for each premix in a capped Eppendorf tube. Store at 4°C protected from light for a maximum of 14 days.
6. Use CD5-HV450, CD19-PECy7, and CD45-PacO according to Tables 3 and 4.

2.3. PMT Voltage Setup

1. Cytometer Setup and Tracking Beads (CS&T beads, BD). Dilute 1 drop in 0.5 mL BD FACS Flow solution.
2. 8-peak Rainbow bead calibration particles. Obtain EuroFlow tested lots from Cytognos, Salamanca, Spain. Dilute 1 drop in 1 mL of distilled water.

2.4. Light Scatter and Compensation Setup

1. PBS/NaN_3: See Subheading 2.1, item 2.
2. FACS Lysing Solution (BD): see Subheading 2.1, item 4.
3. 2 mL PB from a healthy donor.
4. Antibodies as described in Table 5.

3. Methods

3.1. Staining for MRD Flow in CLL

An initial washing step removes serum including possible soluble antigens as well as cell-membrane bound immunoglobulins followed by a stain-lyse-wash protocol (see Note 1). Take necessary bio-hazard precautions and dispose of all NaN_3 containing waste according to EU and national regulations.

1. Determine the white blood cell count (WBC) using a 100 μL sample aliquot and an automatic cell counter (e.g., Sysmex KX-21, Sysmex, Norderstedt, Germany). Pipette the sample volume corresponding to 15×10^6 leukocytes into a 50 mL tube.

Table 1
Reagents and sample preparation for MRD flow in CLL

Tube	Susp. Vol. (μl)	Antibody (μl) FITC	PE	PerCP Cy5.5	APC	PBS NaN_3 (μl)	Lysing sol. (μl)	Final vol. (μl)
1	50	–	–	–	–	–	1,000	200
2	50	CD8 (10)	CD56 (2.5)	CD4 (5)	CD3 (0.5)	7	1,000	200
3	50	κ/λ light chains (5)		CD19 (5)	CD5 (5)	10	1,000	200
5	200	CD20 (20)	CD38 (40)	CD19 (20)	CD5 (20)	0	4,000	500
7	200	CD81 (10)	CD22 (40)	CD19 (20)	CD5 (20)	10	4,000	500
9	200	CD43 (40)	CD79b (20)	CD19 (20)	CD5 (20)	0	4,000	500

All volumes given correspond to a single test. If moabs and PBS/NaN_3 are used in premixes, volumes have to be multiplied by test number

Susp. vol.: Suspension volume of patient sample after adjustment, FITC: fluorescein isothiocyanate , PE: phycoerythrin, PerCP Cy5.5: peridinin chlorophyll protein-cyanin5.5 , APC: allophycocyanin, Lysing sol.: FACS Lysing solution, Final vol.: volume in which the suspension is acquired by the cytometer

Table 2
Antibodies and distributors for MRD flow in CLL

Antibody	Label	Clone	Distributor
CD8	FITC	SK1	BD
κ/λ light chains	FITC/PE	TB 28-2/1-155-2	BD
CD20	FITC	L27	BD
CD81	FITC	JS-81	BD
CD43	FITC	1G10	BD
CD56	PE	B159	BD
CD38	PE	HB-7	BD
CD22	PE	HIB22	BioLegend
CD79b	PE	SN8	BD
CD4	PerCP Cy5.5	SK3	BD
CD19	PerCP Cy5.5	SJ25C1	BD
CD3	APC	SK7	BD
CD5	APC	UTCH2	BD

Table 3
Reagents and sample preparation for MRD flow in MCL

Tube	Susp. Vol. (μl)	Antibody (μl) HV 450	PacO	FITC	PE	PerCP Cy5.5	PE Cy7	APC	PBS NaN_3 (μl)	Lysing sol. (μl)	Final vol. (μl)
1	50	–	–	–	–	–	–	–	–	1,000	200
2	50	–	–	–	–	–	CD19 (5)	–	–	1,000	200
3	200	CD5 (20)	–	κ/λ light chains (20)		LAIR (10)	CD19 (20)	CD200 (5)	105	4,000	500
5	200	CD5 (20)	–	CD10 (60)	CD21 (40)	LAIR (10)	CD19 (20)	CD200 (5)	25	4,000	500
7	50	CD5 (5)	CD45 (1)	CD8 (10) CD20(5)	CD56 (2.5)	CD3 (10)	CD19 (5)	CD23 (2.5) CD4 (2.5)	2.5	1,000	200

HV450: Horizon V450, PacO: Pacific Orange

Table 4
Antigens and distributors for MRD flow in MCL

Antibody	Label	Clone	Distributor
CD5	HV450	L17F12	BD
CD45	PacO	Hi 30	Invitrogen
CD10	FITC	ALB1	Beckman Coulter
CD8	FITC	SK1	BD
CD20	FITC	L27	BD
κ/λ light chains	FITC/PE	TB 28-2/1-155-2	BD
CD56	PE	B159	BD
CD21	PE	B-ly4	BD
LAIR	PerCP Cy5.5	NKTA255	BioLegend
CD3	PerCP Cy5.5	SK7	BD
CD4	APC	SK3	BD
CD23	APC	EBVCS-5	BD
CD200	APC	OX104	eBioSciences
CD19	PE Cy7	J3-119	Beckman Coulter

2. Add approximately 45 mL PBS/BSA/NaN_3, mix well, centrifuge for 5 min at 540×*g*.
3. Remove the supernatant using a vacuum system or a Pasteur pipette carefully avoiding to disturb the pellet.
4. Add approximately 10 mL PBS/BSA/NaN_3, mix well, transfer the suspension to a 15 mL plastic tube and centrifuge for 5 min at 540×*g*.
5. Remove the supernatant carefully avoiding to disturb the pellet. In order to achieve a high leukocyte concentration the supernatant should be removed completely at this stage. Vortex thoroughly.
6. Determine the WBC using a 100 μL aliquot of the pellet. Dilute the pellet with PBS/BSA/NaN_3 if necessary to obtain a WBC of 10,000 leukocytes/μL.
7. Label 5 mL tubes (suitable for sample acquisition for your flow cytometer) with numbers 1 through 9 and with the laboratory code of the sample. Add the appropriate volumes of the cell suspensions to tubes 1 through 3, 5, 7, and 9 (Table 1) while putting aside tubes 4, 6, and 8 (see Note 2).

Table 5
Reagents and sample preparation for compensation setups

Tube	Susp. Vol. (μl)	Antibody (μl) HV 450	PacO	FITC	PE	PerCP Cy5.5	PE Cy7	APC	PBS NaN_3 (μl)	Lysing sol. (μl)	Final vol. (μl)
1	50	–	–	–	–	–	–	–	50	1,000	200
2	50	CD5 (5)	–	–	–	–		–	45	1,000	200
3	50	–	CD45 (1)	–	–	–	–	–	49	1,000	200
4	50	–	–	CD8 (10)	–	–	–	–	40	1,000	200
5	50	–	–	–	CD5 (2.5)	–	–	–	47.5	1,000	200
6	50	–	–	–	–	CD3 (7.5)	–	–	42.5	1,000	200
7	50	–	–	–	–	–	CD19 (5)	–	45	1,000	200
8	50	–	–	–	–	–	–	CD4 (5)	45	1,000	200

The following reagents are used for compensation: CD5-HV450, BD, clone L17F12; CD45-PacO, Invitrogen, clone Hi30; CD8-FITC, BD, clone SK1; CD5-PE, BD, clone UCHT2; CD3-PerCPCy5.5, BD, clone SK7; CD19-PECy7, Beckman Coulter, CD4-APC, BD, clone SK3

Table 6
Sequential gating steps for MRD flow in CLL

Region name	Subpopulation of	Defined in	Applicable to tube	Figures
No_Doublets	All events	FSC-H vs. FSC-A dot-plot	1–3, 5, 7, 9	9.1a
Leukocytes	No_Doublets	FSC-A vs. SSC-A dot-plot	1–3, 5, 7, 9	9.1b
Lymphocytes	Leukocytes	FSC-A vs. SSC-A dot-plot	1–3, 5, 7, 9	9.1c
Neg	Lymphocytes	PerCP Cy5.5 histogram of tube 1 (autofluorescence)	1, 3, 5, 7, 9	9.1d, e
CD19+	Lymphocytes	CD19 PerCP Cy5.5 histogram, CD19 PerCP Cy5.5 vs. CD5 APC dot-plot	3, 5, 7, 9	9.1e, f
CD20CD5BG	CD19+	CD20 FITC vs. CD5 APC dot-plot	5	9.2a–c, 9.3a–c, 9.4a–c
CD20CD38BG	CD19+	CD20 FITC vs. CD38 PE dot-plot	5	9.2a–c, 9.3a–c, 9.4a–c
CD20CD5CD38BG	CD20CD5BG CD20CD38BG	Boolean "AND" operation of CD20CD5BG and CD20CD38BG	5	9.2a–c, 9.3a–c, 9.4a–c
CD22CD5BG	CD19+	CD22 PE vs. CD5 APC dot-plot	7	9.2d–f, 9.3d–f, 9.4d–f
CD81CD22BG	CD19+	CD81 FITC vs. CD22 PE dot-plot	7	9.2d–f, 9.3d–f, 9.4d–f
CD81CD22CD5BG	CD22CD5BG CD81CD22BG	Boolean "AND" operation of CD22CD5BG and CD81CD22BG	7	9.2d–f, 9.3d–f, 9.4d–f
CD79bCD5BG	CD19+	CD79b PE vs. CD5 APC dot-plot	9	9.2g–i, 9.3g–i, 9.4g–i
CD43CD5BG	CD19+	CD43 FITC vs. CD5 APC dot-plot	9	9.2g–i, 9.3g–i, 9.4g–i
CD43CD79bCD5BG	CD79bCD5BG CD43CD5BG	Boolean "AND" operation of CD79bCD5BG and CD43CD5BG	9	9.2g–i, 9.3g–i, 9.4g–i

The subpopulation hierarchy, the parameters used for definition of the gates, the tubes the gates should be applied to and the figures which illustrate the gates are shown

8. Add 25 μL of the appropriate moab premixes according to Table 1 to tubes 2 and 3. Add 100 μL of the appropriate moab premixes according to Table 1 to tubes 5, 7, and 9. Tube 1 is an unstained auto-fluorescence control.
9. Mix well. Incubate with moabs for 15 min in the dark at RT.
10. Add FACS Lysing Solution to tubes 1 through 3, 5, 7, and 9 using the volumes specified in Table 1. Carefully mix the suspension while adding FACS Lysing Solution (see Note 3). Incubate for 10 min at RT protected from light.
11. Centrifuge at 540×*g* for 5 min. Decant the supernatant by turning the tubes upside-down. Remove drops of the supernatant that adhere to the rim of the plastic tubes by gently touching a paper towel.
12. Add 4 mL PBS/BSA/NaN_3 to each of the tubes. Vortex thoroughly to detach the pellet from the bottom of the tube and resuspend the pellet.
13. Centrifuge again at 540×*g* for 5 min. Decant the supernatant by turning the tubes upside-down. Remove drops of the supernatant that adhere to the rim of the plastic tubes by gently touching a paper towel.
14. Add the appropriate volume PBS/BSA to each of the tubes (see Table 1, final volume). Vortex thoroughly to detach the pellet from the bottom of the tube and suspend the pellet.
15. Fill tubes 4, 6, and 8 with 2 mL of PBS. These tubes serve to decrease the risk of contamination caused by carryover between subsequent tubes.
16. Acquire data on a flow cytometer within 60 min (store at 4°C in the dark unless acquired immediately after staining).

3.2. Staining for MRD Flow in MCL

1. Determine the WBC using a 100 μL sample aliquot and an automatic cell counter (e.g., Sysmex KX-21). Pipette the sample volume corresponding to 11×10^6 leukocytes into a 50 mL tube.

Steps 2 through 6 are identical to steps 2 through 6 from Subheading 3.1.

7. Label 5 mL tubes with numbers 1 through 7 and with the laboratory code of the sample. Add the appropriate volumes of the cell suspensions to tubes 1 through 3, 5, and 7 (Table 3) while putting aside tubes 4 and 6.
8. Always add moabs to the samples in the following sequence: CD19-PECy7, CD5-HV450, CD45-PacO followed by the premixed moab cocktail. The appropriate moab volumes are given in Table 3. Add 140 μL of the appropriate moab premix to tubes 3 and 5. Add 35 μL of the moab premix according to Table 7. Tube 1 is an unstained auto-fluorescence control, whereas tube 2 serves to better delineate unspecific CD19 binding and compensation

Table 7
Sequential gating steps for MRD flow in MCL

Region name	Subpopulation of	Defined in	Applicable to tube	Figures
No_Doublets	All events	FSC-H vs. FSC-A dot-plot	1–3, 5, 7	9.1a
Leukocytes	No_Doublets	FSC-A vs. SSC-A dot-plot	1–3, 5, 7	9.1b
Lymphocytes	Leukocytes	FSC-A vs. SSC-A dot-plot	1–3, 5, 7	9.1c
Neg	Lymphocytes	PE Cy7 histogram of tube 1 (autofluorescence)	1–3, 5, 7	9.1d, e
CD19+	Lymphocytes	CD19 PE Cy7 histogram, CD19 PE Cy7 vs. CD5 HV450 dot-plot	3, 5, 7	9.1e, f
CD19-CD5+	Lymphocytes	CD19 PE Cy7 vs. CD5 HV450 dot-plot	3, 5, 7	9.1f
CD200CD5BG	CD19+	CD200 APC vs. CD5 APC dot-plot	3, 5	9.5a
CD200LAIRBG	CD19+	CD200 APC vs. LAIR PerCP Cy5.5 dot-plot	3, 5	9.5b
CD200CD19BG	CD19+	CD200 APC vs. CD19 PE Cy7 dot-plot	3, 5	9.5c
CD200LAIRCD19CD5BG	CD19+	Boolean "AND" operation of CD200CD5BG, CD200LAIRBG and CD200CD19BG	3, 5	9.5a–c

The subpopulation hierarchy, the parameters used for definition of the gates, the tubes the gates should be applied to and the figures which illustrate the gates are shown

artifacts in B-cells. Tube 2 is therefore stained with CD19 only (see Note 4).

9. Mix well. Incubate with moabs for 15 min in the dark at RT.
10. Add FACS Lysing Solution to tubes 1 through 3, 5, and 7 using the appropriate volumes as specified in Table 3. Carefully mix the suspension while adding the Lysing Solution. Incubate for 10 min at RT protected from light.
11. Centrifuge at 540×*g* for 5 min. Decant the supernatant by turning the tubes upside-down. Remove drops of the supernatant that adhere to the rim of the plastic tubes by gently touching a paper towel.
12. Add 4 mL PBS/BSA/NaN_3 to each of the tubes. Vortex thoroughly to detach the pellet from the bottom of the tube and to resuspend the pellet.

13. Centrifuge again at 540 × g for 5 min. Decant the supernant by turning the tubes upside-down. Remove drops of the supernatant that adhere to the rim of the plastic tubes by gently touching a paper towel.
14. Add the appropriate volume PBS/BSA to each of the tubes (see Table 3, final volume). Vortex thoroughly to detach the pellet from the bottom of the tube and resuspend the pellet.
15. Fill tubes 4 and 6 with 2 mL of PBS. These tubes serve to decrease the risk of contamination caused by carryover between subsequent tubes.
16. Acquire on a flow cytometer within 60 min (store at 4°C in the dark unless acquired immediately after staining).

3.3. Standardized Instrument PMT Voltage Setup

Standardized instrument setup and staining allows to directly compare fluorescence intensities of benign and malignant cells over time and between different flow cytometers, thus increasing specificity of MRD flow. For a detailed description of the rationale for the instrument setup and the technical variation determined using this standard operation procedure refer to the manuscript by the EuroFlow consortium (31), for additional information on the setup see "Protocols" at www.euroflow.org (see Note 5).

1. Set up the flow cytometer according to manufacturer's recommendations. Allow for sufficient laser warm-up time (at least 30 min). Perform necessary procedures to degas and clean the flow cell and to set the time-delay in multi-laser instruments according to manufacturer's instructions (e.g., use the CS&T module of the BD FACSCanto instrument) (see Note 6).
2. Create a forward scatter-area (FSC-A) vs. side scatter-area (SSC-A) bivariate dot-plot. Create a gate ("singlets") within this dot-plot. Create an FITC vs. PE dot-plot showing all events of region "singlets." Create a gate "8-peak" within that dot-plot, so that "8-peak" becomes a subpopulation of "singlets." Create 8 histograms each showing one of the fluorescence intensities in channels for PacO, PacificBlue (PacB), FITC, PE, PerCP Cy5.5, PE Cy7, APC, and Allophycocyanin-Hilite7 (APC-H7) for events within the gate "singlets" (see Note 7). Display mean fluorescence and coefficient of variation (CV) for events in "8-peak" for each of the eight fluorescence channels of your instrument.
3. Run diluted rainbow beads at low speed (flow rate: approximately 10 μL/min) at your flow cytometer. Adjust FSC and SSC photomultiplier tube (PMT) voltages to see rainbow beads on scale. Move the "singlets" gate to designate the major bead population showing the lowest FSC and SSC values, thus excluding debris (very low FSC/SSC) and doublets (higher FSC/SSC).

4. Designate the brightest bead population in the FITC vs. PE dot-plot with "8-peak." Check that FITC and PE match target values. Change PMT voltages for the two channels and the position of the "8-peak" gate as necessary.
5. Adjust the PMT voltages for the remaining six fluorescence channels so that the fluorescence intensities of the brightest peak of the rainbow beads match. EuroFlow provides reference values per batch at www.euroflow.org (see Note 8).
6. Record the information for 10,000 beads, check again for match with target values. Adjust PMT voltages as necessary in case of deviation from target values. Check for CVs of the brightest peaks. For optimal instrument performance the CVs of the 8th peak should be below 4% for all but PE Cy7, APC, and APC-H7 channels (here a CV below 6% is required).
7. PMT values from all eight fluorescence channels resulting in optimal match of the brightest peak fluorescence of rainbow beads with EuroFlow targets are recorded and used for all subsequent compensation experiments and MRD assessments (see Note 9).

3.4. Standardized Light Scatter and Compensation Setup

Light scatter is harmonized using lymphocytes from healthy donor PB as standard. Compensation is automatically calculated by dedicated software tools (e.g., integrated into BD FACSDiva) using lymphocytes stained in separate tubes for a brightly expressed antigen for each of the fluorochromes.

1. Label eight 5 mL tubes, add 50 μL of healthy donor blood and incubate with moabs specified in Table 5. Perform the stain-lyse-wash procedure as described in Subheading 3.1.
2. Acquire tube number 1.
3. Gate lymphocytes and display FSC-A and SSC-A mean values for the lymphocyte population. Adjust FSC and SSC PMT voltages so that mean of lymphocyte populations of a healthy donor achieves values of 55,000 (FSC-A) and 13,000 (SSC-A), respectively.
4. Record the PMT values in the FSC-A and SSC-A channels necessary to obtain those values.
5. Run tubes 1 through 8 using FSC-A and SSC-A voltages determined in point 3 and fluorescence detector PMT voltages as determined in Subheading 3.3.
6. Acquire 50,000 cellular events per tube in uncompensated mode (i.e., with compensation turned off) at medium speed (flow rate: approximately 60 μL/min).
7. Identify lymphocytes in a light scatter dot-plot, gate for positive events in each of the fluorescence channel and apply the automatic compensation algorithm of your instrument (e.g., "Calculate compensation" of FACSDiva software) (see Note 10).

3.5. Data Acquisition

1. Start your flow cytometer according to manufacturer's recommendations. Use standard procedures to clean the fluidics system and to remove air bubbles from the cytometer.
2. Allow at least 30 min for the lasers to stabilize.
3. Adjust time delay between lasers according to manufacturer's recommendations (e.g., use the CS&T module of the FACSDiva software on FACSCanto II flow cytometers).
4. Set PMT voltages for scatter and fluorescence channels as determined in Subheadings 3.3 and 3.4. Verify daily that the fluorescence of the rainbow beads match the specified EuroFlow ranges for that lot (see Note 9).
5. Run MRD samples which were stained as described in Subheadings 3.1 and 3.2, respectively. Set the flow rate to medium (approximately 60 mL/min).
6. For CLL MRD flow, record 50,000 cellular events in tubes 1 through 3. For tubes 5, 7, and 9 acquire as many cellular events as possible. For those tubes a time limit is applied so that sample acquisition is automatically stopped before the tube is empty. Tubes 4, 6, and 8 are run in a time-dependent mode for 120 s.
7. For MCL MRD flow, record 50,000 cellular events in tubes 1 and 2. For tubes 3, 5, and 7 acquire as many cellular events as possible. For those tubes a time limit is applied so that sample acquisition is automatically stopped before the tube is empty. Tubes 4 and 6 are run in a time-dependent mode for 120 s.

3.6. Data Analysis for MRD Flow in CLL

Sequential gating steps are applied to identify B-cells with the characteristic immunophenotype of CLL cells using the gating hierarchy outlined in Table 6. Figures 1, 2, 3, and 4 provide typical examples for gating in BM, in PB without interference from anti-CD20 moabs, and in PB after treatment with anti-CD20 moabs (see Note 11).

Gating for T- and NK-cells (tube 2) provides an excellent internal control. As this follows well-defined standard procedures it is not outlined in this chapter.

1. Identify single cellular events in a bivariate forward scatter-height (FSC-H) vs. FSC-A dot-plot (Fig. 1a) using the gate "No_Doublet." Doublets characteristically show disproportionally high FSC-A when compared to FSC-H.
2. Adjust the gate "Leukocytes" by including lymphocytes, neutrophils, monocytes, and eosinophils (Fig. 1b).
3. Adjust the gate "Lymphocytes" so that it includes all lymphocytes. Doublet cells should not be included into this gate (Fig. 1c). It is important to carefully exclude debris (at lower

left border of this gate) and monocytes (upper right border of this gate) from analyses.

4. Copy gates "No_Doublets," "Leukocytes," and "Lymphocytes" to all tubes of the panel. Double-check for consistency of scatter parameters between tubes. Recheck the lymphocyte gate for inclusion of all B-, T-, and NK-cells (appropriate dot-plots not shown).
5. Use the autofluorescence of the lymphocytes to define the autofluorescence gate "neg" (Fig. 1d). It should include the fluorescence of 95–99% of all lymphocytes in the PerCP Cy5.5 channel.
6. Set the CD19^{+} gate (Fig. 1e, f). This gate must not overlap with the gate "neg." It should be at the nadir of the CD19 distribution as shown in histogram (Fig. 1e). The gate should include all CD19^{+}CD5^{+} B cells as represented in Fig. 9.1f. It should typically exclude CD19^{-}CD5^{+} T-cells (upper left quadrant cells in Fig. 1f (see Note 12).
7. Adjust gates CD20CD5BG and CD20CD38BG (Figs. 2, 3, 4a, b; Boolean "AND" operation of the two gates: Fig. 2, 9.3, and 4c), gates CD22CD5BG and CD81CD22BG (Figs. 2, 3, 4d, e; Boolean "AND" operation of the two gates: Figs. 2, 3, and 4f) as well as gates CD79bCD5BG and CD43CD5BG (Figs. 2, 3, 4d,e; Boolean "AND" operation of the two gates Figs. 2, 3, and 4f) (see Note 13). In bone marrow samples taken more than 9 months after last CD20 moab treatment (Fig. 2) tube 5 is usually the most informative one. Include the CD5^{+}CD20^{low} population into the gate CD20CD5BG (Fig. 9.2a). Adjust CD20CD38BG to incorporate the CD20^{low}CD38^{low} population (Fig. 9.2b). In case of doubts, depict in the CD20 vs. CD38 dot-plot not all CD19^{+} lymphocytes but only those B-cells in the gate "CD20CD5BG." The resulting population (Fig. 9.2c) should be homogeneous with respect to all markers but CD38, which might show a bimodal distribution in some CLL cases. The exact position of all gates is best determined by comparison with a pretreatment sample from the same patient.

   Most often, tube 9 provides the second best discrimination (Fig. 2 g–i). CD79b^{low}CD5^{+} B cells are identified first, followed by adjustment of CD43CD5BG. CD79b, CD5, and CD43 are typically homogeneous within the CLL clone. At this stage CD5 expression levels of the malignant population is compared between tubes 9 and 5 (Fig. 2a vs. g). Borders of the gates with respect to CD5 have to be identical between the tubes. Using the CD5 information obtained, the putative CLL clone is marked in the CD22 vs. CD5 dot-plot (Fig. 2d), followed by adjustment of the gate CD81CD22BG (Fig. 2e).

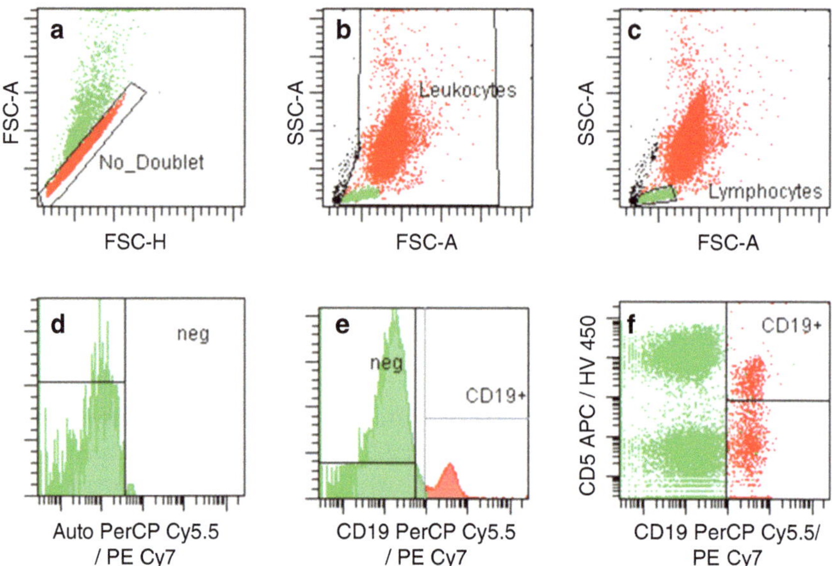

Fig. 1. Common initial steps of a sequential gating strategy for MRD flow in MCL and CLL. (**a**–**c**) First gating steps applicable to all tubes. (**a**) All events, (**b** and **c**) singlets. (**d**) Autofluorescence of unstained lymphocytes in PerCPCy5.5 (CLL) and PECy7 (MCL) channels. The histogram serves to define the range of autofluorescence (region "neg") comprising approximately 99% all cells in those tubes. (**e**) Histogram for CD19 expression of lymphocytes (PerCPCy5.5 in CLL MRD flow, PECy7 in MCL MRD flow). (**f**) Bivariate dot-plots showing CD19 vs. CD5 expressions of lymphocytes (CD5-APC in CLL MRD flow, CD5-HV450 in MCL MRD flow).

All three markers should be expressed homogeneously on the CLL populations.

A very similar strategy is feasible in peripheral blood drawn more than 9 months after the last infusion of a CD20 moab (Fig. 3).

Peripheral blood samples obtained within a 9 months period from the last anti-CD20 treatment lack assessable CD20 (Fig. 4). Due to this it is advisable to first gate within CD22 vs. CD5 and CD81 vs. CD22 dot-plots (Fig. 4d–f), followed by tube 9. The contribution of tube 5 is limited in this situation. Nevertheless, CLL cells can often be identified by lower CD5 than detected in T-cells (Fig. 4a) and low level CD38 (Fig. 4b).

8. Calculate the MRD level. Consider tubes as MRD positive in which more than 20 events with typical CLL immunophenotype can be identified. Those events should form a cluster in all nine dot-plots as shown in Figs. 2, 3, and 4. The MRD level of a single MRD positive tube is calculated as ratio CLL cells in the gates "CD20CD5CD38BG," "CD81CD22CD5BG," and "CD43CD79bCD5BG," respectively, divided by the events in gate "Leukocytes" of the same tube. A sample is considered MRD positive if at least 2 out of 3 tubes are MRD positive. The MRD level for a sample is the mean of the MRD levels of all positive tubes. If 20 or more CLL cells can only be identified

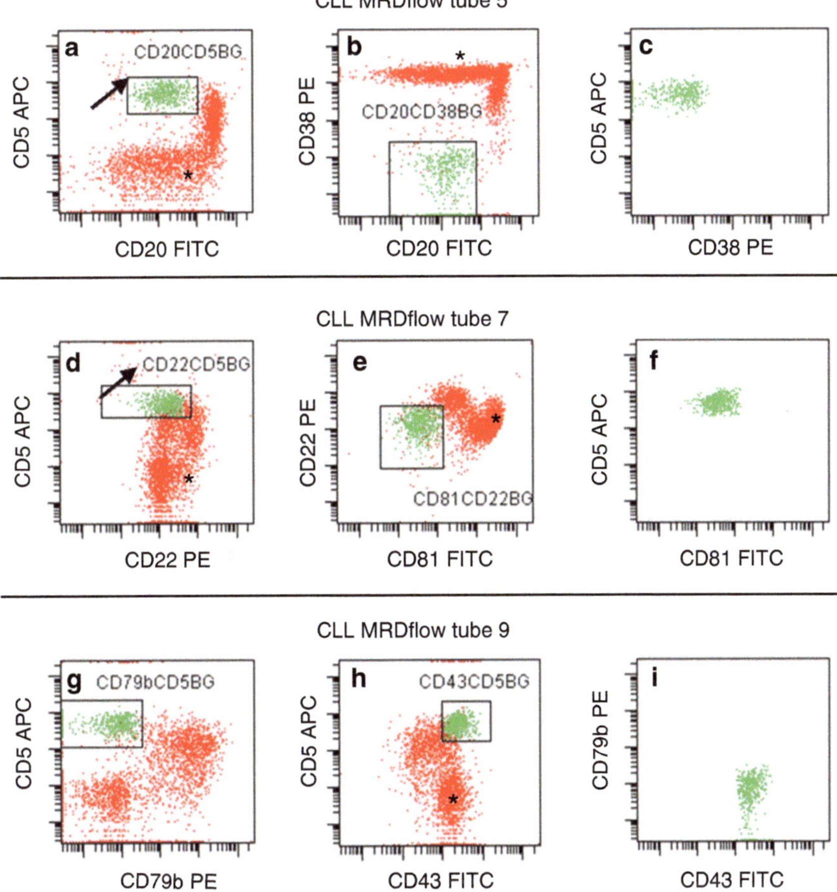

Fig. 2. Example for identification of CLL MRD in regenerating bone marrow. MRD level: 5.2×10^{-4}. (**a**, **b**, **d**, **e**, **g**, **h**) All B lymphocytes. (**c**) Lymphocytes included into both "CD20CD5BG" and "CD20CD38BG." (**f**) Lymphocytes included into both "CD22CD5BG" and "CD81CD22BG". (**i**) Lymphocytes included into both "CD79bCD5BG" and "CD43CD5BG." Benign B-cells are shown in *dark-red*, CLL cells *light green*. *Arrows points* toward contaminating T-cells, *asterisks* marks B-cell progenitors (hematogones).

in one tube or in none of the three tubes, the sensitivity of the approach is calculated as 20 divided by events in gate "Leukocytes" for the tube with the second highest number of leukocytes (see Note 14).

3.7. Data Analysis for MRD Flow in MCL

Sequential gating steps are applied to identify B-cells with the characteristic immunophenotype of MCL cells. The gating hierarchy is outlined in Table 7. Figures 1 and 5 provide a typical example for gating in PB. Similar to CLL MRD flow MCL cells are primarily identified by their aberrant immunophenotype. The correct gating is further confirmed by light chain restriction assessments.

Gating for T- and NK-cells (tube 7) provides an excellent internal control. As this follows standard procedures it is not outlined in this chapter. Tube 7 can also be used to distinguish

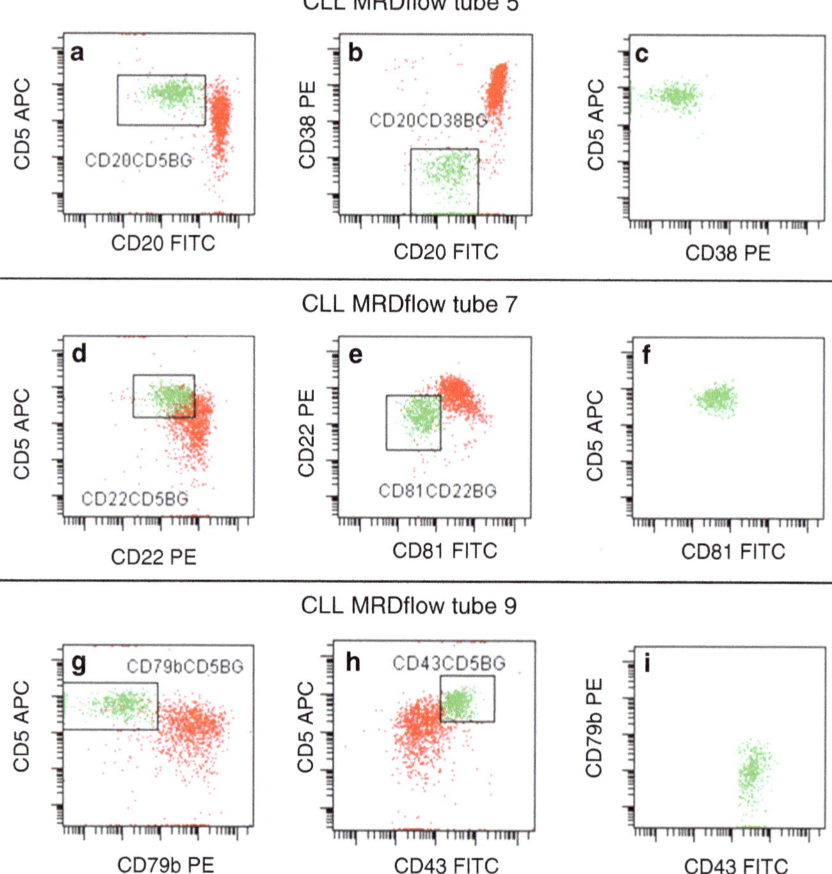

Fig. 3. Example for identification of CLL MRD in regenerating peripheral blood. MRD level: 1.0×10^{-3}. (**a**, **b**, **d**, **e**, **g**, **h**) All B lymphocytes. (**c**) Lymphocytes included into both "CD20CD5BG" and "CD20CD38BG." (**f**) Lymphocytes included into both "CD22CD5BG" and "CD81CD22BG." (**i**) Lymphocytes included into both "CD79bCD5BG" and "CD43CD5BG." Benign B-cells are shown in *dark-red*, CLL cells *light green*. Please note high level expression of CD5 in benign regenerating B-cells.

erythroid precursurs from leukocytes (using CD45) and to further characterize CD5 positive B-cells using CD23 (helpful for characterization of a suspected MCL population).

Steps 1–4 are identical to gating for CLL MRD flow (described in Subheading 3.6).

5. Use the autofluorescence of the lymphocytes to define the autofluorescence gate "neg" (Fig. 1d). It should include the fluorescence of 95–99% of all lymphocytes in the PE Cy7 channel.
6. Set the CD19$^+$ gate (Fig. 1e, f). This gate must not overlap with the gate "neg." It should be at the nadir of the CD19 distribution as shown in histogram (Fig. 1e). The gate should include all CD19$^+$CD5$^+$ B cells as represented in Fig. 1f.
7. Start identification of MCL cells using CD200-APC vs. CD5-HV450 dot-plot showing all B-cells (Fig. 5a). The gate

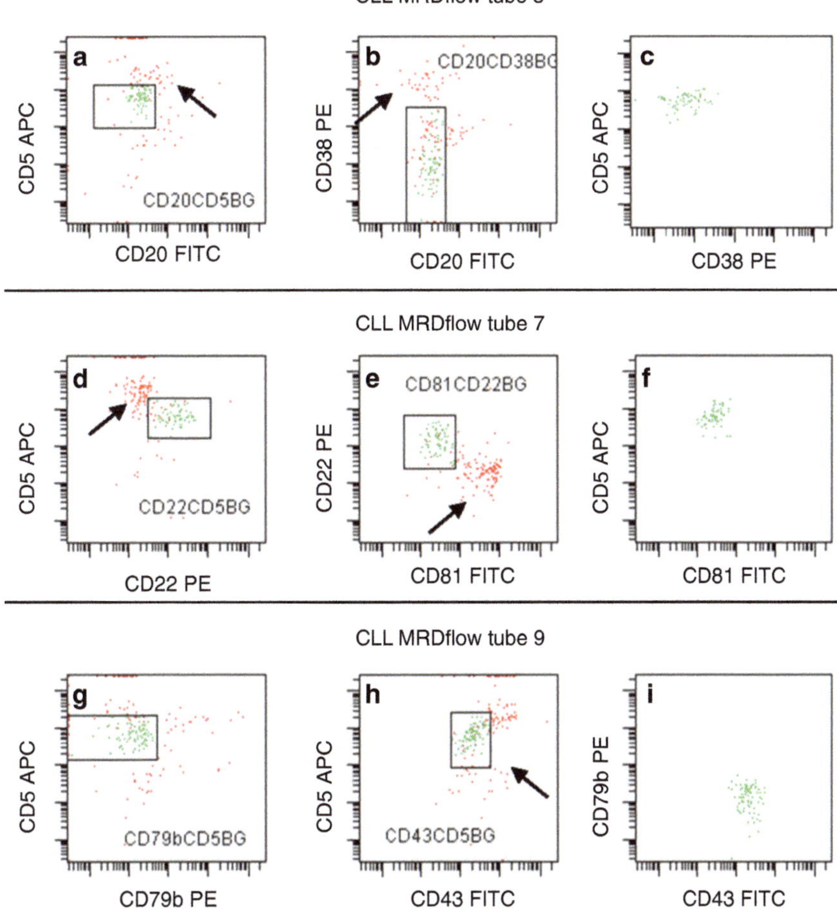

Fig. 4. Example for identification of CLL MRD in peripheral blood within 9 months after anti-CD20 moab treatment. MRD level: 1.4×10^{-4}. (**a**, **b**, **d**, **e**, **g**, **h**) All B lymphocytes. (**c**) Lymphocytes included into both "CD20CD5BG" and "CD20CD38BG." (**f**) Lymphocytes included into both "CD22CD5BG" and "CD81CD22BG." (**i**) Lymphocytes included into both "CD79bCD5BG" and "CD43CD5BG." Benign cells are shown in *dark-red*, CLL cells *light green*. *Arrows point* toward contaminating T-cells. Note the complete lack of benign B-cells.

"CD200CD5BG" encompasses $CD200^{low}CD5^{+}$ B-cells. Start with large gates CD200LAIRBG and CD200CD19BG that include all B-cells (Fig. 5b, c). In this situation the only discriminator for the combined gate CD200LAIRCD19CD5BG is the gate CD200CD5BG.

8. Identify T-cells as $CD19^{-}CD5^{+}$ in tube 3 (Fig. 1f). Depict kappa and lambda light chain levels of T-cells (Fig. 5d) and adjust the quadrant marker so that 97% of all T-cells are included in the quadrant negative for both kappa and lambda. Copy that quadrant marker to a dot-plot that depicts those B-cells only which are included into the gate CD200LAIRCD19CD5BG (Fig. 5e).
9. Check whether or not the cells in the gate CD200LAIRCD19CD5BG are light chain restricted (normal range for κ/λ ratio: $0.38 < \kappa/\lambda$ ratio < 6) (see Note 15). Check the light chain restriction again

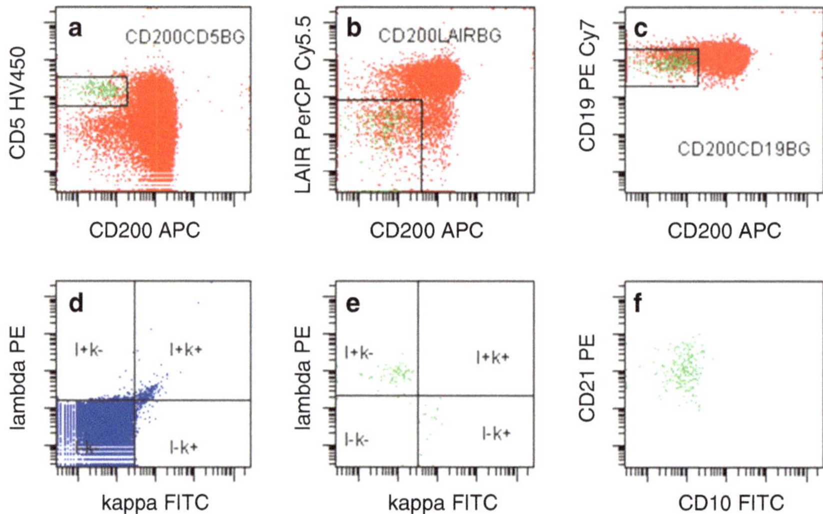

Fig. 5. Example for identification of MCL cells in peripheral blood. MRD level: 2.9×10^{-4}. Benign B-cells are shown in *dark-red*, MCL cells *light green*, T-cells are *dark blue*. (**a–c**) All B lymphocytes. (**d**) CD19-CD5+ lymphocytes (T-cells) are used to set the borders for kappa and lambda light chain expression. (**e** and **f**) Only B-cells included in all three gates CD200CD5BG, CD200LAIRBG, and CD200CD19BG are depicted.

after moving the CD200CD5BG to the CD200^{low}CD5^{-} B-cell subpopulation. Include as many light chain restricted B-cells in the gate CD200CD5BG as possible. The gate might also encompass CD5^{low} or CD5^{-} cells.

10. Try to obtain more skewed light chain ratios by sequentially decreasing the size of the gates CD200LAIRBG and CD200CD19BG. MCL cells typically, but variably show lower levels of CD19 and/or LAIR in comparison to mature B-cells.
11. Copy the gates CD200CD5BG, CD200LAIRBG, and CD200CD19BG to tube 5. The presence of MCL cells can additionally be confirmed by the demonstration that cells included in the gate CD200LAIRCD19CD5BG are CD10^{-}CD21^{low} when compared to benign mature B-cells. Compare the expression levels in MRD cells to expression levels determined prior to therapy.
12. A tube is considered MRD positive when more than 20 cells are included in the gate CD200LAIRCD19CD5BG. Both tubes 3 and 5 have to contain more than 20 cells in the respective gates CD200LAIRCD19CD5BG to consider a sample MRD positive. In addition, the κ/λ ratio needs to be below 0.38 or above 6 within that gate (tube 3). The MRD level for a tube is calculated as ratio of cells in gate CD200LAIRCD19CD5BG divided by the number of events in the gate "Leukocytes." The MRD level of a sample is the mean MRD level from tubes 3 and 5. The sensitivity of a negative test depends on the number of leukocytes in the tube 3 or 5 with fewer leukocytes. This sensitivity is calculated as 20 divided by the number of total leukocytes from this tube.

4. Notes

1. The International harmonization approach for MRD flow in CLL (27) did not require a particular staining or instrument setup protocol, which in turn precluded the direct comparison of staining intensities between different laboratories. To improve standardization we have been using the EuroFlow standard operating procedure for staining and instrument setup in all our analyses, including MRD flow, for the last 4 years (www.euroflow.org).

   Different sample preparation methods have been investigated by the EuroFlow consortium and the technique presented herein was found to be most reliable with respect to practicability, scatter properties, and preservation of fluorescence intensities (31). The addition of a protein source (we recommend BSA) to the washing solution is essential for reduced cell loss. NaN_3 prevents capping of moabs bound to vital cells and reduces the risk of bacterial contamination in protein-containing solutions. As it is poisonous, the final resuspension step omits that reagent so that the flow cytometer is not contaminated. Washing of cell suspensions prior to incubation with moabs is strictly speaking essential only for staining with anti-immunoglobulin-moabs. We use a bulk-wash approach for the sample as a whole in order to improve the comparability of the staining conditions in the individual tubes.

   Ammonium chloride bulk lysis prior to staining was used by some laboratories participating in the International standardization project for CLL MRD flow (27). While this approach might result in a higher concentration of cells in the staining suspension and thus might economize on moab usage, in our evaluations a higher moab concentration was frequently required and some target epitopes appeared sensitive to ammonium chloride. Pending further standardization efforts we currently do not recommend bulk lysis.

2. With the exception of the CD5 moab all moabs were selected from the reagents recommended by the International standardization project (27). The reagents specified in this chapter show low lot-to-lot variation (relative differences between consecutive lots <30%) when checked in our laboratory. Readers should expect to obtain the same staining intensities as shown in this chapter if they use these moabs together with the standardized staining and instrument setup described herein.

3. FACS Lysing solution contains a fixative, that might increase doublet formation when added to cells that adhere to each other. We therefore add this solution mL by mL while vortexing the cell suspension.

4. Fluorochrome choice per channel follows EuroFlow recommendations, aiming at stability, high staining index, and low compensation requirements (31). We found the sequence of incubation steps with the different moabs important for the staining results and therefore standardized it. Clone and moab selection is based on in-house experiments and our experience in four-color MRD flow for MCL (23). Premixing PE Cy7 conjugates is strongly discouraged, as this tandem dye becomes instable when diluted. Pending further experiments it is currently uncertain whether or not HV450- and PacO-labeled moabs are stable in dilutions.
5. PMT voltages are adjusted using rainbow beads that emit fluorescence light of stable intensity in all channels of modern flow cytometers. Voltages are tuned so that the fluorescence of those beads match target ranges on each of the fluorescence channels, resulting in highly similar optical output of all instruments for identical molecule numbers per cell. This is a prerequisite for direct comparison of fluorescence intensities measured at different instruments and over time at the same cytometer. The target values of bead fluorescence emissions were initially created by analyzing many different flow cytometers at all EuroFlow consortium laboratories. These standard values were chosen to minimize electronic noise and at the same time allow the on-scale analysis of all antigens used in clinical flow cytometry. For subsequent masterlots of beads these reference values are cross-calibrated at a EuroFlow reference laboratory. Please note that the last masterlots of EuroFlow rainbow beads were brighter than the original ones, with the eighth peak off scale for some fluorochromes. This forced the consortium to define target MFI values and target CVs for the seventh peak of the beads (i.e., the second brightest bead population)—for details see www.euroflow.org.
6. EuroFlow setup and compensation standard operation procedures have been extensively tested on FACSCanto II (BD), BD LSR II (BD), and CyAn ADP (Dako/Beckman Coulter) cytometers by the consortium (31). The experience of our own group is restricted to FACSCanto II machines.
7. There are many alternative fluorochromes for PacO and PacB from various suppliers available. For MRD flow in MCL we use HV450 instead of EuroFlows original PacB. Please note that while PMT voltage setup remains the same, each alternative fluorochrome requires specific compensation.
8. Reference values are provided for a 262,144 channel scale and have to be converted in case your instrument uses a different resolution (e.g., 4,096 channels at CyAn ADP cytometer, Dako/Beckman Coulter).
9. Bead emissions have to match optimal target values +/−15%. Set your instrument once a month at optimal values. Check

daily for adherence to target values before acquiring samples. In case of a sudden deterioration of CV or loss in fluorescence intensity most often problems in the fluidics system are causative. This can usually be overcome by vigorous cleaning of the instrument and removal of air bubbles. Gradual decline in instrument performance usually requires the help of a service engineer from the manufacturer of your cytometer.

10. Compensation and light scatter values can be used for a month as long as the PMT values for the rainbow beads are within target values. If an adjustment of PMT values of the fluorescence channels becomes necessary or maintenance at fluidics or optical systems of your instrument is required a new setup of compensation and light scatter is strongly recommended.

11. The gating approach follows the general principles of the International standardization project on MRD flow in CLL (27). Staining for CD45/CD14/CD19/CD3 is recommended by this consortium to identify B-T-cells-doublets, assess the contamination of the B-cell gate with T-cells and to refer CLL numbers to $CD45^+$ leukocytes. While the discrimination against T-cells is crucial for MRD flow, we found a high specificity when comparing flow cytometry to ASO primer PCR without using this test. Moreover, identifying nucleated cells by light scatter allowed us to demonstrate a high concordance of MRD levels measured by flow and PCR (25). Consequently, we prefer to omit this CD45/CD14/CD19/CD3 test in order to save moabs. According to our own (22, 25) and the experience of an International consortium (27), the most important factor for the sensitivity of MRD flow in CLL is the total number of acquired cells. We therefore use the total cell number to calculate sensitivity of a test (25) and recommend not to analyze MRD negative samples in which less than 200,000 nucleated cells are measured per tube when a sensitivity of 10^{-4} is required. (14). Data from our group demonstrated that CD20 cannot be assessed up to 9 months after the last treatment with the CD20 moab, rituximab (25). Mature benign B-cells are more sensitive to rituximab treatment than CLL cells and absent from PB for similar periods of time (25). We therefore provide examples (Figs. 2, 3, and 4) separately for patients who received anti-CD20 treatment and those who did not. It is important not only to identify putative CLL-cells but also to identify the position of mature benign B-cells, of benign B-cell precursors and of contaminating T-cells.

    The specificity of MRD flow is greatly enhanced by comparison with the expression levels of all antigens at presentation, made possible by standardized sample preparation and instrument setup. Individual CLL patients show underexpression of the B-cell antigens CD22, CD81, and CD79b to a different extent. Knowledge of the exact initial immunophenotype is particularly helpful for interpretation of CD81 vs. CD22 dot-plots (Figs. 2e, 3e, and 4e). In our experience all B-cell antigens included into the MRD flow panel are stable in the MRD situation relative to the initial presentation. The only exception is CD20 that is not assessable up to 9 months after anti-CD20 treatment. In our own work in 530 samples from CLL patients a sensitivity of at least 10^{-4} and an excellent specificity of this MRD flow approach were demonstrated by comparison to *IGH* RQ-PCR (25).

12. The exclusion of T-cells using this gate is important, while all B-cells have to be included. In case of doubt, set the gate to deliberately include some T-cells (move to the left), identify the position of T-cells in dot-plots as shown in Figs. 2, 3, and 4, and exclude T-cells using e.g., CD22. The CD5 expression of T-cells is usually higher than on CLL and benign B-cells.
13. Mature benign B-cells can show considerable upregulation of CD5 after therapy, so that up to 90% of benign B-cells express this marker (27, 32).
14. Using comparative analysis to PCR, we found a minimal requirement of more than 20 cells in at least 2 tubes sufficient evidence of MRD (25), while the International consortium set this limit at 50 cells per tube (27).
15. There is no generally accepted definition for light chain restriction. The definition provided was established in our work on 4-color MRD flow in MCL (23).

Acknowledgments

We are grateful to E. Harbst, L.Falck, J. Hanani, H. Hinrichs, and L. Henseleit for excellent technical support with establishing the protocols. We thank the members of the EuroFlow and ERIC consortiums for longstanding collaborations. We thank A.W. Langerak for helpful comments on the manuscript.

References

1. Pott C, Schrader C, Gesk S et al (2006) Quantitative assessment of molecular remission after high-dose therapy with autologous stem cell transplantation predicts long-term remission in mantle cell lymphoma. Blood 107:2271–2278
2. Pott C, Hoster E, Delfau-Larue MH et al (2010) Molecular remission is an independent predictor of clinical outcome in patients with mantle cell lymphoma after combined immunochemotherapy: a European MCL intergroup study. Blood 115:3215–3223
3. Ladetto M, De Marco F, Benedetti F et al (2008) Prospective, multicenter randomized GITMO/IIL trial comparing intensive (R-HDS) versus conventional (CHOP-R)

chemoimmunotherapy in high-risk follicular lymphoma at diagnosis: the superior disease control of R-HDS does not translate into an overall survival advantage. Blood 111:4004–4013

4. Lopez-Guillermo A, Cabanillas F, McLaughlin P et al (1998) The clinical significance of molecular response in indolent follicular lymphomas. Blood 91:2955–2960
5. Colombat P, Salles G, Brousse N et al (2001) Rituximab (anti-CD20 monoclonal antibody) as single first-line therapy for patients with follicular lymphoma with a low tumor burden: clinical and molecular evaluation. Blood 97:101–106
6. Rambaldi A, Lazzari M, Manzoni C et al (2002) Monitoring of minimal residual disease after CHOP and rituximab in previously untreated patients with follicular lymphoma. Blood 99:856–862
7. Moreton P, Kennedy B, Lucas G et al (2005) Eradication of minimal residual disease in B-cell chronic lymphocytic leukemia after alemtuzumab therapy is associated with prolonged survival. J Clin Oncol 23:2971–2979
8. Rawstron AC, Kennedy B, Evans PA et al (2001) Quantitation of minimal residual disease levels in chronic lymphocytic leukemia using a sensitive flow cytometric assay improves the prediction of outcome and can be used to optimize therapy. Blood 98:29–35
9. Moreno C, Villamor N, Colomer D et al (2006) Clinical significance of minimal residual disease, as assessed by different techniques, after stem cell transplantation for chronic lymphocytic leukemia. Blood 107:4563–4569
10. Lamanna N, Jurcic JG, Noy A et al (2009) Sequential therapy with fludarabine, high-dose cyclophosphamide, and rituximab in previously untreated patients with chronic lymphocytic leukemia produces high-quality responses: molecular remissions predict for durable complete responses. J Clin Oncol 27:491–497
11. Ysebaert L, Gross E, Kuhlein E et al (2010) Immune recovery after fludarabine-cyclophosphamide-rituximab treatment in B-chronic lymphocytic leukemia: implication for maintenance immunotherapy. Leukemia 24:1310–1316
12. Castro JE, James DF, Sandoval-Sus JD et al (2009) Rituximab in combination with high-dose methylprednisolone for the treatment of chronic lymphocytic leukemia. Leukemia 23:1779–1789
13. Maloum K, Settegrana C, Chapiro E et al (2009) IGHV gene mutational status and LPL/ADAM29 gene expression as clinical outcome predictors in CLL patients in remission following treatment with oral fludarabine plus cyclophosphamide. Ann Hematol 88:1215–1221
14. Bottcher S, Ritgen M, Fischer K et al (2012) Minimal residual disease quantification is an independent predictor of progression free and overall survival in chronic lymphocytic leukemia. A multivariate analysis from the randomized GCLLSG CLL8 trial. J Clin Oncol 30:980–988
15. Bakkus MH, Bouko Y, Samson D et al (2004) Post-transplantation tumour load in bone marrow, as assessed by quantitative ASO-PCR, is a prognostic parameter in multiple myeloma. Br J Haematol 126:665–674
16. Ladetto M, Pagliano G, Ferrero S et al (2010) Major tumor shrinking and persistent molecular remissions after consolidation with bortezomib, thalidomide, and dexamethasone in patients with autografted myeloma. J Clin Oncol 28:2077–2084
17. Paiva B, Vidriales MB, Cervero J et al (2008) Multiparameter flow cytometric remission is the most relevant prognostic factor for multiple myeloma patients who undergo autologous stem cell transplantation. Blood 112:4017–4023
18. Rawstron AC, Davies FE, Dasgupta R et al (2002) Flow cytometric disease monitoring in multiple myeloma: the relationship between normal and neoplastic plasma cells predicts outcome after transplantation. Blood 100:3095–3100
19. San Miguel JF, Almeida J, Mateo G et al (2002) Immunophenotypic evaluation of the plasma cell compartment in multiple myeloma: a tool for comparing the efficacy of different treatment strategies and predicting outcome. Blood 99:1853–1856
20. Dreger P, Dohner H, Ritgen M et al (2010) Allogeneic stem cell transplantation provides durable disease control in poor-risk chronic lymphocytic leukemia: long-term clinical and MRD results of the German CLL Study Group CLL3X trial. Blood 116:2438–2447
21. Ritgen M, Bottcher S, Stilgenbauer S et al (2008) Quantitative MRD monitoring identifies distinct GVL response patterns after allogeneic stem cell transplantation for chronic lymphocytic leukemia: results from the GCLLSG CLL3X trial. Leukemia 22:1377–1386
22. Bottcher S, Ritgen M, Pott C et al (2004) Comparative analysis of minimal residual disease detection using four-color flow cytometry, consensus IgH-PCR, and quantitative IgH PCR in CLL after allogeneic and autologous stem cell transplantation. Leukemia 18:1637–1645

23. Bottcher S, Ritgen M, Buske S et al (2008) Minimal residual disease detection in mantle cell lymphoma: methods and significance of four-color flow cytometry compared to consensus IGH-polymerase chain reaction at initial staging and for follow-up examinations. Haematologica 93:551–559
24. van der Velden VH, Cazzaniga G, Schrauder A et al (2007) Analysis of minimal residual disease by Ig/TCR gene rearrangements: guidelines for interpretation of real-time quantitative PCR data. Leukemia 21:604–611
25. Bottcher S, Stilgenbauer S, Busch R et al (2009) Standardized MRD flow and ASO IGH RQ-PCR for MRD quantification in CLL patients after rituximab-containing immunochemotherapy: a comparative analysis. Leukemia 23:2007–2017
26. Sarasquete ME, Garcia-Sanz R, Gonzalez D et al (2005) Minimal residual disease monitoring in multiple myeloma: a comparison between allelic-specific oligonucleotide real-time quantitative polymerase chain reaction and flow cytometry. Haematologica 90:1365–1372
27. Rawstron AC, Villamor N, Ritgen M et al (2007) International standardized approach for flow cytometric residual disease monitoring in chronic lymphocytic leukaemia. Leukemia 21:956–964
28. Rawstron AC, Orfao A, Beksac M et al (2008) Report of the European Myeloma Network on multiparametric flow cytometry in multiple myeloma and related disorders. Haematologica 93:431–438
29. Rawstron AC, de Tute R, Jack AS et al (2006) Flow cytometric protein expression profiling as a systematic approach for developing disease-specific assays: identification of a chronic lymphocytic leukaemia-specific assay for use in rituximab-containing regimens. Leukemia 20:2102–2110
30. Mateo G, Montalban MA, Vidriales MB et al (2008) Prognostic value of immunophenotyping in multiple myeloma: a study by the PETHEMA/GEM cooperative study groups on patients uniformly treated with high-dose therapy. J Clin Oncol 26: 2737–2744
31. Kalina T, Flores J, van der Velden VH et al (2012) EuroFlow standardization of flow cytometer instrument settings and immunophenotyping protocols. Leukemia 26(9): 1986–2010
32. Bomberger C, Singh-Jairam M, Rodey G et al (1998) Lymphoid reconstitution after autologous PBSC transplantation with FACS-sorted CD34+ hematopoietic progenitors. Blood 91:2588–2600

Chapter 10

MRD Detection in B-Cell Non-Hodgkin Lymphomas Using Ig Gene Rearrangements and Chromosomal Translocations as Targets for Real-Time Quantitative PCR

Christiane Pott, Monika Brüggemann, Matthias Ritgen, Vincent H.J. van der Velden, Jacques J.M. van Dongen, and Michael Kneba

Abstract

Minimal residual disease (MRD) diagnostics is of high clinical relevance in patients with indolent B-cell Non-Hodgkin lymphomas (B-NHL) and serves as a surrogate parameter to evaluate treatment effectiveness and long-term prognosis. MRD diagnostics performed by real-time quantitative PCR (RQ-PCR) is the gold-standard and currently the most sensitive and the most broadly applied method in follicular lymphoma (FL) and mantle cell lymphoma (MCL). RQ-PCR analysis of the junctional regions of the rearranged immunoglobulin heavy-chain gene (IgH) serves as the most broadly applicable MRD target in B-NHL (~80%). Chromosomal translocations as t(14;18) translocation in FL and t(11;14) translocation in MCL can be used in selected lymphoma subtypes. In patients with B-cell chronic lymphocytic leukemia, both flow-cytometry as well as RQ-PCR are equally suitable for MRD assessment as long as a sensitivity of $\leq 10^{-4}$ shall be achieved.

MRD diagnostics targeting the IgH gene is complex and requires extensive knowledge and experience because the junctional regions of each lymphoma have to be identified before the patient-specific RQ-PCR assays can be designed for MRD monitoring. Furthermore, somatic mutations of the IgH region occurring during B-cell development of germinal center and post-germinal center lymphomas may hamper appropriate primer binding leading to false negative results. The translocations mentioned above have the advantage that consensus forward primers and probes, both placed in the breakpoint regions of chromosome 18 in FL and chromosome 11 in MCL, can be used in combination with a reverse primer placed in the IgH joining region of chromosome 14. RQ-PCR-based methods can reach a good sensitivity ($\leq 10^{-4}$). This chapter provides all relevant background information and technical aspects for the complete laboratory process from detection of the clonal IgH gene rearrangement and the chromosomal translocations at diagnosis to the actual MRD measurements in clinical follow-up samples of B-NHL. However, it should be noted that MRD diagnostics for clinical treatment protocols has to be accompanied by regular international quality control rounds to ensure the reproducibility and reliability of the MRD results.

Key words: Minimal residual disease (MRD), B-cell Non-Hodgkin lymphomas (B-NHL), Chronic lymphocytic leukemia (CLL), Immunoglobulin heavy-chain gene (IgH), t(14;18) translocation, t(11;14) translocation, Follicular lymphoma (FL), Mantle cell lymphoma (MCL), Real-time quantitative PCR (RQ-PCR), Quantitative range, Sensitivity

Ralf Küppers (ed.), *Lymphoma: Methods and Protocols*, Methods in Molecular Biology, vol. 971,
DOI 10.1007/978-1-62703-269-8_10, © Springer Science+Business Media, LLC 2013

1. Introduction

Several studies have shown that detection of minimal residual disease (MRD) in indolent B-cell Non-Hodgkin lymphomas (B-NHL) is clinically relevant (1–14). Based on these results, MRD diagnostics is currently implemented in clinical treatment protocols and future studies plan to stratify patients according to MRD levels in bone marrow (BM) or peripheral blood (PB) samples obtained after induction therapy. In most of these MRD-based stratification studies, MRD diagnostics is performed by real-time quantitative PCR (RQ-PCR) analysis of immunoglobulin heavy-chain gene rearrangements (IgH) or chromosomal translocations in B-NHL subtypes [e.g., in follicular lymphoma (FL) and mantle cell lymphoma (MCL)] (9, 10, 14–17).

After rearrangement of the V, (D), and J gene segments of the Ig gene complexes in early B-cell differentiation each lymphocyte obtains a specific combination of V–(D)–J segments that codes for the variable domains of Ig molecules. The random insertion and deletion of nucleotides at the junction sites of V, (D), and J gene segments make the junctional regions of Ig genes into "fingerprint-like" sequences, which are most probably different in each lymphocyte and thus also in each lymphoid malignancy. Therefore, junctional regions of malignant lymphoma cells can be used as tumor-specific targets for MRD-PCR analysis. Such targets can be identified by PCR analysis at initial diagnosis in ~80% of B-NHL (18, 19). Subsequently, the precise nucleotide sequence of the junctional regions can be determined. This sequence information allows the design of junctional region-specific oligonucleotides (either probes or primers), which like in acute lymphoblastic leukemias (ALL) can be used for sensitive detection of low frequencies of malignant cells, down to one malignant cell in 10^4–10^5 normal cells.

Alternatively, in certain histologic subtypes of B-NHL recurrent cytogenetic aberrations as the chromosomal translocations t(14;18) in follicular lymphoma and t(11;14) in MCL can serve as MRD markers. In both translocations the *IGH* locus at 14q32.3 is involved and the *BCL2* gene from 18q21 or the *BCL1* region from 11q13 is translocated into the *IGH* locus. Both translocations involve the process of VDJ recombination and one of the six germline JH gene segments is closely opposed to the *BCL2* gene or the *BCL1* region. As a consequence of the translocation process during VDJ recombination, random insertion of nucleotides at the junction sites of *BCL2* and *BCL1* and (D) J gene segments takes place and generates junctional regions of the translocations that can be used as "fingerprint-like" sequences.

However, chromosomal translocations as MRD markers are not only restricted to selected subtypes of B-NHL but are also detectable only in a subset of patients, about 60% of FL and 30% of MCL (20–22). This is mainly due to scattered breakpoint regions on the derivative chromosomal regions involved in the translocations of chromosomes 18 or 11 that are not accessible by PCR in 100% of translocations.

In this chapter, we describe RQ-PCR-based MRD detection using IgH gene rearrangements and the chromosomal translocations t(14;18) and t(11;14) as MRD-PCR targets in B-NHL patients. Several of the methods for MRD detection using Ig gene rearrangements have already been published by van der Velden and van Dongen (23). Because there is a relevant methodical overlap of MRD analysis in acute leukemias and B-NHL, several of the methods described here, including clonality assessment and sample processing, have been adopted or modified for the application in B-NHL. The approaches described here identify MRD-PCR targets in >80% of B-NHL patients and allow MRD monitoring in the majority of B-NHL patients.

2. Materials

2.1. Bone Marrow and Peripheral Blood Sample Processing at Diagnosis

1. Ficoll Hypaque™ Plus (GE Healthcare BioSciences AB, Uppsala, Sweden). Store in the dark at room temperature.
2. Cell storage medium: 10% dimethyl sulfoxide (DMSO), 20% fetal calf serum (FCS), 70% RPMI-1640 medium.

2.2. Detection of Clonal Ig Targets or Chromosomal Translocations at Diagnosis

2.2.1. Agarose Gel Electrophoresis and Gene Scan Analysis

1. 1× TBE buffer: 90 mM Tris–HCl, 90 mM H_3BO_3, 2.5 mM EDTA, in Milli-Q water.
2. Bromophenol Blue (BPB) gel loading buffer: 30% (v/v) glycerol, 0.25% (w/v) BPB (electrophoresis purity reagent), 0.25% (w/v) Xylene Cyanol FF (electrophoresis purity reagent), in Milli-Q water.
3. Ethidium bromide 1%.
4. $MgCl_2$ (25 mM).
5. Deoxynucleotide (dNTP) mix (25 mM each).
6. DNA mass ladder.
7. AmpliTaqGOLD (5 U/μl) (Applied Biosystems, Foster City, USA).
8. PE buffer II (10×) (Applied Biosystems).
9. Agarose gel (2%).
10. Forward and reverse primers for IgH, t(14;18) or t(11;14) translocation PCR, reverse primers labeled with 5′ FAM 100 μM each (Tables 1 and 2).
11. Electrophoresis units.

Table 1
Primer combinations for detection of IgH rearrangement, t(14;18) and t(11;14) translocation

Target	Forward primer	Reverse primer	Positive control	Product size (bp)	Protocol[a]	References
IGH	VH1/7	JH	Nalm6	300–400	A	(19)
	VH2	JH	Patient DNA	300–400	A	(19)
	VH3	JH	REH	300–400	A	(19)
	VH4/6	JH	ROS16	300–400	A	(19)
	VH5	JH	VH5-TL	300–400	A	(19)
t(14;18)						
MBR	MBR forward primer	JH	DOHH2	150–450	B	(19)
3′ MBR	3′ MBR forward primer 1–4	JH	K231	130–250	B	(19)
MCR	MCR forward primer 1–2	JH	OZ	150–300	B	(19)
5′ MCR	5′ MCR forward primer 1	JH	SC1	260–490	B	(19)
t(11;14)						
	BCL1 MTC forward primer G	JH	JVM-2	100–300	C	(22)
	BCL1 MTC forward primer B	JH	JVM-2	100–300	C	(22)

[a]*See* Table 4 for detailed information about PCR mixture per protocol

12. ABI Prism® 310 Genetic Analyzer with POP-4™ or ABI Prism® 3500 Genetic Analyzer with POP-7™ polymer (Applied Biosystems).
13. Deionized Formamide (Sigma) for ABI Prism® 310 Genetic Analyzer or HIDI-Formamide (ABI P/N 4311320) for ABI Prism® 3500 Genetic Analyzer.
14. ROX500 size standard according to manufacturer's instructions (ABI); from our experience a 1:2 dilution is sufficient.
15. Cathode and anode buffer for the respective Genetic Analyzer.
16. DNA isolated from PB mononuclear cells (MNC) obtained from 5 to 10 healthy individuals (100 ng/μl).
17. DNA from cell lines known to be positive for IgH gene rearrangements or t(14;18) or t(11;14) translocation (100 ng/μl).

2.2.2. PCR-Heteroduplex Analysis

1. 1× TBE buffer: 90 mM Tris–HCl, 90 mM H_3BO_3, 2.5 mM EDTA, in Milli-Q water.
2. BPB loading buffer: 30% (v/v) glycerol, 0.25% (w/v) BPB (electrophoresis purity reagent), 0.25% (w/v) Xylene Cyanol FF (electrophoresis purity reagent), in Milli-Q water.

Table 2
Primer sequences for t(14;18), t(11;14), and IgH PCR

Locus	Primer	Sequence
IGH	FR1-VH1–2	5′-GGCCTCAGTGAAGGTCTCCTGCAAG-3′
	FR1-VH2nw	5′-GTCTGGTCCTACGCTGGTGAAACCC-3′
	FR1-VH3–7	5′-CTGGGGGGTCCCTGAGACTCTCCTG-3′
	FR1-VH4–4	5′-CTTCGGAGACCCTGTCCCTCACCTG-3′
	FR1-VH5–51	5′-CGGGGAGTCTCTGAAGATCTCCTGT-3′
	FR1-VH6	5′-TCGCAGACCCTCTCACTCACCTGTG-3′
	FR2-VH1–2	5′-CTGGGTGCGACAGGCCCCTGGACAA-3′
	FR2-VH2–5	5′-TGGATCCGTCAGCCCCCAGGAAGG-3′
	FR2-VH3–7	5′-GGTCCGCCAGGCTCCAGGGAA-3′
	FR2-VH4–4	5′-TGGATCCGCCAGCCCCCAGGGAAGG-3′
	FR2-VH5–51	5′-GGGTGCGCCACATGCCCGGGAAAGG-3′
	FR2-VH6	5′-TGGATCAGGCAGTCCCCATCGAGAG-3′
	FR3-VH1–2	5′-TGGAGCTGAGCAGCCTGAGATCTGA-3′
	FR3-VH2–5	5′-CAATGACCAAACATGGACCCTGTGGA-3′
	FR3-VH3	5′-TCTGCAAATGAACAGCCTGAGAGCC-3′
	FR3-VH4–4	5′-GAGCTCTGTGACCGCCGCGGACACG-3′
	FR3-VH5–51	5′-CAGCACCGCCTACCTGCAGTGGAGC-3′
	FR3-VH6	5′-GTTCTCCCTGCAGCTGAACTCTGTG-3′
	FR3-VH7	5′-CAGCACGGCATATCTGCAGATCAG-3′
Chr 18 *BCL2*	5′ MBR forward	5′-GCAATTCCGCATTTAATTCATGGTATTCAGGAT-3′
	3′ mbr1 681	5′-GCACCTGCTGGATACAACACTG-3′
	3′ mbr2 1413	5′-GGTGACAGAGCAAAACATGAACA-3′
	3′ mbr3 1745	5′-GTAATGACTGGGGAGCAAATCTT-3′
	3′ mbr4 2676	5′-ACTGGTTGGCGTGGTTTAGAGA-3′
	5′ MCR	5′-CCTTCTGAAAGAAACGAAAGCA-3′
	MCR 1	5′-TAGAGCAAGCGCCCAATAAATA-3′
	MCR 2	5′-TGAATGCCATCTCAAATCCAA-3′
Chr 11 *BCL1*	BCL1—G	5′-GGA GCA TAA TTG CTG CAC TGC-3′
	BCL1-B—B	5′-TTC GGT TAG ACT GTG ATT AGC-3′
IGH	JH consensus[a]	5′-CTTACCTGAGGAGACGGTGACC-3′

[a]For gene scanning this primer is labeled with 5′ FAM

3. Ethidium bromide 1%.
4. $MgCl_2$ (25 mM).
5. dNTP mix (25 mM each).
6. Bovine serum albumin (BSA; 20 mg/ml): 20% (w/v) BSA (Fraction V) in Milli-Q water.
7. Orange G: 20% Ficoll, 10 mM Tris–HCl pH 7.6, 1 mg/ml Orange G, in Milli-Q water.
8. DNA ladders.
9. TaqGOLD (5 U/μl) (Applied Biosystems).

10. PE buffer II (10×) (Applied Biosystems).
11. Agarose gel (2%).
12. Criterion Precast gels (BioRad, Veenendaal, The Netherlands).
13. Forward and reverse primers for the various Ig gene rearrangements or translocations.
14. Electrophoresis units.
15. DNA isolated from PB MNC obtained from 5 to 10 healthy individuals (100 ng/μl).
16. DNA from cell lines or patients known to be positive for certain Ig gene rearrangements or translocations (100 ng/μl).

2.2.3. Sequencing of Clonal Rearrangements

1. 96% ethanol.
2. 2 M NaAc (pH 5.6).
3. 70% ethanol.
4. Millipore Multiscreen-HTS-PCR 96-well plate.
5. Vacuum manifold (Millipore).
6. Sephadex G-50, 6% (w/v) fine DNA grade in Milli-Q water.
7. 10× buffer with EDTA (Applied Biosystems).
8. Elution buffer: 0.5 M ammonium acetate, 10 mM magnesium acetate, 1 mM EDTA, 0.1% SDS, in Milli-Q water.
9. Big Dye® Terminator cycle sequencing ready reaction kit version 3.1 (Applied Biosystems). Store stock at −15 to −25°C; store working dilution at 4°C and protected from light.
10. Forward and reverse primers for the various Ig gene rearrangements or chromosomal translocations.
11. ABI Prism® 310 or 3100 Genetic Analyzer with POP-4™ or POP-6™ polymer (Applied Biosystems).

2.3. RQ-PCR Sensitivity Testing

2.3.1. Design of Allele-Specific Oligonucleotide Primers

1. 1× TE buffer: 10 mM Tris–HCl pH 7.6, 1 mM EDTA, in Milli-Q water.
2. Spectrophotometer (Nanodrop, Wilmington, USA).
3. OLIGO6.0 software (Molecular Biology Insights, Inc., Cascade, USA).

2.3.2. RQ-PCR Analysis of Dilutions of Diagnostic Sample

1. BSA (5%).
2. Universal Master mix (2×) (Applied Biosystems).
3. Forward primer (20 pmol/μl).
4. Reverse primer (20 pmol/μl).
5. TaqMan probe (20 pmol/μl).
6. DNA isolated from PB MNC obtained from 5 to 10 healthy individuals (60 and 100 ng/μl).

2.4. Bone Marrow and Peripheral Blood Sample Processing During Follow-Up

See Subheading 2.1 for materials needed.

2.5. MRD Analysis of Follow-Up Samples

See Subheading 2.3.2 for materials needed.

3. Methods

3.1. Bone Marrow and Peripheral Blood Sample Processing at Diagnosis

1. Perform Ficoll density centrifugation of the BM or PB sample obtained at diagnosis to obtain MNC.
2. Assess the percentage of FL, MCL, and CLL cells in the MNC fraction by flow cytometric immunophenotyping because this information is needed for correction of the tumor load in the standard curve (see Subheading 3.3 below) (see Note 1).
3. For reasons of standardization and prediction of DNA recovery, use a fixed number of 10×10^6 MNC for DNA extraction (see Note 2 and 15).
4. Determine the concentration of the DNA stock solution by analyzing the optical density at 260 and 280 nm in duplicate on the Nanodrop (see Note 3).
5. Prepare 150 μl of a DNA working solution of 60 ng/μl in sterile Milli-Q water; this working dilution will be used for the PCR analyses and RQ-PCR analyses (see below).
6. Check the DNA concentration by analyzing the optical density at 260 and 280 nm in duplicate on the Nanodrop.
7. Store the stock solution and working dilution at 4°C (or −20°C) for short periods of time (<3 months), or at −20°C for long-term storage.
8. Store the remaining MNC at diagnosis in cell storage medium for later studies; store in liquid nitrogen or at −80°C.

3.2. Detection of Clonal Ig Gene Rearrangements and Determination of Chromosomal Translocations at Diagnosis

For MRD marker screening IgH PCR is performed in all B-NHL subtypes including CLL. According to histologic subtype, additionally t(14;18) PCR in FL and t(11;14) PCR in MCL are performed. To determine clonality of IgH gene rearrangements, PCR products have to be further processed by either PCR heteroduplex analysis or gene scanning. This is especially important in malignant lymphomas, where besides the clonal lymphoma cell population polyclonal B cells are mostly present and are amplified by IgH PCR as well. By applying gene scan analysis or PCR heteroduplex analysis with appropriate primer-sets, clonal IgH gene rearrangements can be visualized and distinguished from a polyclonal B-cell background. In gene scanning single-stranded PCR products are separated in a

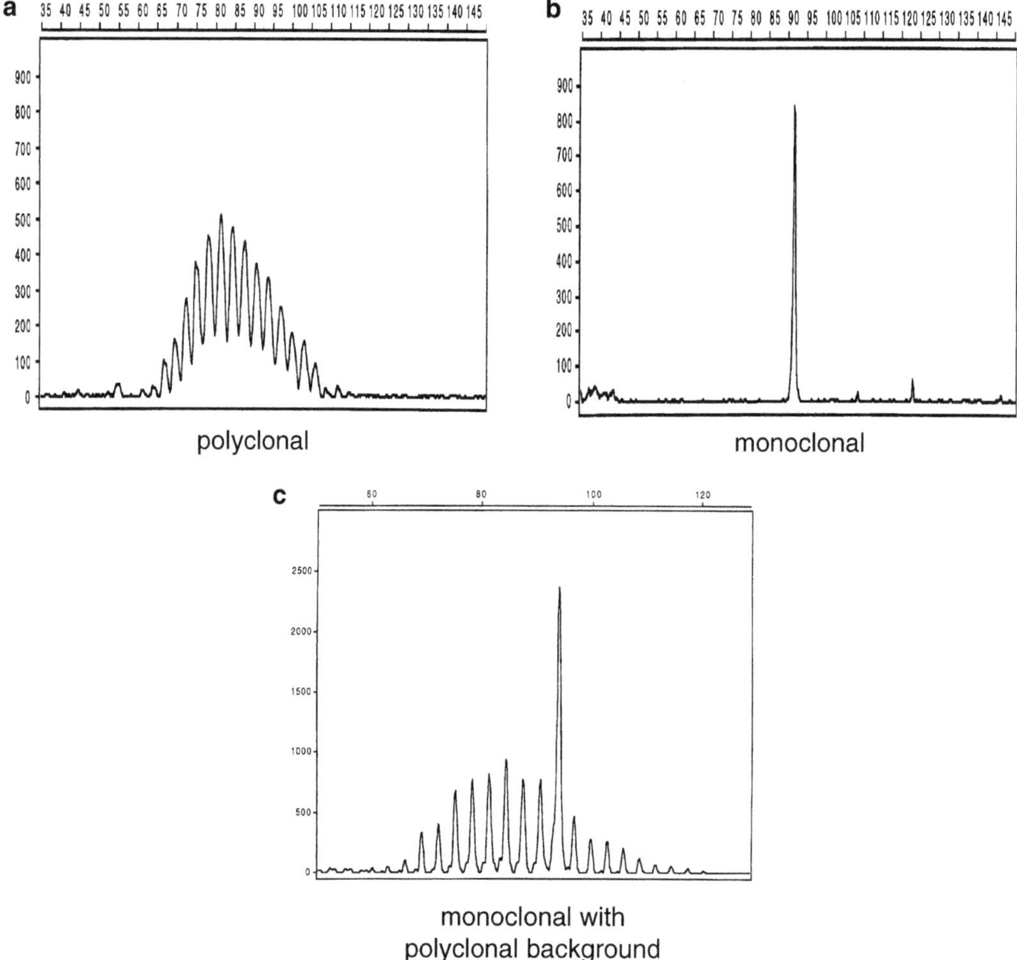

Fig. 1. Gene scan analysis of IG gene rearrangements. (**a**) Mononuclear cells from healthy donors show a polyclonal pattern, (**b**) monoclonal peak in case of a patient with CLL, (**c**) monoclonal peak with polyclonal B-cell background.

high-resolution polymer according to their length only. The PCR primers need to be labeled with a fluorochrome to allow detection of the PCR products with automated sequencing equipment. Figure 1 demonstrates gene scan results of IgH PCR in different settings: (a) a polyclonal gene scan pattern when only polyclonal B cells are present, (b) a monoclonal gene scan pattern in case of a CLL patient, and (c) a monoclonal IgH rearrangement with a polyclonal B-cell background. The latter gene scan pattern is mostly found in nodal FL where besides the lymphoma cell population also polyclonal B cells are present in BM or PB.

In PCR heteroduplex analysis double-stranded PCR products are denaturated at high temperatures followed by rapid random renaturation at low temperatures, preferably +4°C for 1 h (19). This results in case of clonal PCR products in homoduplexes with identical rapid migration speed compared to polyclonal PCR products forming many heteroduplexes with different migration speed when electrophoresis is performed on a 6% polyacrylamide gel.

For the detection of the chromosomal translocations, visualization of the PCR product in an agarose gel is sufficient to detect a t(14;18) or t(11;14) translocation. Here, gene scanning is used to precisely distinguish PCR products from different patients by exact size determination of respective PCR products.

3.2.1. Gene Scan Analysis

1. Perform respective PCR reaction with 100–500 ng diagnostic DNA (Table 1). The reaction mixes are shown in Table 1, the primer sequences are shown in Table 2, and reaction conditions are shown in Tables 3 and 4. Use Milli-Q water as a negative control, normal MNC DNA as a polyclonal control, and an appropriate positive control (Table 1).
2. Analyze PCR products in a 1% analytical agarose gel containing ethidium bromide.
3. For each of the PCR products, positive and negative controls, and the DNA ladder, mix 7 μl of the product with 2 μl gel loading buffer and load these samples on the agarose gel; run for 30 min at 120 V. PCR products are visualized under ultraviolet (UV) light. The positive controls should give a clear band of the expected size, whereas the negative controls should be negative. If this is not the case, repeat the PCR assays with insufficient results.

Table 3
Composition of PCR mixture per protocol

	PCR protocol		
Reaction mix per 50 μl reaction	**A**	**B**	**C**
PE buffer II (10×)	5	5	5
$MgCl_2$ (25 mM)	3	4	4
Milli-Q water	ad 45 μl	ad 45 μl	ad 45 μl[a]
dNTP (25 mM)	0.4	0.4	0.4
BSA (20 mg/ml)	1	1	1
Forward primer (diluted to 10 pmol/μl each)	1	1	1
Reverse primer (diluted to 10 pmol/μl)	1	1	1
TaqGOLD (5 U/μl)	0.2	0.2	0.4
DNA (100 ng/μl)	5	5	5[a]

[a]For the second PCR in t(11;14) seminested PCR 1.5 μl PCR product from the first PCR round is used, so 48.4 μl Milli-Q water and no DNA has to be added

Table 4
PCR reaction conditions per protocol

	PCR protocol					
Cycler protocols	**A**		**B[a]**		**C[b]**	
Initial denaturation	94°C	10 min	94°C	10 min	94°C	10 min
Denaturation	94°C	60 s	94°C	30 s	94°C	60 s
Annealing	60°C	60 s	66°C −0.5°C ramp	30 s	60°C	60 s
Extension	72°C	30 s	–	–	–	–
Final extension	72°C	30 min	–	–	72°C	5 min
Number of cycles	35		23		35	
Initial denaturation			–	–	94°C	10 min
Denaturation			94°C	30 s	94°C	60 s
Annealing			55°C	30 s	60°C	60 s
Extension			72°C	30 s	–	–
Final extension			72°C	10 min	72°C	10 min
Number of cycles			25		35	

[a]The touch down protocol does not require another pipetting step and the second cycler protocol is run directly after the first

[b]The seminested PCR protocol requires a second pipetting step where 1.5 μl PCR product of PCR 1 is used as template

4. For gene scanning select samples and put 1 μl PCR product with 0.5 μl ROX standard and 12 μl formamide (HIDI-formamide if a ABI Prism® 3100 genetic analyzer is used) in a 96-well plate. Centrifuge the plate shortly and heat the covered plate (rubber seal) at 92°C for 10 min in a PCR machine.
5. Place samples in the respective genetic analyzer and start gene scan run.
6. Analyze data according to manufacturer's instructions with the gene mapper program and identify clonal rearrangements.
7. Interpret the results (Fig. 1). One peak of expected size indicates one clonal rearrangement; one dominant peak with an additional polyclonal background pattern indicates the presence of polyclonal B cells and a polyclonal B-cell pattern indicates polyclonal rearrangements only.

3.2.2. PCR-Heteroduplex Analysis

1. Perform PCR reaction with 100–500 ng diagnostic DNA (Table 1). The reaction mixes and primers are shown in Tables 1 and 2. For PCR heteroduplex analysis use unlabelled

primers only. Use Milli-Q water as a negative control, normal MNC DNA as a polyclonal control, and an appropriate positive control (Table 1).

2. For each of the positive and negative controls and the DNA ladder, mix 7 μl of the control with 2 μl gel loading buffer and load these samples on the agarose gel; run for 30 min at 15 V/cm. The positive controls should give a clear band of the expected size, whereas the negative controls should be negative. If this is not the case, repeat the PCR assays with insufficient results.
3. Heat the PCR products of the patient samples tested for 5 min at 94°C in a PCR machine.
4. Cool the PCR products for 60 min at 4°C in a PCR machine.
5. Use Criterion Precast gels (see Note 4); remove the white strip at the bottom of the cassette and place the Criterion Precast gel in the electrophoresis cell.
6. Fill the buffer chamber of the gel cassette with 60 ml TBE buffer (1×) and fill both sides of the electrophoresis cell with 1× TBE buffer till the indicated bar (approximately 800 ml).
7. Remove the comb.
8. Mix 10 μl denaturated and renaturated PCR product with 2 μl loading buffer in a 96-well plate.
9. Pipet the mixture immediately in one of the gel slots.
10. Load the DNA ladder on the gel.
11. Cover the electrophoresis cell, place the electrodes, and run the gel for 12 h at 3 V/cm (the exact time depends on the size of the PCR products that are loaded on the gel).
12. After turning off the power, remove the cover and carefully take out the gel cassette.
13. Open the gel cassette and rinse the gel of the cassette using 100 ml Milli-Q water.
14. Incubate the gel for 2–5 min in 100 ml Milli-Q water with 10 μl ethidium bromide on a rocker platform.
15. Wash the gel for 2–5 min in 100 ml Milli-Q water on a rocker platform.
16. Analyze the gel using UV light, and make a photograph of the gel.
17. Interpret the gel. One band of expected size indicates one clonal rearrangement; two homoduplex bands plus two heteroduplex bands indicate two clonal rearrangements (biallelic or bi-clonal); >4 bands ("ladder pattern") indicate oligoclonal rearrangements; a smear indicates polyclonal rearrangements. Weak or very weak bands may indicate subclonal rearrangements.

3.2.3. PCR Product Purification and Sequencing of Clonal Rearrangements

After identification of clonal IgH gene rearrangements or the chromosomal translocations t(11;14) in the DNA sample of the patient at diagnosis, the precise sequences have to be determined. By comparing the obtained sequences with the sequences of the germline gene segments, the exact composition of the patient-specific junctional regions can be obtained. Clonal t(14;18) rearrangements can be quantified by RQ-PCR with a consensus primer–probe system without prior sequencing by using a t(14;18) positive cell line for quantification.

Monoallelic rearrangements and PCR products of t(14;18) and t(11;14) rearrangements can generally be sequenced directly from the PCR product. Use 1 μl PCR product for the sequencing reaction.

1. Prior to sequencing, PCR products have to be purified to remove primer dimers, excess primers, and dNTPs.
2. PCR products are purified by putting the 50 μl PCR products each in a Millipore Multiscreen-HTS-PCR 96-well plate. Put the plate on a vacuum device (Millipore) and apply vacuum for max. 10 min until the membranes are dry.
3. Put 25 μl Milli-Q water in each well and shake for 5–10 min.
4. Transfer the solution in fresh Eppendorf cups.
5. To estimate the concentration of DNA, 1/10 of the DNA is run on a 1% analytical agarose gel containing ethidium bromide together with a DNA mass ladder which allows the estimation of the DNA concentration by comparing the strength of the PCR product band with the strength of the DNA fragments of the mass ladder.
6. Perform sequencing using the BigDye® Terminator (BDT) v3.1 Cycle Sequencing Kit in a total reaction volume of 20 μl. The reaction contains template DNA (PCR product; see step 4), 1 μl BDT, 3.5 μl sequencing buffer, 1.3 μl primer (2.5 pmol/μl), and Milli-Q water (see Note 5). In case of multiplex sequencing reactions, use 0.5 μl of each primer (2 pmol/μl).
7. Use the same primers as used for the PCR reactions, but consider that primers should be unlabelled when used for sequencing (Table 2). In case of multiple primers in the first PCR reaction, use all forward or reverse primers for sequencing.
8. The amount of template DNA should be the following: singleplex PCR, strong band: 1 μl; weak band: 2 μl; very weak band: 3 μl; eluted homo- or heteroduplex band, strong: 2 μl; weak: 3 μl; very weak: 4 μl.
9. The PCR conditions are the following: 96°C, 1 min; 25 cycles of 96°C 10 s, 50°C 5 s, 60°C 3 min.
10. After the sequencing reaction purify the samples, e.g., by using a Millipore Multiscreen-HTS-PCR-96-well plate and vacuum manifolder.

11. Load and run the purified samples on the sequencing instrument, e.g., an ABI3100 or 3500.
12. In order to obtain a reliable junctional region sequence, sequence the clonal PCR product always from both directions (i.e., using both a forward as well as a reverse primer). In case of doubt, a second (independent) clonal PCR product should be sequenced.
13. Analyze the sequence data using Sequence Navigator or comparable software programs.

In case of biallelic rearrangements or polyclonal background heteroduplex analysis is used to isolate the clonal bands for sequencing. For this purpose follow the instructions in Subheading 3.2.2 for PCR heteroduplex analysis. Instead of 10 μl PCR product as recommended in step 11 mix 3×10 μl denaturated and renaturated PCR product with 2 μl BFB loading buffer each in a 96-well plate and pipet the mixture immediately in three lanes of 10 μl each of the gel slots. Then continue with the procedure as in Subheading 3.2.2, step 10.

1. Interpret the gel. In case of biallelic rearrangement or a clonal PCR product with polyclonal background the clonal band has to be excised.
   (a) In case of biallelic rearrangements, the relevant homoduplex bands need to be isolated. If the two homoduplexes are not well separated, the relevant heteroduplex bands should be isolated. The bands of interest are cut from the gel (on a UV lamp), and each put in a microcentrifuge tube. 100 μl aqua dest. is added to each tube and the tubes are incubated overnight at 37°C. The eluate is subsequently transferred to a new tube and 10 μl NaAc and 200 μl ice-cold 96% EtOH are added. After overnight incubation at -20°C (or 2×45 min on dry ice), the tube is centrifuged for 20 min at 10,000 rpm. The pellet is washed with 200 μl 70% EtOH, followed by centrifugation for 20 min at 10,000 rpm. After removal of the supernatant, the pellet is air-dried and dissolved in 10 μl Milli-Q water. Generally, 2 μl PCR product can be used for the sequencing reaction. If the recovery of the excised DNA is low, a higher volume may however be necessary.
   (b) Alternatively, 50 μl aqua dest. is added to each tube. After vortexing and short centrifugation the tubes are incubated overnight at 4°C. The next day the tubes are vortexed again and 7 μl of the solution is directly used for sequencing.
   (c) Exceptionally, if the above approaches are not successful, one might decide to clone the PCR products, followed by sequencing of 10–15 clones.

3.2.4. Sequence Interpretation

The junctional regions of rearranged IgH genes can be considered as "DNA fingerprints" of the lymphoma cells which are used as MRD-PCR targets. Also chromosomal translocations t(14;18) and t(11;14) involving the IgH gene on chromosome 14 bear a unique junctional region. Therefore, it is crucial to correctly assess the precise composition of the junctional region. To this end, the IgH sequence obtained should be aligned with the germline V, D, and/or J gene segments so that the exact number of deleted and inserted nucleotides can be determined. Alignment can be done using several databases freely available on the Internet or using hard copies. Since some variations can be seen between different databases, it is advised to check the alignment of the sequence in at least two ways.

1. Identify sequences using DNAPLOT software by searching for homology with all known human germline sequences obtained from the VBASE directory of human Ig genes (http://www.mrc-cpe.cam.ac.uk/DNAPLOT.php?menu=901) and IMGT (http://imgt.cines.fr) for human Ig and TCR genes. Sequences can be confirmed by using BLAST sequence similarity searching tool (National Center for Biotechnology Information: http://www.ncbi.nlm.nih.gov/BLAST/ or http://www.ncbi.nlm.nih.gov/igblast/).
2. First, try to assign the appropriate V, D, and/or J gene segments to the sequence. For recognition of D segments, a minimum length of 1/3 of the complete germline segment is required, with a minimum length of five base pairs (see Note 6).
3. Analysis of D and/or J gene segment can be done by VBASE directory of human Ig genes (http://www.mrc-cpe.cam.ac.uk/DNAPLOT.php?menu=901) and IMGT (http://imgt.cines.fr) for human Ig and TCR genes.
4. After having identified the appropriate V, D, and/or J gene segments of a clonal IgH rearrangement determine the number of deletions at both sides of all joining sites, that is, the number of nucleotides present in the germline V, D, and/or J gene segments that are lost in the patient's sequence. The remaining parts of the germline V, D, and/or J gene segments should give a 100% alignment with the patient's sequence. Lack of complete alignment may be due to somatic mutations or due to polymorphisms. Since such variations may impact RQ-PCR primer and/or probe binding, these variations should clearly be marked in the patient's sequence.
5. Subsequently, determine the number of insertions in the joining site(s).
6. For chromosomal translocations: Try to assign the chromosomal breakpoints on chromosome 18 or 11 by using published sequences. The identification of the junctional region with D

and J gene segments can be done manually from published sequences or by using electronic databases. The usage of VBASE directory of human Ig genes (http://www.mrc-cpe.cam.ac.uk/DNAPLOT.php?menu=901) IMGT (http://imgt.cines.fr) for the exact identification of human Ig and TCR genes requires some modification. Because most of the programs require a minimum sequence length (about 150 bp) and the junctional region and the JH gene segment involved in the translocation region are too short, an "artificial" VH-germline sequence can be added manually 5′ of the junctional N-DH-N-JH sequence of the respective rearrangement instead of the chromosome 18 or 11 sequence. With this artificial extension sequence proper identification of DH and JH gene segments involved in the translocation can be performed.

3.2.5. Selection of MRD-PCR Targets

In order to limit the risk of false-negative MRD results preferably two MRD-PCR targets should be used for each lymphoma patient if available. These MRD-PCR targets should be selected based on expected sensitivity. A sensitivity of at least 10^{-4} should be obtained. When using IgH gene rearrangements, somatic mutations may hamper RQ-PCR sensitivity, especially in germinal center/post-germinal center B-cell malignancies (f.e. FL) since RQ-PCR primer and/or probe binding may be constrained. Therefore, these variations should be carefully marked in the patient's sequence and should be considered for the selection of MRD-PCR targets. In patients with chromosomal translocations as t(14;18) translocation or t(11;14), these translocations can serve as the preferred MRD target because they are highly stable and the JH-region involved in the translocation is generally unmutated (24, 25).

The sensitivity of MRD-PCR analysis of junctional regions is dependent on several factors, including the size of the junctional region, the "background" of normal lymphoid cells with comparable polyclonal IgH gene rearrangements, and the frequency of somatic IgH mutations (26). If a relatively high proportion of lymphoma cells and a low background of polyclonal B cells are present as f.e. in CLL at diagnosis, the MRD level can generally reliably be quantified from the diagnostic sample. In case of low MRD levels in peripheral blood and bone marrow, f.e. in primary nodal lymphomas as FL or MCL, the assay becomes less accurate. A prerequisite for quantification from diagnostic material is therefore the assessment of the percentage of lymphoma cells in the MNC fraction by flow cytometric immunophenotyping.

Within the European Study Group on MRD detection in ALL (ESG-MRD-ALL; now EURO-MRD group see Note 7) it therefore was decided to define the "Quantitative range" and the "sensitivity" (27). The "Quantitative range" reflects the part of the standard curve in which the MRD levels can be quantified reproducibly and accurately, whereas the "sensitivity" reflects the lowest

MRD level that still can be detected, although not reproducibly and accurately. These definitions are also used for the quantification of MRD levels in malignant lymphomas.

Generally, IgH gene-based MRD assessment results in higher sensitivities in lymphoma entities with a lower frequency of somatic mutations f.e unmutated or low mutated MCL and CLL. In germinal center/post-germinal center B-cell malignancies with a high mutation load a completely sequence-based design of primers and probes might be necessary to achieve the required sensitivity of 10^{-4}. Alternatively, light chain rearrangements of rearranged Ig kappa (Igκ) are less frequently mutated and may serve as alternative MRD targets (28).

3.3. RQ-PCR Sensitivity Testing

3.3.1. Design of Allele-Specific Oligonucleotide Primers

The following instructions have been developed within the EURO-MRD network for the design of allele-specific oligonucleotide (ASO) primers in general and have been successfully applied for MRD assessment in ALL, CLL, and MCL. For the application in B-NHL the same guidelines can be followed and are therefore cited here as already published (27).

For RQ-PCR analysis generally the TaqMan (hydrolysis) probe approach is being used (see Note 8). Several primer–probe sets for RQ-PCR-based detection of tumor-specific *IGH* have been designed and successfully applied in patients with ALL (29, 30). Although both an ASO probe approach and ASO primer approach can be used, the ASO primer approach has the advantage that it is much cheaper since no patient-specific probes need to be ordered. In cases with frequent somatic mutations of IgH gene rearrangements, a completely sequence-based design of primers and probes is recommended for RQ-PCR analysis.

The germline primer and probe sets should be used in combination with a patient-specific ASO primer. This ASO primer should be positioned within and around the junctional region in such a way that it provides maximal specificity.

1. Enter the sequence of the IgH rearrangement or the t(11;14) translocation in Primer Express software.
2. Select an appropriate germline primer and probe.
3. Mark the position of the germline primer and probe in the sequence.
4. Design an ASO primer using Primer Express and OLIGO6.0 software according to the following guidelines.
   (a) Place the 3′ end of the primer in or just over the junctional region; maximum overlap with germline sequence: 2–6 nucleotides. In case of *IGH* gene rearrangements, the design of an ASO primer within the D_H–J_H junctional region has the advantage of high specificity. In case of chromosomal translocations, the ASO primer can be

placed either completely in the D_H–J_H junctional region or overlapping the breakpoint region on the respective chromosome and the D_H–J_H junctional region.

(b) Melting temperature (Tm) 57–60°C, according to the Primer Express nearest neighbor algorithm.

(c) The GC content should be in the range of 20–80%.

(d) Check in OLIGO 6.0:

- Internal stability graph (if possible no more than 2 G/Cs in the last five bases on 3′ end according to Primer Express): Strong binding in middle part of primer; lower binding at 3′ end ($\Delta G > -10$).
- 3′-Dimers forward–forward and forward–reverse primer (<5 kcal/mol).
- Dimers forward–forward and forward–reverse primer (preferably <10 kcal/mol).
- Hairpin stems 3′ ends (no hairpin with Tm within 10°C from Tm primer).

5. Order the primers (several companies available).
6. Upon receipt, dissolve the primer in 1× TE buffer; prepare a concentration of approximately 100 pmol/μl. Check the concentration of the primer with the Nanodrop® (ND-1000 spectrophotometer) by measuring 1.0 μl primer (in duplicate).
7. Prepare 300 μl of a primer working dilution in Milli-Q water with a concentration of 30 pmol/μl. Store the stock and working solution at −20°C.

3.3.2. Application of Consensus Primers and Probes for t(14;18) RQ-PCR

For RQ-PCR analysis of the t(14;18) translocation forward primer and probe are placed in chromosome 18 in combination with a reverse primer placed in the consensus JH region on chromosome 14. Because only cells bearing the translocation can be amplified, the assay is specific without the necessity to use an ASO primer. With this approach the same primers and probes can be universally used for different patients making this approach much cheaper and faster since no ASO primers have to be designed. A further advantage of this approach is the possible use of t(14;18)-bearing cell lines as diagnostic samples. This is especially important in FL which frequently does not have a leukemic presentation and the degree of infiltration in the BM is unknown. Beyond this, the usage of a cell line as diagnostic sample provides the advantage that different patients can be quantified with the same cell line as diagnostic samples. Primers and probes for this approach are listed in Table 2.

However, there are several restrictions for the usage of t(14;18) RQ-PCR assays. The t(14;18) breakpoints cluster in at least four different regions: the major breakpoint region (MBR), the minor clustering region (mcr), the 3′-MBR, and the 5′-mcr (31, 32).

A number of variant translocations have been described that cannot be usefully assessed by PCR (33). Chromosome 14 breakpoints are almost exclusively located within one of the six germline JH segments allowing PCR amplification of the translocation. Only MBR, and to a lesser extent mcr, has been extensively exploited for MRD analysis. Thus, only a subset of FL patients, ranging from 50 to 65%, can be assessed using the BCL2-IgH PCR (18, 34).

It is part of the current activities of the EURO-MRD group to develop and validate RQ-PCR assays for other breakpoints than MBR and this is work in progress. Therefore, the procedures described here will only focus on patients with a t(14;18) MBR translocation.

Besides the breakpoint location on chromosome 18 the size of the respective RQ-PCR product is of great importance since PCR products excessing 300 bp in size are not reliably quantified due to a lower amplification efficiency and to a possibly lower amplification efficiency in comparison with the diagnostic cell line.

3.3.3. RQ-PCR Analysis of Dilution Series of Diagnostic Sample

In order to determine the sensitivity of the RQ-PCR assay and the specificity of a designed ASO primer, serial dilutions of the diagnostic sample should be used.

1. Based on the morphological (BM) or preferably immunophenotypically determined lymphoma cell percentage (PB and BM) in the diagnostic MNC preparation, the diagnostic DNA sample (60 ng/μl) should be diluted in normal control MNC DNA (60 ng/μl) to a lymphoma cell percentage of 10% (see Notes 9 and 10). For t(14;18) quantification cell line DNA (f.e. DOHH2) is used. Prepare 150 μl of this 10^{-1} dilution (see Note 11).
2. Prepare a 10^{-2} dilution by mixing 15 μl of the 10^{-1} dilution with 135 μl normal control MNC DNA (60 ng/μl). Comparably, prepare 150 μl of a 10^{-3}, 10^{-4}, and 10^{-5} dilution. Prepare a 5×10^{-4} dilution by mixing 7 μl of the 10^{-2} dilution with 133 μl normal control MNC DNA (60 ng/μl).
3. Perform an RQ-PCR reaction using the 10^{-1}–10^{-5} dilutions in duplicate as well as the normal control MNC DNA in sixfold (see Note 12) and Milli-Q water in duplicate. 600 ng DNA should be used per well (equivalent to approximately 90,000 cells, required to obtain a theoretical sensitivity of 10^{-5}). Reaction mixes are shown in Table 5.
4. Thermal cycler conditions are the following: 50°C, 2 min; 95°C, 10 min; 50 cycles of 95°C, 15 s, 60°C, 1 min.
5. Analyze the RQ-PCR data using appropriate software.
6. The threshold of the RQ-PCR assay should always be set in the region of exponential amplification across all amplification plots. This region is depicted in the log view of amplification plots as the portion of the curve that is linear. Often the threshold

automatically determined by the instrument software can be used. However, if the threshold appears to be positioned outside the linear part of (some of) the amplification curves, adjustments may be made (26, 27).

7. By plotting the logarithmic value of the dilution series against the cycle threshold (C_T) obtained, a standard curve can be obtained.

3.3.4. RQ-PCR Data Interpretation: Quantitative Range and Sensitivity

For defining the sensitivity, several criteria (including reproducibility of the measurement, the difference between specific and nonspecific amplification, slope and correlation coefficient of the standard curve) should be taken into account (27) (see note 18). Furthermore, in order to compare data between different studies and/or different laboratories, it is essential to have uniform guidelines for RQ-PCR data interpretation. For Ig/TCR-based MRD data interpretation in ALL, such guidelines have been developed within the ESG-MRD-ALL (now EURO-MRD consortium) and have been applied for MRD assessment in malignant lymphoma by the same group (see Notes 13 and 14).

3.4. Bone Marrow and Peripheral Blood Sample Processing During Follow-Up

1. Obtain a total of 2.5–3 ml BM in one first aspirate or use 10 ml PB.
2. Perform Ficoll density centrifugation for obtaining MNC.
3. Extract DNA from 10×10^6 MNC using a DNA extraction column or any other DNA extraction method, generally resulting in 30–40 μg of DNA (27). This is sufficient for standard RQ-PCR analysis (applying two MRD-PCR targets and one control gene) (see Note 16).
4. Make an RQ-PCR solution of 60 ng/μl and store at 4°C (or −20°C).
5. Store the remaining MNC of the BM follow-up samples in cell storage medium for later studies; store in liquid nitrogen or as a dry pellet at −80°C.

3.5. MRD Analysis of Follow-Up Samples

The actual MRD analysis includes three main steps: RQ-PCR analysis of a control gene for checking the quality and quantity of the DNA, RQ-PCR analysis of the IgH targets or chromosomal translocations, and finally interpretation of the data obtained.

3.5.1. Control Gene RQ-PCR Analysis

In order to check for the quantity and quality of the DNA, RQ-PCR analysis of a control gene is required. For this purpose, the albumin (*ALB*) gene on chromosome 4 can be used (35).

1. Make dilutions of normal control MNC DNA (100 ng/μl) in Milli-Q water so that solutions of 50, 10, and 1 ng/μl are obtained.
2. Analyze the four normal control DNA samples (i.e., 100, 50, 10, and 1 ng/μl), the (undiluted) diagnostic DNA (60 ng/μl),

and the follow-up samples (60 ng/μl) in duplicate. Use Milli-Q water as a negative control (in duplicate). Use the reaction mix as indicated in Table 10.5.

3. Thermal cycler conditions (for ABI 7000; other platforms may require different optimal temperatures and times): 50°C, 2 min; 95°C, 10 min; 95°C, 15 s, 60°C, 1 min (50×).
4. Analyze the RQ-PCR data using appropriate software. It is advised to use a common threshold setting for all control gene RQ-PCR reactions, since this allows an easier comparison of data obtained in different RQ-PCR runs. The C_T values obtained for the standard curve wells can be used to monitor the performance of the RQ-PCR reaction over time.
5. Construct a standard curve by plotting the C_T values of the four normal control MNC DNA against their C_T values.
6. Determine the DNA content of the diagnostic and follow-up DNA samples using their C_T value and the obtained standard curve (see Note 16).

3.5.2. MRD-PCR Target RQ-PCR Analysis

For the actual analysis of MRD in the follow-up samples, RQ-PCR analysis is performed for the selected targets.

1. For each MRD-PCR target, analyze the dilution series (10^{-1}–10^{-5}; see Subheading 3.3.2) of the diagnostic DNA (60 ng/μl) in duplicate, the follow-up samples (60 ng/μl) in triplicate, and normal control MNC DNA in sixfold. Use Milli-Q water as a negative control (in triplicate). Use the reaction mix as indicated in Table 5.
2. Of importance, the control gene PCR and MRD-PCR target PCRs should be performed using exactly the same DNA sample in identical reaction volumes and using identical amounts of DNA.

Table 5
RQ-PCR reaction mixtures

Reaction mix per 25 μl reaction	Final concentration	μl/well
Universal master mix (2×)	1×	12.5
Forward primer (20 pmol/μl)	300 nM	0.38
Reverse primer (20 pmol/μl)	300 nM	0.38
TaqMan probe (20 pmol/μl)	100 nM	0.13
BSA (5%)	0.04%	0.2
Milli-Q water		5.4
DNA (100 ng/μl)		6

3. Thermal cycler conditions (for ABI 7000; other platforms may require different optimal temperatures and times): 50°C, 2 min; 95°C, 10 min; 95°C, 15 s, 60°C, 1 min (50×) (see Note 17).
4. Analyze the RQ-PCR data using appropriate software.
5. The threshold of the RQ-PCR assay should always be set in the region of exponential amplification across all amplification plots. This region is depicted in the log view of amplification plots as the portion of the curve that is linear. Often the threshold automatically determined by the instrument software can be used. However, if the threshold appears to be positioned outside the linear part of (some of) the amplification curves, adjustments may be made (26, 27).

3.5.3. RQ-PCR MRD Data Interpretation

For interpretation of the MRD data, the guidelines of the ESG-MRD-ALL (27) should be used.

For MRD status interpretation, the MRD-PCR target with the highest MRD level should be regarded as the most reliable target. So, always the highest MRD level per time point should be reported. The optimal marker should have a quantitative range of $\leq 10^{-4}$ and a sensitivity of 10^{-5}; less sensitive markers with a quantitative range of $\leq 10^{-3}$ and a sensitivity $\leq 10^{-4}$ are acceptable. The acceptance of an MRD marker is dependent on the protocol and purpose of a clinical trial.

3.6. Concluding Remarks

In contrast to MRD diagnostics in ALL the clinical implications of PCR-based MRD diagnostics in malignant lymphoma have only evolved recently. As in ALL patients, it is also clear that the PCR-based MRD diagnostics via Ig genes and chromosomal translocations is complex, requires extensive knowledge about gene rearrangement processes, and has to consider the specific characteristics of the heterogeneous histological subtypes. Therefore, usage of standardized assays and internationally accepted guidelines for interpretation of RQ-PCR results as well as regular internal and external quality control rounds are a prerequisite for MRD-based treatment modification.

4. Notes

1. Ideally the diagnostic material for generating the standard curves contains 70–100% tumor cells. However, in the majority of MCL and especially nodal FL as well as in a subset of CLL patients infiltration of BM or PB is less than 70%. Therefore, flow cytometric immunophenotyping allows the determination of the percentage of tumor cells and with this the correction of the tumor load in the diagnostic sample.

While in MCL and CLL the percentage of lymphoma cells is almost equal in PB and BM, circulating lymphoma cell level is about one log higher in BM compared to PB. Therefore in FL it is useful in many cases to use BM for the standard curve. The use of samples with low percentages of lymphoma cells (<10%) may hamper the reliable quantification.

2. For DNA extraction different column-based and other methods give comparable good results. The usage of fixed number of MNC facilitates reproducibility of PCR analysis because extraction of DNA from 10×10^6 MNC will generally result in 30–40 μg of DNA, which is sufficient for all relevant PCR analyses at diagnosis including the preparation of a standard curve.
3. For assessment of DNA concentration, the Nanodrop method is preferred over analysis using another type of spectrophotometer, since the results are more robust and accurate.
4. Instead of commercially available gels, also homemade 6% PAGE gels can be used.
5. Alternatively, comparable commercially available sequencing kits can be used.
6. The recognition of (second) D segments is often not done by software. Therefore, if the N region consists of >10 bp check for D segment by hand (using hard copies) or use only the sequence around the N region for analysis in databases.
7. The EuroMRD group (formerly ESG-MRD-ALL) under the umbrella of ESHLO (European Scientific Foundation of Laboratory Hemato-Oncology) is a consortium of international MRD-PCR laboratories, coordinated by J.J.M. van Dongen. The main aims of the group are the organization of a quality control program twice per year, the collaborative development and clinical evaluation of new MRD strategies and techniques, and the development of guidelines for the interpretation of RQ-PCR-based MRD data.
8. For RQ-PCR analysis, other probe formats (e.g., hybridization probes) can also be used. The use of nonspecific dyes, like SYBR Green I, should be avoided, since this limits the sensitivity.
9. MNC obtained from PB can be used. For Ig gene rearrangements, a pool of at least five healthy individuals should be used. When performing t(14;18) PCR, normal control MNC DNA should be tested for the presence of t(14;18) translocation because t(14;18)-positive cells can be present in healthy donors without lymphoma. Alternatively, DNA from cell lines without the t(14;18) translocation can be used.
10. In many cases the degree of infiltration is less than 10% in the diagnostic material. To our experience, it is possible to generate

reliable standard curves from diagnostic samples with a minimum infiltration of 1%. This implies that the standard curve starts at 10^{-2} dilution as the lowest dilution of the standard curve.

11. If only a limited amount of diagnostic DNA is available, one might decide to skip the 10^{-1} dilution and to use the 10^{-2} dilution as the lowest dilution of the standard curve.
12. Since nonspecific amplification is generally only detected at low levels and outside the reproducible range of the RQ-PCR, nonspecific amplification controls should be run at least in sixfold in each RQ-PCR analysis for each Ig or TCR marker.
13. The background of the RQ-PCR assay, i.e., the nonspecific amplification of comparable Ig gene rearrangements present in normal cells or t(14;18)-positive cells not belonging to the lymphoma, is defined by the lowest C_T value obtained in the nonspecific amplification controls (normal MNC DNA).
14. If the quantitative range and/or sensitivity of the RQ-PCR test is not sufficient, i.e., has nonspecific amplification, the annealing temperature can be increased (up to 70°C) in order to improve specificity. If this is not successful, one can decide to design an alternative ASO primer.
15. At least 2×10^6 MNC should be used for DNA extraction since at least 5–8 μg of DNA should be available for RQ-PCR analysis.
16. If the RQ-PCR of the control gene shows a lower amount of template than expected based on physical measurements (e.g., Nanodrop measurement), special caution is needed since this lower value can be the result of inhibition, which can be found in a substantial number of BM or PB samples (5–10%). Addition of BSA prevents inhibition, and the EURO-MRD therefore recommends the inclusion of 0.04% BSA in all RQ-PCR reactions (27). Since the addition of less template will result in loss of sensitivity, the control gene values of all samples need to be within predefined ranges, that is, between 250 and 1,000 ng/reaction. If the DNA amount appears to be higher, the follow-up sample should be diluted and retested because too high DNA concentrations may also result in PCR inhibition. If the DNA amount is lower, correct interpretation may not be possible. Consequently, reanalysis is recommended after adaptation of the amount of DNA (27).
17. The annealing temperature can be different, dependent on the results of the sensitivity testing.
18. Logically, very low MRD levels (below the "quantitative range") should always be judged with caution; especially if only one well of the three replicates is positive. In such a case, reanalysis of the doubtful sample(s) may be performed, but one should be aware that by definition the results will often not be reproducible.

Acknowledgements

We gratefully acknowledge Ariane Stuhr, Karin Olsen, and Petra Chall for their excellent technical support.

References

1. Brown JR, Feng Y, Gribben JG, Neuberg D, Fisher DC, Mauch P et al (2007) Long-term survival after autologous bone marrow transplantation for follicular lymphoma in first remission. Biol Blood Marrow Transplant 13:1057–1065
2. Goff L, Summers K, Iqbal S, Kuhlmann J, Kunz M, Louton T et al (2009) Quantitative PCR analysis for Bcl-2/IgH in a phase III study of Yttrium-90 Ibritumomab Tiuxetan as consolidation of first remission in patients with follicular lymphoma. J Clin Oncol 27:6094–6100
3. Hardingham JE, Kotasek D, Sage RE, Gooley LT, Mi JX, Dobrovic A et al (1995) Significance of molecular marker-positive cells after autologous peripheral-blood stem-cell transplantation for non-Hodgkin's lymphoma. J Clin Oncol 13:1073–1079
4. Hirt C, Schuler F, Dolken G (2003) Minimal residual disease (MRD) in follicular lymphoma in the era of immunotherapy with rituximab. Semin Cancer Biol 13:223–231
5. Ladetto M, Magni M, Pagliano G, De Marco F, Drandi D, Ricca I et al (2006) Rituximab induces effective clearance of minimal residual disease in molecular relapses of mantle cell lymphoma. Biol Blood Marrow Transplant 12:1270–1276
6. Moos M, Schulz R, Martin S, Benner A, Haas R (1998) The remission status before and the PCR status after high-dose therapy with peripheral blood stem cell support are prognostic factors for relapse-free survival in patients with follicular non-Hodgkin's lymphoma. Leukemia 12:1971–1976
7. Moreton P, Kennedy B, Lucas G, Leach M, Rassam SMB, Haynes A et al (2005) Eradication of minimal residual disease in B-cell chronic lymphocytic leukemia after Alemtuzumab therapy Is associated with prolonged survival. J Clin Oncol 23:2971–2979
8. Pott C, Hoster E, Beldjord K, Macintyre E, Böttcher S, Asnafi V et al (2010) R-CHOP/R-DHAP prior to autologous stem cell transplantation induces higher rates of molecular remission than R-CHOP in younger patients with MCL: results of the *MCL younger* intergroup trial of the *European MCL Network*. Blood 116: abstract 528
9. Pott C, Hoster E, Delfau-Larue MH, Beldjord K, Bottcher S, Asnafi V et al (2010) Molecular remission is an independent predictor of clinical outcome in patients with mantle cell lymphoma after combined immunochemotherapy: a European MCL intergroup study. Blood 115:3215–3223
10. Pott C, Schrader C, Gesk S, Harder L, Tiemann M, Raff T et al (2006) Quantitative assessment of molecular remission after high-dose therapy with autologous stem cell transplantation predicts long-term remission in mantle cell lymphoma. Blood 107:2271–2278
11. Provan D, Bartlett-Pandite L, Zwicky C, Neuberg D, Maddocks A, Corradini P et al (1996) Eradication of polymerase chain reaction-detectable chronic lymphocytic leukemia cells is associated with improved outcome after bone marrow transplantation. Blood 88: 2228–2235
12. Geisler CH, Kolstad A, Laurell A, Andersen NS, Pedersen LB, Jerkeman M et al (2008) Long-term progression-free survival of mantle cell lymphoma after intensive front-line immunochemotherapy with in vivo-purged stem cell rescue: a nonrandomized phase 2 multicenter study by the Nordic Lymphoma Group. Blood 112:2687–2693
13. Ladetto M, De Marco F, Benedetti F, Vitolo U, Patti C, Rambaldi A et al (2008) Prospective, multicenter randomized GITMO/IIL trial comparing intensive (R-HDS) versus conventional (CHOP-R) chemoimmunotherapy in high-risk follicular lymphoma at diagnosis: the superior disease control of R-HDS does not translate into an overall survival advantage. Blood 111:4004–4013
14. Rambaldi A, Carlotti E, Oldani E, Della SI, Baccarani M, Cortelazzo S et al (2005) Quantitative PCR of bone marrow BCL2/IgH positive cells at diagnosis predicts treatment response and long-term outcome in Follicular non Hodgkin's Lymphoma. Blood 105:3428–3433
15. Andersen NS, Donovan JW, Zuckerman A, Pedersen L, Geisler C, Gribben JG (2002)

Real-time polymerase chain reaction estimation of bone marrow tumor burden using clonal immunoglobulin heavy chain gene and bcl-1/JH rearrangements in mantle cell lymphoma. Exp Hematol 30:703–710

16. Andersen NS, Pedersen LB, Laurell A, Elonen E, Kolstad A, Boesen AM et al (2009) Pre-emptive treatment with rituximab of molecular relapse after autologous stem cell transplantation in mantle cell lymphoma. J Clin Oncol 27:4365–4370
17. Rambaldi A, Lazzari M, Manzoni C, Carlotti E, Arcaini L, Baccarani M et al (2002) Monitoring of minimal residual disease after CHOP and rituximab in previously untreated patients with follicular lymphoma. Blood 99:856–862
18. Evans PAS, Pott C, Groenen PJTA, Salles G, Davi F, Berger F et al (2006) Significantly improved PCR-based clonality testing in B-cell malignancies by use of multiple immunoglobulin gene targets. Report of the BIOMED-2 Concerted Action BHM4-CT98-3936. Leukemia 21:207–214
19. van Dongen JJ, Langerak AW, Bruggemann M, Evans PA, Hummel M, Lavender FL et al (2003) Design and standardization of PCR primers and protocols for detection of clonal immunoglobulin and T-cell receptor gene recombinations in suspect lymphoproliferations: report of the BIOMED-2 Concerted Action BMH4-CT98-3936. Leukemia 17:2257–2317
20. Kneba M, Bergholz M, Bolz I, Hulpke M, Batge R, Schauer A et al (1990) Heterogeneity of immunoglobulin gene rearrangements in B-cell lymphomas. Int J Cancer 45:609–613
21. Kneba M, Eick S, Herbst H, Willigeroth S, Pott C, Bolz I et al (1991) Frequency and structure of t(14;18) major breakpoint regions in non-Hodgkin's lymphomas typed according to the Kiel classification: analysis by direct DNA sequencing. Cancer Res 51:3243–3250
22. Pott C, Tiemann M, Linke B, Ott MM, von Hofen M, Bolz I et al (1998) Structure of Bcl-1 and IgH-CDR3 rearrangements as clonal markers in mantle cell lymphomas. Leukemia 12:1630–1637
23. van der Velden V, van Dongen J (2009) MRD detection in acute lymphoblastic leukemia patients using Ig/TCR gene rearrangements as targets for real-time quantitative PCR. In: Walker JM (ed) Lymphoma: methods and protocols. Humana Press, p. 115–150
24. Olsson K, Gerard CJ, Zehnder J, Jones C, Ramanathan R, Reading C et al (1999) Real-time t(11;14) and t(14;18) PCR assays provide sensitive and quantitative assessments of minimal residual disease (MRD). Leukemia 13:1833–1842
25. Pott C, Schrader C, Trautmann H, Irmer S, Desgranges C, Ritgen M et al (2003) VH mutational status and VH gene usage define molecular heterogeneity in mantle cell lymphoma. Onkologie 26(suppl 5):162
26. van der Velden VH, Hochhaus A, Cazzaniga G, Szczepanski T, Gabert J, van Dongen JJ (2003) Detection of minimal residual disease in hematologic malignancies by real-time quantitative PCR: principles, approaches, and laboratory aspects. Leukemia 17:1013–1034
27. van der Velden VHJ, Cazzaniga G, Schrauder A, Hancock J, Bader P, Panzer-Grumayer ER et al (2007) Analysis of minimal residual disease by Ig//TCR gene rearrangements: guidelines for interpretation of real-time quantitative PCR data. Leukemia 21:604–611
28. Payne K, Wright P, Grant JW, Huang Y, Hamoudi R, Bacon CM et al (2011) BIOMED-2 PCR assays for IGK gene rearrangements are essential for B-cell clonality analysis in follicular lymphoma. Br J Haematol 155:84–92
29. Bruggemann M, Droese J, Bolz I, Luth P, Pott C, von Neuhoff N et al (2000) Improved assessment of minimal residual disease in B cell malignancies using fluorogenic consensus probes for real-time quantitative PCR. Leukemia 14: 1419–1425
30. Donovan JW, Ladetto M, Zou G, Neuberg D, Poor C, Bowers D et al (2000) Immunoglobulin heavy-chain consensus probes for real-time PCR quantification of residual disease in acute lymphoblastic leukemia. Blood 95: 2651–2658
31. Akasaka T, Akasaka H, Yonetani N, Ohno H, Yamabe H, Fukuhara S, Okuma M (1998) Refinement of the BCL2/immunoglobulin heavy chain fusion gene in t(14;18)(q32;q21) by polymerase chain reaction amplification for long targets. Genes Chromosomes Cancer 21:17–29
32. Buchonnet G, Lenain P, Ruminy P, Lepretre S, Stamatoullas A, Parmentier F et al (2000) Characterisation of BCL2-JH rearrangements in follicular lymphoma: PCR detection of 3′ BCL2 breakpoints and evidence of a new cluster. Leukemia 14:1563–1569
33. Albinger-Hegyi A, Hochreutener B, Abdou MT, Hegyi I, Dours-Zimmermann MT, Kurrer MO et al (2002) High frequency of t(14;18)-translocation breakpoints outside of major breakpoint and minor cluster regions in follicular lymphomas: improved polymerase chain reaction protocols for their detection. Am J Pathol 160:823–832

34. van Krieken JHJM, Langerak AW, Macintyre EA, Kneba M, Hodges E, Sanz RG et al (2006) Improved reliability of lymphoma diagnostics via PCR-based clonality testing: [mdash] Report of the BIOMED-2 Concerted Action BHM4-CT98-3936. Leukemia 21:201–206

35. Pongers-Willemse MJ, Verhagen OJ, Tibbe GJ, Wijkhuijs AJ, de Haas V, Roovers E et al (1998) Real-time quantitative PCR for the detection of minimal residual disease in acute lymphoblastic leukemia using junctional region specific TaqMan probes. Leukemia 12:2006–2014

Chapter 11

Enrichment of Methylated DNA by Methyl-CpG Immunoprecipitation

Miriam Sonnet, Constance Baer, Michael Rehli, Dieter Weichenhan, and Christoph Plass

Abstract

Normal DNA methylation is an epigenetic modification required for proper development. Aberrant DNA methylation, in contrast, is frequently observed in many different malignancies including leukemias and lymphomas. Global DNA methylation profiling addresses the methylated sequences (methylome) of patient genomes to identify disease-specific methylation patterns. Workload in methylome analyses can be considerably reduced by methylome enrichment using proteins or antibodies with high affinity to methylated DNA. Methyl-CpG Immunoprecipitation (MCIp) employs an immobilized recombinant human methyl-CpG binding domain protein 2, MBD2, which binds methylated CpGs in double-stranded DNA. Elution with increasing salt concentrations allows the fractionated enrichment of different degrees of methylation.

Key words: Epigenetics, DNA methylation, Methyl-CpG-Immunoprecipitation, MBD2

1. Introduction

1.1. Aberrant DNA Methylation in Disease

Methylation of cytosine at its C-5 atom (5mC) in CpG dinucleotides is the most abundant and best studied epigenetic DNA modification. DNA modifications other than 5mC comprise 5-hydroxymethylcytosine (1, 2), 5-formylcytosine and 5-carboxycytosine (3, 4). Though their function is still largely elusive, initial evidence points to roles in DNA demethylation and tissue-specific regulation of genes.

Accumulations of CpGs, so-called CpG islands, are often but not exclusively found in promoter regions (5). Aberrant DNA methylation in promoters is often associated with changes in gene

Ralf Küppers (ed.), *Lymphoma: Methods and Protocols*, Methods in Molecular Biology, vol. 971,
DOI 10.1007/978-1-62703-269-8_11, © Springer Science+Business Media, LLC 2013

expression which contribute to diseases. As a hallmark of many malignancies, gain of promoter methylation leads to downregulation of tumor suppressor genes and, as a consequence, carcinogenesis. Global DNA hypomethylation, on the other hand, is associated with genome instability and chromosomal rearrangements which are characteristic for various cancer types (6).

CpG methylation is a stable DNA modification and hence, can be reliably analyzed in archived biological specimen such as paraffin-embedded tissue slices. Sites of aberrant DNA methylation are used as biomarkers for diagnosis and prognosis of a malignancy, while epigenetically deregulated genes serve as druggable targets for cancer therapy (7). Accordingly, the identification of epigenetic aberrations on a genome-wide scale bears the prospect for a better understanding of disease development and novel therapeutic regimens.

1.2. Methods of Methylome Analysis

Bisulfite treatment of DNA allows the direct readout of the methylation state at the nucleotide level by altering unmethylated cytosines to uracil while methylated cytosines remain unchanged. Whole genome bisulfite treatment followed by next generation sequencing (NGS) is the gold standard of methylome analysis, yet still an exception, because it is not affordable as a routine method. Enrichment of methylated DNA prior to microarray analysis or NGS provides a much cheaper alternative. During the last 5 years, affinity capture techniques such as methylated DNA immunoprecipitation (8), MethylCap (9), Methylated-CpG island recovery assay (10), or Methyl-CpG Immunoprecipitation (MCIp) (11, 12) have been published and have even become commercially available. Though similar in the final outcome, the methods are different in their flexibility and efficiency to enrich moderately and highly methylated DNA stretches (13, 14).

MCIp uses a recombinant human methyl-CpG binding domain protein 2 (MBD2), a MBD2-Fc fusion protein, which is produced in cell culture by *Drosophila melanogaster* S2 Schneider cells (11). Via its immunoglobulin Fc part, the protein is immobilized to protein A-coated sepharose or magnetic bead particles. Upon incubation with fragmented sample DNA, MBD2 selectively captures methylated sequence stretches which are subsequently eluted in fractions of increasing degree of methylation using buffers with increasing salt concentrations. To monitor enrichment efficiency, quantitative polymerase chain reaction (qPCR) is performed targeting an imprinted gene such as *SNRPN* or *Mest*. In addition to a detailed MCIp protocol (3.1–3.4), we also present a concise protocol for the production of the MBD2-Fc protein (Appendix).

2. Materials

2.1. Sonication of Genomic DNA

1. 2.5 μg DNA.
2. Sonicator and tubes for sonication (e.g., Bioruptor NextGen and TPX Diagenode tube, Diagenode, Liège, Belgium).
3. 1.5% agarose gel.
4. 100 base pair (bp) DNA size ladder.
5. Ethidium bromide.
6. Gel documentation station.

2.2. Immobilization of MBD2-Fc Protein to Protein A-Coated Magnetic Beads

1. MBD2-Fc protein.
2. Protein A-coated magnetic beads (e.g., Diagenode).
3. Magnet (e.g., magneto Pure micro, Chemicell, Berlin, Germany).
4. 8-well strip, suitable for magnet.
5. 10× Tris-buffered saline (TBS): 500 mM Tris–HCl, 1.5 M NaCl, 10 mM EDTA; adjust to pH 7.4. Dilute 1:10 with water.

2.3. Capture of Methylated DNA by Immobilized MBD2-Fc Protein

1. Buffer A: 300 mM NaCl, 1×TME, 0.1% Nonidet P-40 (NP-40).
2. Buffer B: 400 mM NaCl, 1×TME, 0.1% Nonidet P-40 (NP-40).
3. Buffer C: 500 mM NaCl, 1×TME, 0.1% Nonidet P-40 (NP-40).
4. Buffer D: 550 mM NaCl, 1×TME, 0.1% Nonidet P-40 (NP-40).
5. Buffer E: 1 M NaCl, 1×TME, 0.1% Nonidet P-40 (NP-40).
6. Buffer X: 300 mM NaCl, 1×TME, 0.1% Nonidet P-40 (NP-40) after addition of DNA solution (see Note 1).
7. 10×TME: 0.2 M Tris–HCl, 0.02 M $MgCl_2$, 0.005 M EDTA; adjust to pH 8.
8. Magnet (see Subheading 2.2).
9. 8-well strips, suitable for magnet.

3. Methods

3.1. Sonication of Genomic DNA

1. Dissolve 2.5 μg DNA (see Note 2) in a suitable volume of H_2O depending on the sonicator in the next step.
2. Sonicate the DNA to a size range appropriate for subsequent downstream applications (see Note 3).
3. Analyze the result of sonication on a 1.5% agarose gel.

3.1.1. Example: Use of the Bioruptor Next Gen (Diagenode)

1. Perform sonication at 4°C using either a thermostate or an ice bath for cooling.
2. Transfer 125 µl DNA solution (20 ng/µl) to the bottom of a 1.5 ml TPX tube (Diagenode) (see Note 4).

   Place the tube in the tube adapter suited for 1.5 ml tubes. Be sure that all positions are engaged by equally filled tubes. Fix the tubes, insert the adapter, and sonicate with the following settings (see Note 5):

   Low energy

   Per cycle, 30 s ON, 30 s OFF

   30 cycles (centrifuge briefly after 15 cycles to recollect the DNA solution at the bottom of the tube
3. Analyze 10 µl of the sonicated DNA on a 1.5% agarose gel together with a 100 bp ladder as size marker. The DNA smear should range from 100 to 600 bp (Fig. 1).

If a considerable portion of the smear extends to a larger size, repeat sonication using 6–10 more cycles.

3.2. Immobilization of MBD2-Fc Protein to Protein A-Coated Magnetic Beads

1. Per DNA sample, transfer 40 µl thoroughly homogenized suspension of protein A-coated magnetic beads to a well of an 8-well strip containing 160 µl TBS/well. Wash the beads by 10× pipetting up and down (see Note 6).
2. Transfer the strip to an appropriate magnet, wait for about 30 s until all beads stick to the side of the tube, and discard supernatant.
3. Remove the strip from the magnet, resuspend and wash the beads with 200 µl TBS by 20× pipetting up and down.
4. Repeat step 2.
5. Remove the strip from the magnet, resuspend beads in 200 µl TBS containing 60 µg MDB2-Fc protein. Close the strip (see Note 7).

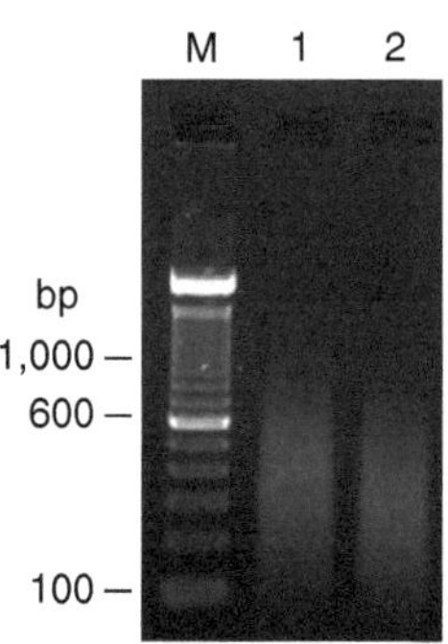

Fig. 1. Sonicated DNA. M, 100 bp ladder; 1 and 2, DNA samples.

6. To immobilize the MBD2-Fc protein to the beads by affinity capture, rotate the strip at 4°C overnight.

3.3. Capture of Methylated DNA by the Immobilized MBD2-Fc Protein and Stepwise Elution of Methylated DNA

1. After overnight rotation, transfer the strip to the magnet, wait for about 30 s, and discard supernatant.
2. Remove the strip from the magnet and resuspend the beads with 200 μl buffer A by 20× pipetting up and down.
3. Repeat steps 1 and 2 once, transfer the strip to the magnet, wait for about 30 s, and discard supernatant.
4. Remove the strip from the magnet and resuspend the beads with 50 μl buffer X. Add 100 μl sonicated DNA and close the strip.
5. To capture the methylated DNA by the immobilized MBD2-Fc protein, rotate the strip at 4°C for 2 h.
6. After rotation, transfer the strip to the magnet, wait for about 30 s, and transfer the supernatant to a collection tube labeled A: eluate A (300 mM NaCl). Store eluate A on ice.
7. Remove the strip from the magnet and resuspend the beads with 150 μl elution buffer B by 20× pipetting up and down (see Note 8).
8. Transfer the strip to the magnet, wait for about 30 s, and transfer the supernatant to a collection tube labeled B: eluate B (400 mM NaCl). Store eluate B on ice.
9. Repeat steps 7 and 8 with elution buffers C, D, and E (500, 550, 1 M NaCl), each time collecting 150 μl eluates C, D, and E in tubes labeled C, D, and E, respectively. Store on ice until desalting.
10. Purify DNA in eluates (e.g., with MinElute PCR Purification Kit, Qiagen) (see Note 9).

See Note 10.

3.4. Quantification of Enrichment

Enrichment of methylated DNA can be tested by qPCR quantifying the unmethylated and the methylated allele of an imprinted gene in the different elution fractions. The gene *SNRPN* is used for human, the gene *Mest* is used for mouse and rat DNA (see Table 1) (see Note 11). The unmethylated allele enriches in eluate A, the methylated allele enriches in eluate E (Fig. 2).

To continue with downstream applications, it is recommended to determine DNA concentrations, e.g., with Qubit (Invitrogen, Darmstadt, Germany) (see Note 12).

Table 1
Primers for qPCR

Gene	Forward primer (5′-3′)	Reverse primer (5′-3′)
SNRPN (human)	TACATCAGGGTGATTGCAGTTCC	TACCGATCACTTCACGTACCTTCG
Mest (mouse)	CAGACGCCACCTCCGATCC	GGCCGCATTATCCCATGCC
Mest (rat)	GCCATCCCCGATCCTGTACC	GGCCGCATTATCCCATGCC

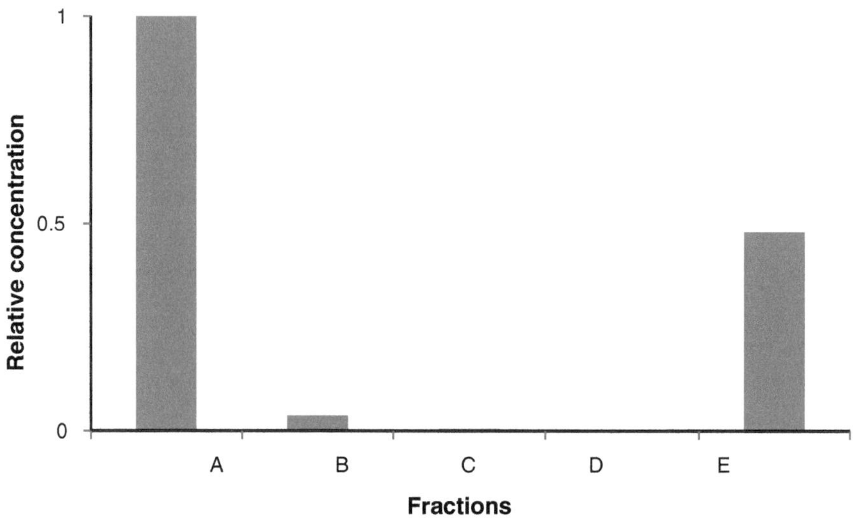

Fig. 2. Relative concentrations of alleles from an imprinted gene. The unmethylated allele enriches in eluate A, while the methylated allele enriches in eluate E. Almost no enrichment is present in eluates B, C and D. The calculated relative concentration in fraction A was set 1.

4. Notes

1. The mixture of buffer X and DNA solution corresponds to buffer A. Upon DNA elution, this mixture provides eluate A.
2. DNA isolated with Trizol is inappropriate for MCIp. Use of a DNA isolation kit like the DNeasy Blood and Tissue Kit (Qiagen) or careful phenol/chloroform extraction avoiding phenol/chloroform carryover is recommended.
3. A size range of 100–600 bp applies best for microarray analyses while a range of 50–250 bp is best suited for NGS on an Illumina or Solexa platform.
4. Other collection tubes may be used for the Bioruptor sonication as well, yet require additional optimization.

5. Different DNA samples may require different numbers of sonication cycles.
6. Use of a multichannel pipette facilitates work if more than a single DNA sample is processed.
7. Smaller amounts of protein A-coated magnetic beads and MBD2-Fc protein may be sufficient.
8. Eluting DNA in two steps with each 75 μl of the respective elution buffer may provide higher yields.
9. After desalting, the final fraction E should be eluted to special tubes with low DNA binding capacity (e.g., Low Binding Tubes, Eppendorf, Hamburg, Germany).
10. Instead of manual MCIp, automated MCIp offering the advantage of high reproducibility using a laboratory robot (e.g., SX-8 G IP-Star, Diagenode) may be considered. Moreover, commercial kits for the enrichment of methylated DNA (e.g., Epimark DNA enrichment Kit, New England Biolabs, Ipswich, USA) are alternatives to MCIp, yet require individual optimization as well.
11. Depending on the source tissue and/or tumor, genes other than *SNRPN* or *Mest* may be tested in the qPCR to properly prove enrichment of methylated DNA.
12. Usually, only eluate E corresponding to highly methylated DNA is used in downstream applications such as NGS or CpG island/promoter microarray analysis. Starting with an input DNA amount of about 2.5 μg, the yield in eluate E usually ranges between 15 and 50 ng, depending on the source tissue. While an amount of 15 ng is sufficient to obtain at least 20 million mappable reads in NGS, a number well suited for the identification of aberrantly methylated regions in tumor samples, an amount of 30 ng is sufficient for proper labeling of DNA in array hybridization analysis.

Appendix: Production of MBD2-Fc protein

1. Materials

1.1. Cell Culture

1. *Drosophila melanogaster* S2 cells, stably transfected with pMT-Bip/MBD-Fc and pCoHygro (11).
2. Cell culture flasks 75 cm^2 (Greiner Bio-One, Frickenhausen, Germany).

3. 2 L cell culture roller bottles (Corning, Lowell, MA, USA).
4. Insect-Xpress medium, protein-free with L-glutamine (Lonza, Cologne, Germany).
5. Penicillin/Streptomycin (e.g., Pen/Strep, Invitrogen).
6. Hygromycin B (50 mg/ml, Clontech, Saint-Germain-en-Laye, France).
7. Microscope and Neubauer counting chamber.
8. 100 mM $CuSO_4$, sterile filtered; working concentration is 0.5 mM.
9. Cooling incubator (e.g., Thermo Fisher Scientific, Karlsruhe, Germany) with roller equipment (e.g., Cellroll, Integra Biosciences, Fernwald, Germany).
10. Centrifuge tubes (e.g., 200 ml one way conical tubes, BD Biosciences, Heidelberg, Germany).
11. Benchtop cooling centrifuge (e.g., 4 K15, Sigma, Osterode, Germany) with swinging bucket rotor and buckets including adapters for 250 ml conical centrifuge bottles.

1.2. Protein Isolation

1. Dialysis tubing (e.g., Spectra/Por MWCO 6–8,000, Serva, Heidelberg, Germany) and suitable tubing clamps.
2. rProtein A-sepharose fast flow (e.g., GE Healthcare, München, Germany).
3. High-speed centrifuge with fixed angle rotor for 250 ml centrifuge bottles (e.g., Beckman J-25 with rotor JA 14, Beckmann Coulter, Krefeld, Germany) and suitable 250 ml reusable polypropylene centrifuge bottles.
4. Chromatography column (e.g., XK 16, GE Healthcare).
5. Peristaltic pump (e.g., RP-1, Rainin, Emeryville, CA, USA).
6. TBS, pH 7.4.
7. TBS, pH 7.4, 500 mM NaCl.
8. TBS, pH 7.4, 20% ethanol.
9. 3 M KCl.
10. 0.1 M citric acid, pH 3.
11. 1.5 M Tris–HCl, pH 8.8.
12. UV spectrophotometer (e.g., BioPhotometer, Eppendorf) and UV-compatible cuvettes.
13. Dialysis cassette (e.g., Slide-A-Lyzer, 0.5–3 ml capacity, 3,500 MWCO, Thermo Fisher Scientific).
14. Protease inhibitor cocktail (e.g., Roche, Mannheim, Germany).
15. 100% glycerol.

2. Methods

An overview of cell numbers required for the different steps of protein production is given in Table 2 (2.4). A short summary of the protein production procedure is provided in Table 3 (2.4).

2.1. Cell Culture of Drosophila S2 Cells

1. For initial expansion in a cell culture flask, inoculate Xpress medium containing hygromycin and Pen/Strep from a frozen stock of S2 cells at a titer of 5–10 × 10^5 cells/ml and incubate at 21°C (see Note 1).
2. Split the culture if the cell number reaches 5 × 10^6 cells/ml. Do not dilute below 5 × 10^5 cells/ml.

2.2. Induction of MBD2-Fc Protein Expression

1. Seed two cultures of 300 ml Xpress medium containing 2 × 10^6 cells/ml, Pen/Strep and Hygromycin at 21°C in 2 L roller bottles until the cell number reaches approximately 1 × 10^7 cells/ml which usually takes 3 days (see Note 2). Set the roller to a speed of 6 rounds/min.
2. At about 1 × 10^7 cells/ml, distribute the 2 × 300 ml cultures to four conical one way centrifuge bottles and centrifuge using a swinging bucket rotor at 15°C for 8 min, 300 × *g*.
3. To induce the MBD2-Fc gene, discard the supernatant and resuspend the cells in Xpress medium containing Pen/Strep and 0.5 mM $CuSO_4$ but no hygromycin (see Note 3).
4. Retransfer the four suspensions to the two roller bottles and continue incubation at 21°C for 4 days (see Note 4) in which the cell number usually reaches 2.5–3 × 10^7 cells/ml.
5. Distribute the 2 × 300 ml cell cultures to four conical centrifuge bottles and centrifuge using a swinging bucket rotor at 15°C for 8 min, 300 × *g*.
6. Transfer the supernatants without cellular carryover to four 250 ml centrifuge bottles and centrifuge in a fixed angle rotor at 4°C for 60 min, 15,000 × *g* (see Note 5).

2.3. MBD-Fc Protein Isolation by Affinity Chromatography on rProtein A-Sepharose Fast Flow

1. Dialyze the cleared supernatants from the previous step at 4°C against 5 L TBS, pH 7.4, with 3–4 buffer changes (see Note 6).
2. Pack the chromatography column with 5 ml rProtein A-sepharose fast flow and equilibrate the sepharose matrix with 80 ml cold TBS (see Note 7).
3. Load the dialyzed culture supernatant from Subheading 2.3 item 1 (Appendix) onto the matrix.
4. Wash the matrix with 80 ml TBS containing 500 mM NaCl, then with 80 ml TBS (see Note 8).

5. Elute the MBD-Fc protein with 0.1 M citric acid, pH 3 in 20 fractions of 1 ml each into Eppendorf tubes containing 50 μl 1.5 M Tris–HCl, pH 8.8 (see Note 9).
6. Identify the fractions with the highest protein content by measuring absorbance at 280 nm using a UV spectrophotometer. Keep fractions with absorbance higher than 2 (see Note 10).
7. Combine these fractions and transfer the mixture to a dialysis cassette. Dialyze against 1 L cold (4°C) TBS with two buffer changes.
8. Transfer dialysate to a 50 ml tube and add 10% NaN_3 to a final concentration of 0.02%, 4 μl protease inhibitor cocktail per ml dialysate and 100% glycerol to a final concentration of 20%. Homogenize carefully but thoroughly and store in 0.5–1 ml aliquots at −20°C.

See Note 11.
See Note 12.
See Note 13.

2.4. Overview of Cell Culturing and Protein Production

See Tables 2 and 3.

Table 2
Culturing *Drosophila* S2 cells

Condition	Description
Morphology	Round singly growing cells
Medium	Xpress medium, Hygromycin (0.4 mg/ml), Pen/Strep (50 μ/l Penicillin, 50 μg/l Streptomycin)
Subculture	Seed out after thawing at about 1×10^6 cells/ml, optimal split ratio of about 1:5 every 5 days, maintain at 2×10^6 cells/ml
Temperature	21°C
$CuSO_4$ induction	Seed about 1×10^7 cells/ml without Hygromycin (recommended volume: 2×300 ml)
Harvest	2.5–3×10^7 cells/ml
Storage	Frozen in liquid nitrogen with 50% Xpress-Insect medium containing Pen/Strep, 40% FCS, 10% DMSO at about 1×10^7 cells/ml

FCS fetal calf serum, *DMSO* dimethyl sulfoxide

Table 3
MBD-Fc protein production

Day	Task
1	Induction of protein expression
4	Purification and dialysis of supernatant
5	Affinity purification of MBD-Fc protein
	Dialysis of MBD-Fc protein
	Restoration of column matrix

3. Notes

1. The initial expansion medium should contain 20–30% conditioned medium, i.e., supernatant of a previous culture.
2. Starting with a titer of more than 1×10^7 cells/ml may hamper optimal protein production though may be tolerable. At this step, a weekly production cycle starts which is continued 3 days later with induction of protein expression.
3. $CuSO_4$ induces the expression of the MBD-Fc gene via the metallothionein promoter.
4. During these 4 days of incubation, the MBD-Fc protein is expressed and excreted into the medium via an excretion signal engineered in the expression vector.
5. Cells can be recycled: Resuspend the cells in one of the conical centrifuge bottles with 50 ml Xpress medium containing Pen/Strep and Hygromycin and start a next round of production (refer to subheading 2.2 item 1 (Appendix)). The weekly protein production cycle may be continued until a considerable decrease in the amount of produced protein becomes obvious.
6. Use a funnel to facilitate transfer to dialysis tubing. Dialyze in cold room using a 10 L plastic bucket placed on a magnetic stirrer. Add 2.5 g sodium azid (NaN_3) to 5 L TBS to prevent fungal growth.
7. Set the speed of the peristaltic pump such that 80 ml buffer or dialysate passes the sepharose matrix within about 60 min.
8. Presence of 500 mM NaCl at this step proved very efficient to remove possible *Drosophila* DNA contamination derived from the cell culture.
9. 1.5 M Tris–HCl, pH 8.8 neutralizes the acidic elution buffer and, hence, stabilizes the eluted protein.

10. Fractions with absorbance higher than 2 usually start with fractions 7–9 and end with fractions 11–13.
11. The protein concentration may be determined by any established protein quantitation method, e.g., the Pierce Protein Assay. MBD-Fc protein yield from 600 ml S2-cell culture usually ranges between 5 and 15 mg. Protein purity may be tested by SDS PAGE. Protein size is approximately 40 kDa.
12. Avoid repeated freezing and thawing since this may harm protein stability.
13. To restore the rProtein A-sepharose column matrix, wash with 80 ml 3 M KCl, then with 80 ml TBS and finally with TBS containing 20% ethanol. Keep the column with sepharose matrix at 4°C until next use. The column matrix may be reused four to five more times. It should be replaced if the yield drops considerably or the amount of contaminating protein as observed by SDS-PAGE considerably increases.

References

1. Tahiliani M, Koh KP, Shen Y, Pastor WA, Bandukwala H, Brudno Y, Agarwal S, Iyer LM, Liu DR, Aravind L, Rao A (2009) Conversion of 5-methylcytosine to 5-hydroxymethylcytosine in mammalian DNA by MLL partner TET1. Science 324:930–935
2. Kriaucionis S, Heintz N (2009) The nuclear DNA base 5-hydroxymethylcytosine is present in Purkinje neurons and the brain. Science 324:929–930
3. Inoue A, Shen L, Dai Q, He C, Zhang Y (2011) Generation and replication-dependent dilution of 5fC and 5caC during mouse preimplantation development. Cell Res 21:1670–1676
4. Ito S, Shen L, Dai Q, Wu SC, Collins LB, Swenberg JA, He C, Zhang Y (2011) Tet proteins can convert 5-methylcytosine to 5-formylcytosine and 5-carboxylcytosine. Science 333:1300–1303
5. Takai D, Jones PA (2002) Comprehensive analysis of CpG islands in human chromosomes 21 and 22. Proc Natl Acad Sci USA 99:3740–3745
6. Jones PA, Baylin SB (2002) The fundamental role of epigenetic events in cancer. Nat Rev Genet 3:415–428
7. Rodriguez-Paredes M, Esteller M (2011) Cancer epigenetics reaches mainstream oncology. Nat Med 17:330–339
8. Weber M, Davies JJ, Wittig D, Oakeley EJ, Haase M, Lam WL, Schubeler D (2005) Chromosome-wide and promoter-specific analyses identify sites of differential DNA methylation in normal and transformed human cells. Nat Genet 37:853–862
9. Brinkman AB, Simmer F, Ma K, Kaan A, Zhu J, Stunnenberg HG (2010) Whole-genome DNA methylation profiling using MethylCap-seq. Methods 52:232–236
10. Rauch T, Pfeifer GP (2005) Methylated-CpG island recovery assay: a new technique for the rapid detection of methylated-CpG islands in cancer. Lab Invest 85:1172–1180
11. Gebhard C, Schwarzfischer L, Pham TH, Schilling E, Klug M, Andreesen R, Rehli M (2006) Genome-wide profiling of CpG methylation identifies novel targets of aberrant hypermethylation in myeloid leukemia. Cancer Res 66:6118–6128
12. Schilling E, Rehli M (2007) Global, comparative analysis of tissue-specific promoter CpG methylation. Genomics 90:314–323
13. Robinson MD, Stirzaker C, Statham AL, Coolen MW, Song JZ, Nair SS, Strbenac D, Speed TP, Clark SJ (2010) Evaluation of affinity-based genome-wide DNA methylation data: effects of CpG density, amplification bias, and copy number variation. Genome Res 20:1719–1729
14. Nair SS, Coolen MW, Stirzaker C, Song JZ, Statham AL, Strbenac D, Robinson MW, Clark SJ (2011) Comparison of methyl-DNA immunoprecipitation (MeDIP) and methyl-CpG binding domain (MBD) protein capture for genome-wide DNA methylation analysis reveal CpG sequence coverage bias. Epigenetics 6:34–44

Chapter 12

Gene Expression Profile Analysis of Lymphomas

Katia Basso and Ulf Klein

Abstract

Through the genome-wide characterization of a cell type's transcriptome, gene expression profile analysis provides a potent tool for analyzing the pathogenesis of lymphomas and has had a major impact on the understanding of lymphoid neoplasia. The analysis of gene expression patterns of lymphomas and normal lymphocytes permits (1) the definition of molecular subtypes of lymphoma; (2) the identification of the normal cellular counterpart of a lymphoma subtype; (3) the identification of diagnostic markers and therapeutic targets; and (4) the identification of signaling pathways affected by the oncogenic transformation. This chapter presents an approach to accomplish these goals.

Key words: DNA microarrays, Genechips, Lymphomas, Gene expression profiling, Gene expression data analysis, Molecular classification of lymphomas, Lymphomagenesis

1. Introduction

1.1. General Concepts

The invention of DNA microarrays, on which virtually all genes encoded in the genome are represented, allowed the fast and comprehensive identification of the genes expressed in a cell type or a tissue. The comparison of a large number of samples that potentially leads to the identification of gene expression differences among samples is commonly referred to as gene expression profile (GEP) analysis. Microarrays (also known as gene chips) contain DNA molecules that can either represent cDNAs or DNA oligonucleotides and that are complementary to their corresponding mRNA. For both approaches, the mRNA derived from the respective cells or tissue biopsy samples needs to be labeled, thus allowing the detection of the mRNA hybridization to the DNA on the microarray and the quantification of the mRNA expression levels. cDNA microarrays require the hybridization of two differently labeled mRNAs, i.e., the test sample with one fluorochrome, and

Ralf Küppers (ed.), *Lymphoma: Methods and Protocols*, Methods in Molecular Biology, vol. 971,
DOI 10.1007/978-1-62703-269-8_12, © Springer Science+Business Media, LLC 2013

the control with a second one eliciting a different color. Therefore, they provide a relative measurement of gene expression levels. Conversely, oligonucleotide microarrays require labeling with only one fluorochrome and provide an absolute measurement of the concentration of the respective mRNA represented on the microarray. The most common type of the commercially available microarrays is the oligonucleotide microarray; instead, cDNA microarrays are usually generated in-house by spotting cDNAs on glass slides. There are two types of oligonucleotide microarrays which are generated (1) by in situ synthesis of oligonucleotides using a photolithographic approach (1) and (2) by attaching the oligonucleotides to special matrices. Both approaches yield high-density microarrays on which virtually all known genes are represented, thus making them powerful tools for a global GEP analysis of different cells or biopsies. While the majority of commercially available oligonucleotide-based DNA microarrays typically comprise sequences corresponding to 3′ untranslated regions, recently a new generation of microarrays has been developed which display probes targeting all exons, thus allowing the detection of alternative splice variants. Oligonucleotide microarrays now show low variability and thus high reproducibility. It is important to be aware, however, that data generated using different microarrays cannot be compared in the same data analysis (2). In the near future, the rapidly decreasing costs of deep sequencing technologies are expected to ultimately enable the fast and comprehensive sequencing of the whole transcriptome, including mRNAs and noncoding RNAs. The clear advantage of this approach, that provides a snapshot of all expressed transcript variants, is currently hampered by the need of sophisticated analysis methods and specific computational expertise.

1.2. Data Analysis

Following microarray hybridization and scanning, the resulting normalized GEP data of a panel of samples are comparatively analyzed in order to determine gene expression differences among the samples (Fig. 1). Evidently, this approach generates an enormous amount of data points, necessitating the development of suitable biostatistical methods that enable the extraction of useful information from these experiments. Such biostatistical analysis methods employ hierarchical clustering algorithms, also referred to as pattern discovery algorithms, that allow the identification of gene expression patterns among a large number of samples. One distinguishes two analysis methods that are both based on clustering algorithms, namely, (1) unsupervised clustering that is used to identify cell types which have not been classified a priori and (2) supervised analysis that facilitates the identification of differentially expressed genes between samples defined a priori according to criteria such as cell type, genotype, clinical prognosis, or patient survival. Supervised analysis can also be used to generate classifiers, or predictors, that allow the assignment of cases from an unrelated

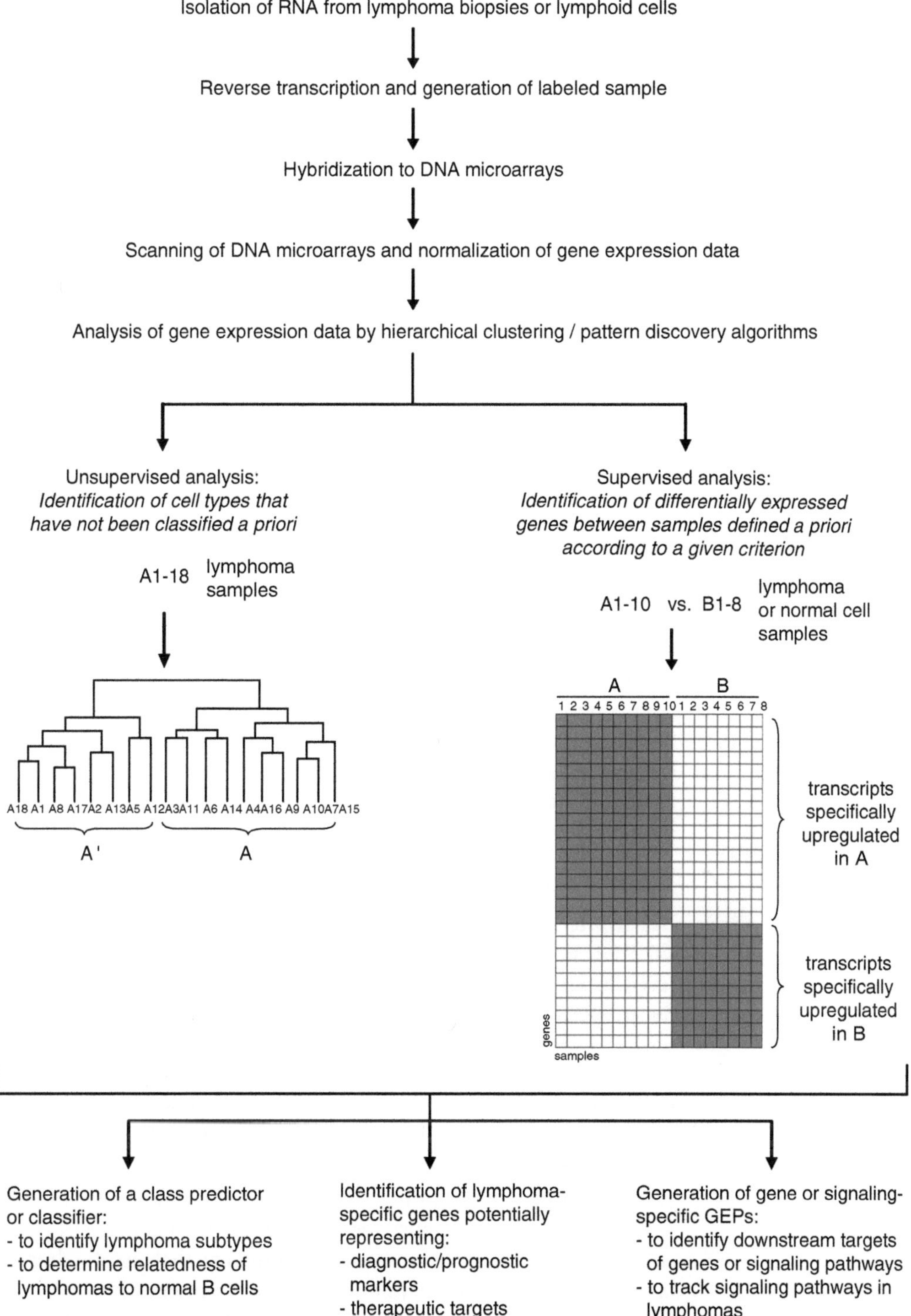

Fig. 1. Strategy for the identification of GEP-based molecular subtypes and differentially expressed genes in lymphomas.

panel of samples into subgroups defined by GEP analysis. The outputs of unsupervised and supervised GEP analyses are lists of genes associated with a distinct phenotype, genotype, or other characteristics. Such gene sets can be employed in a classification analysis (also referred to as class prediction analysis) where the identity of the genes is secondary. Of particular interest in most cases, however, is the identification of genes associated with a particular phenotype or genotype with biological pathways and functional categories such as specific signaling pathways. Over the past decade, GEP approaches have proved to be extremely useful in the characterization of new lymphoma subtypes and in the identification of the underlying pathology of lymphoid diseases.

1.3. Applications to the Analysis of Lymphomas

GEP analysis can provide insights into the phenotype and cell derivation of lymphomas and allows the identification of diagnostic markers as well as gene products specifically associated with the lymphoma phenotype that may be potential therapeutic targets.

1.3.1. Molecular Classification of Tumors

The first study demonstrating that GEP analysis allows the molecular classification of cancers, through a combination of class discovery and class prediction approaches, was performed on a set of cases comprising both acute lymphoblastic leukemia (ALL) and acute myeloid leukemia (AML). In this analysis, ALL could be distinguished from AML, and a class predictor (also referred to as "classifier") was able to assign unrelated cases to the correct class, i.e., either to ALL or to AML (3). A groundbreaking study that could identify new lymphoma subgroups through a GEP-based approach was aimed at resolving the basis of the clinical heterogeneity observed in a morphologically homogenous lymphoma type, namely, diffuse large B-cell lymphoma (DLBCL) (4). This analysis has led to the now generally recognized characterization of the DLBCL subgroups of germinal center (GCB)- and activated B-cell (ABC)-type DLBCL which, despite their similar morphology, differ in both their underlying biology and key pathogenic mechanisms. Later, class discovery was applied to several subtypes of B-cell tumors where, in direct contrast to the situation of DLBCL, GEP analysis could confirm the molecular homogeneity of several tumor subtypes, including B-cell chronic lymphocytic leukemia (CLL), hairy cell leukemia (HCL), and mantle cell lymphoma (5–8). Especially the finding that CLL showed molecular homogeneity was surprising because there are two known genetically distinct subgroups of this mature B-cell malignancy that are furthermore characterized by differences in their clinical presentation.

The subset of genes, that is specifically expressed or not expressed in a cell type or a particular condition, is generally referred to as a "*signature.*" The correlation of a specific signature with a certain phenotype or other defined characteristics allows performing class prediction. Predictive signatures have been used to assign new samples to previously defined subgroups, as exemplified by a study that could successfully predict survival in DLBCL (9). Ideally, specific GEP associated with good or bad tumor prognosis will allow predicting the outcome of a patient as early as at the time of diagnosis. In addition, it is envisioned that knowing the nature of the genes associated with a bad outcome will provide important information for the development of effective therapeutic strategies.

1.3.2. Identification of the Putative Lymphoma Precursor Cell

Knowing the normal cellular counterpart of a lymphoma is a prerequisite for identifying the molecular mechanisms that contributed to the oncogenic transformation of the lymphoma precursor cell. Frequently, a lymphoid malignancy has no obvious morphologic or immunophenotypic resemblance to a normal lymphocyte subpopulation. Gene expression profiles corresponding to the normal populations markedly enlarge the number of genes associated with a particular stage of development that can be used to track the cell of origin of a lymphoma phenotype. For example, GEP-based analyses that compared B-cell lymphoma types to normal stages of B-cell development have revealed that (1) DLBCL subtypes show phenotypic relatedness to either GC B cells or activated B cells (4) and (2) CLL and HCL are more similar in their GEP to post-GC antigen-experienced memory B cells than to any other of the normal B-cell populations analyzed (5, 8). Once the putative cell of origin has been identified, the gene expression differences between lymphoma and normal cells can be determined by supervised analysis and may reveal, for instance, the aberrant activation of cell signaling pathways in the lymphoma cells.

1.3.3. Identification of Lymphoma-Specific Markers

GEP-based approaches are valuable tools in the identification of lymphoma-associated gene products that could represent novel diagnostic or prognostic markers or potential therapeutic targets. An example of a diagnostic marker identified by GEP analysis is represented by the ANXA1 molecule in HCL. Indeed, HCL cannot always be unambiguously distinguished from other B-cell-derived malignancies. GEP analysis of HCL versus a panel of B-cell non-Hodgkin lymphomas and normal B-cell subpopulations, followed by immunohistochemical validation in a large tumor panel, could identify ANXA1 as a molecule that is specifically upregulated in HCL (8, 10). Immunohistochemical detection of ANXA1 has been developed into a simple and highly specific assay for diagnosis of HCL (10). The comparison of the gene expression signature of a given lymphoma to both normal cells and other lymphoma types can lead to the identification of genes that are specifically associated with the disease and may turn out to be valuable diagnostic markers. The identification of prognostic markers that would allow classifying subsets of lymphomas with similar clinical behavior but different prognosis requires the analysis of GEPs of large panels of cases. An ideal target for therapy would be one that is specifically expressed in the lymphoma under investigation but that is absent in normal B cells. The GEP analysis of CLL has led to the identification of several molecules, including the orphan-receptor ROR1 (5, 6), that are currently being evaluated for targeting by specific therapy. It is important to note that before a gene identified by GEP-based approaches should be further considered as a diagnostic or a prognostic marker, or as a therapeutic target, its expression needs to be further validated in independent panels of cases at the mRNA and protein levels.

1.3.4. Tracking Signaling Pathways

GEP data of different cell populations can be directly assessed for changes in the expression of known downstream targets of signaling pathways. Most importantly, however, GEP-based approaches are especially valuable in identifying new target genes of signaling pathways. Suitable in vitro or in vivo systems can be employed to recapitulate the activation of the signaling pathway under study; for instance, cells can be activated through a cell surface receptor in vitro, or cells made deficient for a gene by knockout technology or gene silencing approaches can be studied for the effects of the inactivation of this gene relative to wild-type cells. Similarly, pharmacological drugs can be tested for their effects on the activation of a particular signaling pathway by large-scale GEP approaches. Such generated gene expression data can then be tracked within the GEPs derived from lymphomas aiming at the identification of the activity of a signaling pathway or transcriptional response in the tumor cells.

2. Materials

The following reagents and materials are necessary for 3′IVT expression analysis performed using the Affymetrix oligonucleotide GeneChip® cartridge arrays.

2.1. Generation of Gene Expression Profiles

2.1.1. RNA Isolation

Homogenizer.

- TRIzol Reagent (Invitrogen, Grand Island, NY, USA).
- RNeasy Mini Kit (QIAGEN, Valencia, CA, USA).
- UV spectrophotometer (e.g., NanoDrop® ND-8000 UV–vis Spectrophotometer; NanoDrop Technologies, Wilmington, DE, USA).
- Gel electrophoresis unit with appropriate buffer.

  Optional:

- Agilent 2100 Bioanalyzer (Agilent Technologies, Santa Clara, CA, USA).
- RNA 6000 Nano LabChip kit (Agilent Technologies).

2.1.2. Double-Strand cDNA and Labeled aRNA Synthesis

- GeneChip® 3′ IVT Express Kit (Affymetrix, Santa Clara, CA, USA).

2.1.3. Hybridization, Staining, Washing, and Scanning

- GeneChip® Hybridization, Wash, and Stain Kit (Affymetrix).
- GeneChip® Arrays (Affymetrix).
- GeneChip® Hybridization Oven 640 (Affymetrix).
- GeneChip® Fluidics Station 450 (Affymetrix).
- GeneChip® Scanner 3000 (or higher) (Affymetrix).

3. Methods

3.1. Sample Selection

The careful selection of adequate biological specimens is a mandatory step toward a successful gene expression profiling experiment. The selection process must aim to reduce as much as possible all variables that are not associated with the biological identity of the lymphoma cells. The major sources of variability that need to be considered include (1) processing and storage of the specimen at the time of collection; (2) percentage of tumor cells; (3) tumor localization; and (4) institution of origin. The gene expression profile of a tissue can be dramatically affected by any stress encountered at the time of removal from the body. Ideally, the specimen should be immediately placed in a cold isotonic solution and transferred from the operating room to the laboratory for processing. To ascertain the optimal preservation of RNA, the specimen should be snap frozen and stored in liquid nitrogen. In reality, this procedure is the exception, and in most cases the tissue is included in OCT and stored at −80°C. Although OCT-embedded samples are suitable for RNA isolation, it is recommended to remove most of the OCT (working on dry ice to avoid thawing of the tissue) before proceeding with the RNA isolation. Unless the tumor cells have been purified from the biopsy specimen at the time of collection, the sample will include a mixture of normal and malignant cells. The fraction of malignant cells must be evaluated and quantified by standard diagnostic procedures, and only specimens containing a similar percentage (preferentially more than 90%) of tumor cells should be considered. This evaluation should be performed on the very same specimen that will be subject to the RNA isolation because different portions of the same biopsy may vary considerably in the content of malignant cells. Moreover, it is highly recommended that the diagnosis and tumor cell content for each sample included in the dataset should be revised by the same pathologist. Since the tumors may reside in different body locations, the normal infiltrating cells may differ in cell number and type across specimens with the same lymphoma diagnosis. Even if nonmalignant infiltrating cells represent only a small percentage, they will impact a comparative analysis in that different specimens are distinguished on the basis of their non-tumor components. Specimens collected at different institutions are most likely to display distinct gene expression profiles reflecting the differences in the collection, processing, and storage of the material.

Overall, although it may not always be possible to eliminate the sources of variability, being aware of their potential influence will allow to monitor the corresponding GEPs for patterns of expression associated with these variables.

3.2. Generation of Gene Expression Profiles

The protocols reported here are for 3′IVT expression analysis performed using the Affymetrix oligonucleotide GeneChip® cartridge arrays (see Note 1).

3.2.1. RNA Isolation and Quality Control

RNA isolation can be performed using column-based approaches offered by a large number of companies or using the TRIzol Reagent (Invitrogen) followed by a cleanup step by the RNeasy columns (Qiagen). In order to obtain good-quality RNA, the isolation must start from optimally preserved fresh or frozen tissue. Fresh tissue must be maintained in a serum-free isotonic solution and processed by a few hours from body removal. Frozen tissue must be stored at −80°C or preferentially in liquid nitrogen; when processing frozen tissue, do not allow the tissue to thaw by keeping it on dry ice. The frozen tissue can be sliced by a cryostat or transferred as whole in Trizol and mechanically homogenized. Following isolation, RNA must be quantified using a spectrophotometer and RNA integrity be checked by agarose gel electrophoresis or with the Agilent 2100 Bioanalyzer and the RNA 6000 Nano LabChip® kit following the manufacturer's instructions. RNA integrity and purity strongly affect the efficiency of the cDNA and aRNA synthesis (see Notes 2 and 3). The purity of the RNA is assessed by the ratio of absorbance readings at 260 and 280 nm, which should range between 1.7 and 2.1, if the RNA is free of contaminating proteins and other molecules associated with the isolation procedure such as phenol, ethanol, and salts (see Note 3). Highly intact RNA is required to grant a proper reverse transcription, since partially degraded mRNA will lead to cDNA which may lack portions of the transcripts that are interrogated by probes on the array. The integrity of the RNA sample can be checked by agarose gel electrophoresis or using the Agilent 2100 Bioanalyzer. RNA resolved on an agarose gel will display rRNA bands and an intact RNA will associate with a 28S rRNA band appearing approximately twice as intense as the 18S rRNA. The integrity of an RNA sample tested on the Agilent 2100 Bioanalyzer will be defined by the RNA Integrity Number (RIN), a metric developed by Agilent Technologies to provide indications on the RNA degradation state.

3.2.2. Double-Strand cDNA and Labeled aRNA Synthesis

The recommended input total RNA amount is 100 ng, with a minimum of 50 ng and a maximum of 500 ng, according to the amount of mRNA present in the tissue of interest. The RNA is reverse-transcribed using oligo-dT primers containing a T7 sequence in order to obtain a double-stranded cDNA that is subsequently used as template for the aRNA synthesis incorporating biotinylated nucleotides. The labeled aRNA is purified, quantified (usually 40–80 μg are obtained), and its quality should be evaluated by gel electrophoresis or using the Agilent 2100 Bioanalyzer system. Good-quality aRNA appears as a smear ranging 250–5,500 nt with the strongest signal between 600 and 1,200 nt. Fifteen micrograms of the biotinylated aRNA are subjected to fragmentation in order to break down full-length aRNA into fragments ranging from 35 to 200 nt with a peak at 100–120 nt. Similarly to the aRNA, the fragmented aRNA sample needs to be checked by gel electrophoresis or using the Agilent 2100 Bioanalyzer system.

All samples of a gene expression profile study must be prepared starting from the same amount of RNA and using exactly the same protocol, keeping the incubation times constant. The detailed protocols are reported in the "3′ IVT Express Kit User Manual" downloadable at http://www.affymetrix.com/support/technical/manuals.affx.

3.2.3. Hybridization, Staining, Washing, and Scanning

The cartridge array is incubated with the pre-hybridization mix for 10 min, then refilled with the hybridization mixture, and incubated for 16 h. After the overnight hybridization, the hybridization cocktail is removed from the array cartridge and can be stored at −20°C and reused for hybridization on another array. In order to proceed with the washing and staining procedures that are performed automatically using the Fluidics Station, the sample needs to be registered in the Affymetrix® GeneChip Command Console (AGCC). The arrays will be then washed and stained with a streptavidin–phycoerythrin (SAPE) solution followed by a signal amplification protocol using anti-streptavidin biotinylated antibodies and SAPE solution. The Fluidics Station takes approximately 90 min to perform automatically the washing and staining procedures. Upon staining, the microarrays can be immediately scanned or stored for several hours at 4°C in the dark. The detailed protocol for the hybridization is reported in the "3′ IVT Express Kit User Manual," while the staining and washing protocols and the procedure for scanning are detailed in the "GeneChip® Expression Wash, Stain and Scan User Manual" downloadable at http://www.affymetrix.com/support/technical/manuals.affx.

3.3. Gene Expression Data Mining

In most cases, a core facility is in charge of the hybridization and scanning procedures and the user receives the raw gene expression data represented by the images acquired by the scanner (.dat and .cel files) and the gene expression signal values obtained by the elaboration of the array images (.chp file). Information on the quality controls and the hybridization procedure are stored in a .rpt file as an additional record.

The signal values reported in the probe set summarization file (.chp) are calculated by integrating information from approximately 11 probe pairs (in the most recent GeneChip® arrays), each of them represented by an oligonucleotide perfectly matching the target sequence and a second one carrying a single mismatch in its middle part, designed to control for any possible unspecific annealing. The provided "signal value" represents a quantitative measurement of the transcripts while the "detection call" defines each probe set as present, absent, or marginal (neither present nor absent), based on a "detection p-value" that estimates the probability of each probe set to be expressed above the background. The Expression Console™ software is provided by Affymetrix to enable probe set summarization and initial data quality examination for Affymetrix expression arrays. Different algorithms can be used to generate a probe set summarization file: Affymetrix initially proposed the MAS algorithms (the current version is called MAS5) followed by the PLIER algorithm, while the microarray user community contributed with the creation of the Robust Multichip Analysis (RMA) algorithm. The MAS5 and the RMA algorithms are currently the most widely used and their main differences are that MAS5 analyzes each array independently and uses the mismatch probes to adjust the perfect match intensity, while RMA is a multichip analysis approach and disregards the information provided by the mismatch probes.

The Expression Console™ software provides a tabular view of all the quality control (QC) metrics. In order to monitor the data quality, it is essential to check each sample for several of these QC parameters. In particular, the following information should be carefully evaluated: signals, detections, and 3′/5′ ratio for spike controls and housekeeping genes (ACTB and GAPDH); percentage of present, absent, and marginal genes; and background and scaling factor. Note that quality control metrics depend on the algorithm used to generate the probe set summarization files and that some thresholds such as the percentage of present/absent genes may largely vary according to the investigated tissue type.

Details on the use of the Expression Console™ software and the different algorithms used to generate the probe set summarization files can be found in the "Affymetrix Expression Console™ software Version 1.0-User guide" downloadable at http://www.affymetrix.com/support/technical/manuals.affx.

3.4. Gene Expression Data Analysis

The analysis of genome-wide expression data requires computational tools that can be found as freely available software or as user-friendly tool packages commercially provided by numerous companies. Rather than discussing specific analysis tools, we will provide the essential concepts for a successful gene expression analysis in lymphomas. For tips on choosing analysis software see Note 4.

3.4.1. Molecular Classification of Hematological Malignancies

Gene expression-based molecular classification relies on the identification of gene expression patterns associated with a clinically and histologically defined malignancy or with its subtypes that may be recognizable only based on their transcriptional signatures. Indeed, the ability to simultaneously interrogate a large number of transcripts provides a valuable resource to molecularly characterize a specimen in addition to its morphological features. In order to investigate the heterogeneity of a tumor type it is necessary to apply an unsupervised method that recognizes differences in the gene expression profiles without the bias of any a priori class definition. The most commonly used unsupervised methods rely on the hierarchical clustering of the samples based on their expression profiles. Although gene expression-based molecular classification can prove extremely powerful, it is essential to monitor the analysis to exclude molecular signatures that are not related to the disease itself. As discussed in Subheading 3.1, numerous variables including specimen handling, RNA isolation procedure, and percentage of tumor cells may adversely affect the transcriptional profiling. The accurate annotation of the subgroups identified by hierarchical clustering will allow controlling for factors unrelated to the disease and identifying the subclusters that are likely to represent biological subtypes of the tumor. The availability of complete information for each sample, including clinical and genetic data, may be essential to associate a specific gene expression pattern with clinical or genetic features. Of note, the analysis of lymphoma GEP data can be strongly affected by the location of the analyzed specimens in the body that may include different types of infiltrating normal cells. Indeed, gene expression data represent the average expression across the mixture of tumor and normal cells represented in the specimen. Several reports suggest that the non-tumor component may contribute to the molecular features of some type of lymphomas. However, it is evident that the specific characterization of GEPs from tumor cells requires the purification of those cells.

The subset of genes (signature) associated with a lymphoma or one of its molecular subtypes can be used as a "class predictor" in order to assign unclassified cases to a defined molecular class. In general, predictive signatures can be employed to assign new samples to previously defined subgroups (e.g., clinical, cytogenetic, and molecular groups). To generate a class predictor, the dataset must include both types of samples "belonging" and "not belonging" to the class of interest. A subset of the available tumor samples (training set) is used to identify genes (class predictor genes) that are specifically associated with the subtype of interest. The remaining samples (testing set) are then tested in order to investigate how efficiently the predictor genes are able to assign the samples to the correct class. The procedure is repeated multiple times using different samples in the training and testing sets. When all samples have been tested, the resulting set of selected genes defines the class predictor. The accuracy of a class predictor is strictly dependent on, and increases with, the number of samples used in the identification of the signature genes.

3.4.2. Identification of the Putative Lymphoma Precursor Cell

In order to identify the putative cell of origin of a lymphoma, it is necessary to generate gene expression profiles from the candidate normal cell populations. This analysis is particularly important in the study of lymphomas which arise from the malignant transformation of B or T cells at different stages of development. During the lymphomagenesis process the tumor cells acquire aberrant morphological and immunophenotypic features which mask the characteristics of their cell of origin. Normal cell populations are purified by cell sorting based on the expression of surface markers, or isolated from the lymphoid tissue by microdissection. GEPs are investigated by a supervised pattern discovery approach aimed at identifying the distinct gene signatures associated with each normal subpopulation. A classification method is usually applied when the purpose is to track the specific signatures obtained from the normal cells within the tumor profiles. Generally, classification methods measure the fraction of the normal cell signature that can be detected in the GEPs of the tumor and assign to each tumor sample a score of relatedness to a normal population.

3.4.3. Identification of Lymphoma-Specific Markers

Although supervised analyses are the methods of choice for the identification of specific features associated with a given class, unsupervised approaches are also commonly used to extract subsets of gene signatures clustering with a certain phenotype. The specificity of a marker is defined by its expression (or lack of it) exclusively in the tumor cells when compared to their normal counterpart and to any other related tumor types. Evidently, the purity of the tumor samples is an important factor toward the identification of genes specifically associated with the lymphoma cells. Candidate lymphoma-specific genes identified by gene expression analysis need to be validated for protein expression, preferentially by immunohistochemistry, to confirm the restricted expression in the lymphoma cells. This validation should be extended to a large panel of biopsies including specimens not represented in the discovery panel.

4. Notes

1. Several options are now available to generate GEPs. Currently, the most common approach is the array-based detection of poly-A transcripts. Here we reported on the use of arrays designed to detect transcripts based on probes recognizing regions in their

3′-end. In order to study alternatively spliced transcripts, arrays that interrogate each exon of a gene (exon arrays) have been designed. Although this approach enables the study of multiple splice variants, the data analysis is more complex and requires computational expertise. RNA-seq is the most recent technology used to generate large-scale expression profiles. Although it has the advantage of sequencing all poly-A transcripts expressed in the sample analyzed, it requires advanced computational expertise in addition to adequate data storage capabilities and analysis power.

2. RNA integrity and purity are the most critical parameters for a successful gene expression analysis. RNA integrity is compromised by inappropriate storage and/or mishandling during the isolation procedure. To optimally preserve the RNA, the biological specimen must be quickly frozen and stored in liquid nitrogen or at −80°C. RNA isolation must be performed in an RNAse-free environment.

3. The *RNA purity* depends on the isolation method. To eliminate phenol contamination carried on using phenol-based isolation approaches, it is suggested to clean up the isolated RNA with columns. Salt contaminations are usually associated with column-based isolation methods and can be reduced by increasing the number of washes. Finally, to eliminate ethanol from the RNA preparation, let the RNA dry for a few minutes before dissolving it in water (the pellet should become transparent, although still visible). If using a column-based isolation procedure, apply, after the washing step (performed using the ethanol-containing buffer), an extra round of centrifugation using a clean collection tube.

4. The data analysis may be the most critical and time-consuming step in a GEP project. The broad availability of software for gene expression data analysis may be confusing. Talk to investigators who have experience with GEP analysis: the core facility providing the hybridization service is usually able to direct you to some computational support. Choose one approach and try to become familiar with your data using that particular software, and then interrogate the same gene expression dataset using other programs. The data analysis may take longer than the generation of the profiles and may in fact never be completely finished since new tools to interpret GEP data become constantly available.

References

1. Lipshutz RJ, Fodor SP, Gingeras TR, Lockhart DJ (1999) High density synthetic oligonucleotide arrays. Nat Genet 21:20–24
2. Kuo WP, Jenssen TK, Butte AJ et al (2002) Analysis of matched mRNA measurements from two different microarray technologies. Bioinformatics 18:405–412
3. Golub TR, Slonim DK, Tamayo P et al (1999) Molecular classification of cancer: class discovery and class prediction by gene expression monitoring. Science 286:531–537

4. Alizadeh AA, Eisen MB, Davis RE et al (2000) Distinct types of diffuse large B-cell lymphoma identified by gene expression profiling. Nature 403:503–511
5. Klein U, Tu Y, Stolovitzky GA et al (2001) Gene expression profiling of B cell chronic lymphocytic leukemia reveals a homogeneous phenotype related to memory B cells. J Exp Med 194:1625–1638
6. Rosenwald A, Alizadeh AA, Widhopf G et al (2001) Relation of gene expression phenotype to immunoglobulin mutation genotype in B cell chronic lymphocytic leukemia. J Exp Med 194:1639–1647
7. Martínez N, Camacho FI, Algara P et al (2003) The molecular signature of mantle cell lymphoma reveals multiple signals favoring cell survival. Cancer Res 63:8226–8232
8. Basso K, Liso A, Tiacci E et al (2004) Gene expression profiling of hairy cell leukemia reveals a phenotype related to memory B cells with altered expression of chemokine and adhesion receptors. J Exp Med 199:59–68
9. Rosenwald A, Wright G, Chan WC et al (2002) The use of molecular profiling to predict survival after chemotherapy for diffuse large B-cell lymphoma. N Engl J Med 346:1937–1947
10. Falini B, Tiacci E, Liso A et al (2004) Simple diagnostic assay for hairy cell leukaemia by immunocytochemical detection of annexin A1 (ANXA1). Lancet 363:1869–1870

Chapter 13

FISH and FICTION to Detect Chromosomal Aberrations in Lymphomas

Maciej Giefing and Reiner Siebert

Abstract

Fluorescence In Situ Hybridization (FISH) is a powerful and robust technique allowing the visualization of target sequences like genes in interphase nuclei. It is widely used in routine diagnostics to identify cancer specific aberrations including lymphoma associated translocations or gene copy number changes in single tumor cells. By combining FISH with immunophenotyping—a technique called Fluorescence Immunophenotyping and Interphase Cytogenetic as a Tool for Investigation Of Neoplasia (FICTION)—it is moreover possible to identify a cell population of interest. Here we describe standard protocols for FISH and FICTION as used in our laboratory in diagnosis and research.

Key words: Fluorescent In Situ Hybridization (FISH), Fluorescence Immunophenotyping and Interphase Cytogenetics as a Tool for Investigation of Neoplasia (FICTION), Lymphoma research, Translocation

1. Introduction

In Situ Hybridization (ISH) is based on the ability of complementary nucleic acids to anneal and form stable DNA–RNA or DNA–DNA structures. It allows the precise localization of a target sequence with the complementary probe—initially an RNA sequence labeled with a radio isotope. In the late 1970s and 1980s of the twentieth century the harmful radio isotopes have been replaced by fluorophores and the detection of the fluorescently labeled hybrids was developed producing the first Fluorescent In Situ Hybridization (FISH) experiments (1–5). The introduction of labeling procedures like nick translation or random priming that are based on the direct incorporation of labeled nucleotides into the nucleic acid further facilitated the preparation of FISH probes. Together with

Ralf Küppers (ed.), *Lymphoma: Methods and Protocols*, Methods in Molecular Biology, vol. 971,
DOI 10.1007/978-1-62703-269-8_13, © Springer Science+Business Media, LLC 2013

the development of chromosome banding techniques FISH allowed for the first time gene mapping with relatively high resolution and resulted in a burst of new applications in cancer research and other fields. Detailed information on the historical development of ISH and FISH is excellently reviewed elsewhere (5–7).

A major advantage of FISH is its ability to visualize target sequences in interphase nuclei without prior cell culture and laborious metaphase preparations. It is of special importance in cancer research where it is difficult to obtain metaphase spreads of high quality. Thus, FISH became a method of choice for fast and reliable identification of chromosomal alterations including copy number abnormalities, genomic breakpoints and the ensuing translocations or oncogenic fusion genes.

With the discovery of recurrent chromosomal aberrations in lymphomas like t(8;14) in Burkitt lymphoma, t(14;18) in follicular lymphoma or t(11;14) in mantle cell lymphoma FISH became widely used in routine diagnostics of lymphatic neoplasms (8). Today, the availability of Bacterial Artificial Chromosomes (BAC) or P1 Artificial Chromosome (PAC) libraries—wherein the entire human genome divided into 50–300 kb fragments per bacterial clone is stored—allows fast probe preparation for the desired genomic region. Using differently labeled BACs/PACs it is possible to design multicolor FISH assays for any combination of genes that aided the identification of secondary aberrations in lymphomas and provided insight into the complex genetic background of lymphomagenesis (Table 1) (8–10). Novel approaches for probe development include oligonucleotide libraries which have moreover the advantage of being repeat free.

FISH analyses of Hodgkin lymphoma were initially hampered mainly for the following reasons: (a) Hodgkin lymphoma is characteristic for its scarcity of the neoplastic Hodgkin and Reed-Sternberg (HRS) cells that usually do not exceed 1% of cell content in the tumor, (b) during the isolation of interphase nuclei for FISH is it necessary to widely remove the cytoplasm what leads to the loss of cellular phenotype and makes the HRS cell nuclei difficult to distinguish from other cell populations (c) visual identification of HRS cells in primary tissue sections is demanding and requires practice. This was overcome by the development of the Fluorescence Immunophenotyping and Interphase Cytogenetics as a Tool for Investigation Of Neoplasia (FICTION) technique that combines fluorescence immunophenotyping and fluorescence in situ hybridization (11–13). The advantage of FICTION over FISH in analyzing Hodgkin lymphoma is that it uses the phenomenon that HRS cells, in contrast to the wide majority of other cell populations infiltrating the tumor of Hodgkin lymphoma, strongly express the CD30 protein. For FICTION, tissue sections or imprints are

Table 1
Commercially available FISH probes for the detection of chromosomal translocations recurrently identified in lymphomas

Lymphoid malignancy	Chromosomal translocation	Genes involved	Available commercial probes[a]
B-cell derived			
Burkitt lymphoma	t(8;14)(q24;q32)	*MYC/IGH*	MYC/IGH Dual Fusion
	t(2;8)(p12;q24)	*IGK/MYC*	MYC Break Apart; IGK Breakapart
	t(8;22)(q24;q11)	*MYC/IGL*	MYC Break Apart; IGL Breakapart
Diffuse large B-cell lymphoma	t(3q27)	*BCL6*	BCL6 Break Apart
	t(14q32)	*IGH*	IGH Break Apart
	t(14;18)(q32;q21)	*IGH/BCL2*	IGH Break Apart; BCL2 Break Apart; IGH/BCL2 Dual Fusion
Follicular lymphoma	t(14;18)(q32;q21)	*IGH/BCL2*	IGH Break Apart; BCL2 Break Apart; IGH/BCL2 Dual Fusion
	t(3;14)(q27;q32)	*BCL6/IGH*	BCL6 Break Apart; IGH Break Apart
MALT lymphoma	t(11;18)(q21;q21)	*API2/MALT1*	BIRC3 (API2)/MALT1 Dual Fusion; MALT1 Break Apart
	t(14;18)(q32;q21)	*IGH/MALT1*	IGH/MALT1 Dual Fusion; MALT1 Break Apart; IGH Break Apart
	t(1;14)(p22;q32)	*BCL10/IGH*	IGH Break Apart
Mantle cell lymphoma	t(11;14)(q13;q32)	*CCND1/IGH*	IGH/CCND1 Dual Fusion; CCND1 Break Apart; IGH Break Apart
T-cell derived			
Anaplastic large cell lymphoma	t(2;5)(p23;q35)	*NPM/ALK*	ALK Break Apart
Precursor T-cell lymphoblastic lymphoma	14q11.2; 7q35; 7p14-15/10q24 or 5q35	*αTCR* and *δTCR; βTCR; γTCR/TLX1* or *TLX3*	TCRAD BAP; TLX1 Break Apart; TLX3 Break Apart

The table was prepared referring to refs. (8, 10, 14)
[a]Suppliers like Abbott Molecular, Cytocell, or Kreatech offer a large collection of hematology FISH probes

immunophenotyped with primary antibodies directed against CD30 and then visualized either by conjugation of the primary or of secondary antibodies with a fluorescent dye. Thereafter, fluorescent in situ hybridization is performed with the desired combination of probes. Using different filters in the fluorescence microscope FICTION allows visualization of the neoplastic HRS cells in the section and simultaneous analysis of the fluorescent signals in the nuclei.

In this chapter we describe the standard protocols for FISH and FICTION used in our laboratory for routine diagnostics of lymphatic neoplasms and cancer research.

2. Materials

2.1. Preparation of Locus-Specific FISH Probes

2.1.1. DNA Preparation from BAC/PAC Clones

Instruments/Web site for Probe Selection

1. http://genome.ucsc.edu/cgi-bin/hgTracks (FISH clones, BAC end pairs, Fosmid end pairs).
2. Centrifuge capable of 15,000×*g* at 2–8°C equipped with racks for 15 ml, 50 ml tubes or alternative for large-scale preparations.
3. Microcentrifuge for 1.5 and 2 ml tubes.
4. Incubation oven with shake option.
5. Water bath.
6. NanoDrop or alternative for DNA concentration and quality measurement.

Reagents and buffers

1. BAC/PAC clones harboring the DNA of interest.
2. Selection antibiotic (see Note 1).
3. Glycerine 50%.
4. LB medium: dissolve 20 g LB Broth Base in 1,000 ml distilled water. Autoclave and store at +4°C for maximum 1 year.
5. PhasePrep™ BAC DNA Kit (Sigma-Aldrich Chemie GmbH Munich, Germany).

Equipment

1. Vented tubes for culturing BACs/PACs.
2. 250 ml conical flasks.
3. 1.5 ml tubes

2.1.2. DNA Labeling by Random Priming

Instruments

1. Water bath (0–99°C).
2. Centrifuge capable of 15,000×*g*.

Reagents and buffers

1. Bioprime DNA Labeling System (Invitrogen GmbH, Darmstadt, Germany).
2. Desoxynucleoside Triphosphate Set (dilute unlabelled dNTPs each 1:10 with distilled water to final concentration of 10 mM).
3. Orange-dUTP or Green-dUTP (Enzo Life Sciences GmbH, Lörrach Germany): 50 nmol of dUTP, 50 μl distilled water.
4. dUTPs labeling mixture: 2.5 μl 10 mM dATP, 2.5 μl 10 mM dCTP, 2.5 μl 10 mM dGTP, 1.25 μl 10 mM dTTP, 12.5 μl 1 mM dUTP (orange or green), 3.75 μl distilled water.

Equipment

1. Amicon Ultra-0.5, Ultracel-30 Membrane, 30 kDa (Millipore, Cork, Ireland).
2. 1.5 ml tubes.

2.1.3. Probe Precipitation

Instruments

1. Centrifuge capable of 15,000×*g*.

Reagents and buffers

1. Cotl DNA.
2. 3 M sodium acetate, pH 5.0.
3. Ethanol abs.
4. 20× SSC: 3 M NaCl, 0.3 M sodium citrate, dissolve in 800 ml distilled water and adjust to 1,000 ml, pH 7.0.
5. Mastermix: 2.5 ml formamide, deionized, 1.25 ml 40% dextran sulfate, 0.5 ml 20× SSC, 0.25 ml distilled water, stored in −20°C. We recommend using a face mask during the preparation of mastermix as dextran sulfate is toxic.
6. Control chromosome enumeration probes (CEP probes) if appropriate (Abbott, Kreatech, Metasystems or alternative).

Equipment

1. 1.5 ml tubes.
2. Face mask.

2.2. Sample Preparation

2.2.1. Preparation of Cell Suspensions Form Cultured Tumor Cells from Bone Marrow or Blood

Instruments

1. Sterile bench.
2. Cell culture incubator with CO_2.
3. Centrifuge with racks for 15 ml tubes.
4. Vacuum pump.
5. Sterile bench.

Reagents and buffers

1. RPMI 1640 medium with 25 mM Hepes, without L-glutamine.
2. Fetal calf serum (FCS).
3. L-glutamine.
4. Penicillin Streptomycin.
5. Conditioned medium:
   - Collect peripheral blood from volunteers using heparin-monovettes
   - Incubate at 37°C in 5% CO_2 for 2–5 h till sediment is formed

- Prepare in a sterile flask: 500 ml RPMI 1640, 125 ml FCS, 6.25 ml L-glutamine, 6.25 ml Penicillin Streptomycin, 6.25 ml phytohemagglutinin
- Aliquot each 9 ml of the medium in 25 cm^2 surface area culture flasks
- Collect the supernatant from the blood samples (in case a leukocytes ring formed mix it with the supernatant) and add 1 ml of the supernatant to the 9 ml of the medium
- Incubate for 72 h in 37°C with 5% CO_2
- Transfer the cultures in 15 ml tubes and centrifuge at $200 \times g$ for 10 min.
- Collect the supernatants in 50 ml tubes and filter the medium using Steriflips with a vacuum pump.
- Remove the Steriflips sterile under the bench und close the tubes with a sterile caps
- Store in −20°C.

6. Phytohemagglutinin.
7. 0,075 M KCl.
8. Ice cold fixative: 50 ml acetic acid, 150 ml absolute methanol.

Equipment

1. Heparin-monovettes.
2. 500 ml flask.
3. 15 and 50 ml tubes.
4. Steriflips (Millipore, Cork, Ireland).
5. Twist Top Vials.

2.2.2. Preparation of FISH Slides Form Cell Suspensions

Instruments

1. Fume hood.
2. Light microscope.

Reagents and buffers

1. Protease 1 solution: 250 mg protease 1, 500 ml VP2000 protease buffer, 0.01 N HCl.
2. Paraformaldehyde.
3. Ethanol abs.

Equipment

1. Pasteur pipette.
2. Glass slides.
3. Diamond pen.
4. Coplin-jars for glass slides.

2.3. Hybridization

Instruments

1. Water bath.
2. Hybridization oven.

Equipment

1. Cover glasses Ø 10 mm.
2. Fixogum rubber cement (Marabu GmbH & Co. KG, Tamm, Germany).
3. Hybridization metal box.

2.4. Washing

Instruments

1. Water bath.

Reagents and buffers

1. 2× SSC.
2. Wash buffer 1: 2 ml 20×SCC, 300 μl Tween20, adjust to 100 ml with distilled water.
3. Wash buffer 2: 10 ml 20×SCC, 100 μl Tween20, adjust to 100 ml with distilled water.
4. DAPI stock solution: 0.2 mg DAPI (4′,6-diamidino-2-phenylindole·2HCl)/ml distilled water, aliquot in 1.5 ml tubes and store in +4°C in the dark.
5. DAPI working solutions: 100 μl DAPI stock solution, 100 ml 2× SSC, the solution can be used up to 1 month.
6. Vectashield Mounting Medium H –1000 (Axxora GmbH, Lörrach, Germany).

Equipment

1. Coplin-jars for glass slides.
2. Cover glasses 124×60 mm.

2.5. Evaluation

1. Fluorescent microscope with appropriate filter set (see Note 2).
2. Documentation software for example Isis-Workstation Metasystems software (Metasystems GmbH, Altlussheim, Germany).

2.6. FISH to Paraffin Embedded Tissue Sections

Instruments

1. Fume hood.
2. Pressure cooker.

Reagents and Buffers

1. Roti-Histol (Carl Roth GmbH, Karlsruhe, Germany).
2. Ethanol abs.

3. 0.01 M citrate buffer: 2.3 g 99.5% citric acid (dehydrated), 800 ml distilled water, 5.4 ml 5 N NaOH, adjust with distilled water to 1,000 ml, pH 6.0, prepare daily new solution.
4. Protease 1 solution: 250 mg protease 1, 500 ml VP2000 protease buffer, 0.01 N HCl.
5. Paraformaldehyde.

Equipment

1. Coplin-jars for glass slides
2. Steel rack for glass slides.

2.7. FICTION

2.7.1. FICTION on Cryo Sections

Reagents and buffers

1. Acetone.
2. PN buffer: 13.8 g sodium hydrogen phosphate ($NaH_2PO_4 \times H_2O$), 14.2 g disodium hydrogen phosphate (Na_2HPO_4), adjust to 1,000 ml with distilled water, pH 8.0, store at room temperature up to 1 year.
3. PNM buffer: 50 g skimmed powdered milk, 0.2 g sodium azide (NaN_3), adjust to 1,000 ml with PN buffer, warm up and stir the buffer for 1–2 h, centrifuge the buffer to pellet the debris for 15 min in 850 g, aliquot the buffer in 50 ml tubes, store at +4°C up to 1 year.
4. Ice cold fixative: 50 ml acetic acid, 150 ml 100% methanol.
5. Paraformaldehyde.
6. Ethanol abs.

Equipment

1. Shandon coverplates and racks (Thermo Scientific, Waltham, USA).
2. Coplin-jars for glass slides.

3. Methods

3.1. Preparation of Locus-Specific FISH Probe

3.1.1. DNA Preparation from BAC/PAC Clones

1. Select appropriate BAC/PAC clones (see Note 3).
2. Transfer 50 μl of the medium in which the clones were shipped to 3 ml LB medium and add 1.5 μl of the appropriate selective antibiotic (200 mg/ml) (see Note 4).
3. Incubate overnight at 37°C, shaking.
4. Prepare several glycerine stocks by adding 500 μl of 50% glycerine to each 500 μl of the overnight culture.
5. In a 250 ml conical flask prepare 100 ml LB medium, add 50 μl (200 mg/ml) of the appropriate antibiotic and 10 μl of the glycerine stock (see Note 5).

6. Incubate overnight at 37°C in an incubation oven, shaking.
7. Isolate BAC/PAC DNA using the PhasePrep™ BAC DNA Kit (Sigma-Aldrich Chemie Gmbh Munich, Germany) according to the instructions of the supplier (see Note 6).
8. Dissolve the isolated DNA for more than 1 h in distilled water at +4°C. Thereafter thoroughly vortex and measure the concentration of DNA. DNA of sufficient quality should show the value of 1.8–2.0 for the absorption ratio A_{260}/A_{280}. Store the DNA at –20°C.

3.1.2. DNA Labeling by Random Priming

1. In a 1.5 ml tube with rubber gasket dissolve 1 μg DNA in a final volume of 24 μl distilled water.
2. Add 20 μl of 2.5× random primer solution.
3. Denature the mixture in boiling water for 5 min.
4. Immediately thereafter cool the mixture down on ice for 5 min.
5. Add 5 μl of diluted dUTPs labeling mixture. From now on try to avoid exposing the probe to light.
6. Add 1 μl Klenow fragment.
7. Vortex carefully and centrifuge. Avoid harsh vortexing and centrifugation that may damage the sensitive Klenow polymerase.
8. Incubate overnight in 37°C.
9. Add 5 μl stop buffer, vortex and centrifuge.
10. Label the Amicon centrifugation filter and insert it into the 1.5 ml collection tube.
11. Add 300 μl distilled water to the labeling mixture and transfer it into the centrifugation filter.
12. Centrifuge at 13,000 × *g* for 10 min.
13. Discard the flowthrough.
14. Add 300 μl distilled water into the centrifugation filter und centrifuge at 13,000 × *g* for 10 min.
15. Add 50 μl distilled water into the centrifugation filter and insert the filter upside down into a new collection tube.
16. Centrifuge at 1,700 × *g* for 5 min.
17. Transfer the purified labeled DNA into the original 1.5 ml tube with rubber gasket. Store at –20°C.

3.1.3. Probe Precipitation

1. Prepare in a 1.5 ml tube the following: 5 μl Cot1 DNA.
2. Add 10 μl of the labeled DNA (larger volumes of the probe, or two or more differently labeled BAC/PAC clones may be jointly used without increasing the volume of Cot1 DNA).
3. Add 1/10 volume 3 M sodium acetate, pH 5.0.
4. Add ×2.5 volume abs. ethanol.

5. Centrifuge at 15,000×*g* or higher for 30 min at room temperature.
6. Carefully remove the liquid with a pipette tip not disturbing the pellet.
7. Let the pellet dry in the dark for 15 min (leave the tube open) (alternatively use a SpeedVac for 10 min).
8. Resuspend the pellet in 10 μl 50% mastermix (or larger volumes if more labeled DNA was used in point 1) and add 1 μl commercially available chromosome enumeration probe (CEP probe) for each 10 μl of the mastermix.
9. Shake the probe in 37°C for at least 1 h.
10. Store at –20°C.

3.2. Sample Preparation

3.2.1. Preparation of Cell Suspensions Form Cultured Tumor Cells from Bone Marrow or Blood

(For steps 1–5 work under a sterile bench)

1. Transfer 1–9 ml of the bone marrow or blood sample into a 15 ml tube and adjust with RPMI 1640 up to 14 ml. Mix gently.
2. Centrifuge at 200×*g* for 10 min.
3. Remove the supernatant.
4. Resuspend the cells in 10 ml RPMI and transfer them to a culture flask.
5. Add 1 ml FCS, 1 ml conditioned culture medium, and 100 μl L-glutamine.
6. Incubate overnight at 37°C in 5% CO_2.
7. Transfer the cell culture into a 15 ml tube and centrifuge at 200×*g* for 10 min.
8. Remove the supernatant (we recommend using a vacuum pump).
9. Add 37°C warm 0.075 M KCl to the cell pellet up to 9 ml and mix.
10. Incubate at 37°C for 30 min.
11. Add few drops of ice cold fixative and mix immediately.
12. Centrifuge at 200×*g* for 10 min.
13. Remove the supernatant and add ice cold fixative up to 10 ml.
14. Repeat step 13–14 at least ×3 till the pellet is white.
15. Resuspend the pellet in 2 ml ice cold fixative and transfer it to Twist Top vials.

3.2.2. Preparation of FISH Slides from Cell Suspensions

1. Mix the cell suspension.
2. Using a Pasteur pipette drop from about 20 cm height one droplet of the cell suspension onto a glass slide (work under a fume hood).

3. Let the slides dry in room temperature for 20 min.
4. Analyze the slides under a light microscope and mark the region of highest cell density using a diamond pen.
5. Digest the slides in protease 1 solution for 5 min in 37°C (see Note 7).
6. Wash the slides in a Coplin-jar with distilled water for 2 min in room temperature.
7. Fix the slides in a Coplin-jar in 1% paraformaldehyde for 2 min in room temperature.
8. Wash the slides in a Coplin-jar with distilled water for 1 min in room temperature.
9. Dehydrate the slides in 70; 85%; abs. ethanol each for 2 min in room temperature.
10. Air dry the glass slides for 10 min in room temperature.

3.3. Hybridization

1. Apply 1.3 μl probe for one hybridization. Pipette the probe onto the region previously marked with the diamond pen having high cell density (avoid exposition to light).
2. Cover the hybridization region with Ø 10 mm cover glass). Avoid formation of air bubbles.
3. Cover the entire hybridization region with Fixogum rubber cement.
4. Transfer the glass slides into a hybridization metal box with lid and denature 7 min in 75°C in a water bath.
5. Hybridize for 12–72 h in 37°C in a hybridization oven.

3.4. Washing (See Note 8)

1. Place a Coplin-jar with wash buffer 1 in 72°C water bath.
2. Using a pincette carefully remove the Fixogum rubber cement and the cover glass.
3. Wash the glass slides in a Coplin-jar with 2× SSC for 1–10 min.
4. Place the slides in 72°C warm wash buffer 1 for 2 min.
5. Transfer the glass slides in wash buffer 2 for 1 min in room temperature.
6. Transfer the glass slides to 2× SSC for 2–10 min in room temperature.
7. Stain the slides in DAPI for 5 min in room temperature.
8. Wash once in 2× SSC for 3 min in room temperature.
9. Put two drops Vectashield Mounting Medium onto a wet glass slide and cover it with 124 × 60 mm cover glass.
10. Store in the dark at room temperature.

3.5. Evaluation

1. Hybridize your probe to 5 normal male controls and analyze the signal quality in at least 100 nuclei each and 10 metaphases each to test for potential cross hybridizations or hybridization to a another chromosome. Only probes giving intensive and not diffused signals at the correct chromosomal position are appropriate for further analysis (see Note 9).
2. Calculate the detection threshold of a probe taking for example the mean % of false positive signals from the 5 controls (100 interphase nuclei each) and add three times the standard deviation calculated for the five controls. Evaluate regions with bright signals and non overlapping nuclei. For break-apart or dual-fusion probes detect a "merge" pattern if the distance between the signals is less than two times the signal diameter (see Note 10).
3. For evaluation of your samples analyze at least 100 nuclei each and 10 metaphases if present. In case of tissue sections analyze cells across the whole hybridization area to detect possible subpopulations of cells. Any signal pattern distinct from the expected should be regarded as abnormal and interpreted. Each sample should be evaluated independently by two observers. Further considerations on FISH probe evaluation may be found in (14).

3.6. FISH to Paraffin Embedded Tissue Sections

(For deparaffinization work under a fume hood)

1. Incubate the slides with the paraffin sections 3×5 min in Roti-Histol in room temperature.
2. Hydrate the slides in abs, 80 and 70% ethanol for 5 min each in room temperature.
3. Place the slides in a steel rack and put it in a Coplin-jar with distilled water.
4. Boil 1 L 0.01 M citrate buffer in a pressure cooker and when boiling place the steel rack with the slides into the cooker; close the lid.
5. Warm up the pressure cooker until it reaches the maximal pressure and let it cook for 160 s.
6. Cool down the pressure cooker under cold tap water and release gradually the steam.
7. Open the lid and carefully pour cold tap water inside. Do not pour the water directly onto the steel rack.
8. Place the steel rack with the slides into a Coplin-jar with distilled water.
9. Digest the slides in protease 1 solution for 30 min in 37°C.
10. Wash the slides in a Coplin-jar with distilled water for 2 min in room temperature.

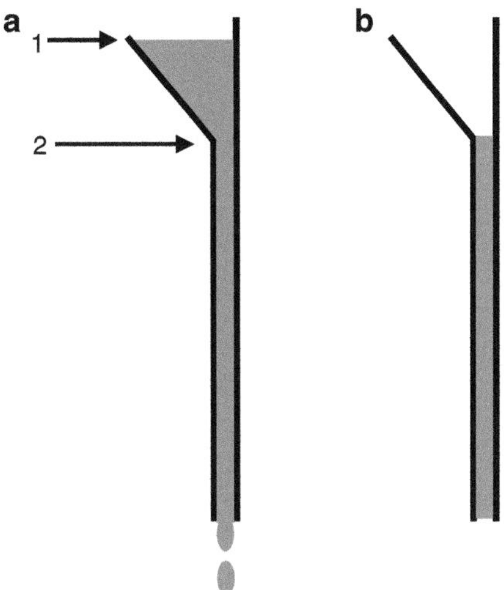

Fig. 1. Shandon coverplates; (**a**) Antibody solution is added to the funnel (position 1) created between the Shandon coverplate and the slide; (**b**) The antibody solution passes between the Shandon coverplate and the slide until the funnel is empty (position 2); the slide remains continuously covered by the antibody solution thanks to capillarity (15).

11. Fix the slides in a Coplin-jar in 1% paraformaldehyde for 2 min in room temperature.
12. Wash the slides in a Coplin-jar with distilled water for 1 min in room temperature.
13. Dehydrate the slides in 70; 85%; abs. ethanol each for 2 min in room temperature.
14. Air dry the slides for 5 min in room temperature.
15. Proceed form Subheading 3.3.

3.7. FICTION on Cryo Sections

1. Air dry the slides for 30 min.
2. Fix the slides for 10 min in acetone.
3. Pour one drop of the PN buffer on the slides and cover it with the plastic Shandon coverplates and attach them in the Shandon racks and pour 100 μl of the PN buffer into the funnel (Fig. 1).
4. After the PN buffer passes through (position 2) add 100 μl of the primary antibody solution for immunohistochemical staining of the section and incubate for 30 min; cover the rack with the attached lid to protect the slides from light (see Note 11).
5. Wash the slides with 100 μl of the PN buffer.
6. After the PN buffer passes through (position 2) add 100 μl of the secondary antibody solution and incubate for 30 min; cover the rack with the attached lid to protect the slides from light (see Note 12).

7. Remove the slides form the Shandon racks and place them in a Coplin-jar containing PN buffer in room temperature, protect the slides form light.
8. Fix the slides in ice cold fixative for 10 min.
9. Wash the slides in distilled water for 1 min.
10. Fix the slides in a Coplin-jar in 1% paraformaldehyde for 2 min in room temperature.
11. Wash the slides in a Coplin-jar with distilled water for 1 min in room temperature.
12. Dehydrate the slides in 70; 85%; abs. ethanol each for 2 min in room temperature.
13. Air dry the slides for 10 min in room temperature.
14. Proceed form Subheading 3.3.

3.8. FICTION on Paraffin Embedded Tissue

1. Proceed with deparaffinization as described in Subheading 3.6, (steps 1–8) and continue with Subheading 3.7 (steps 1–14) (see Note 13).

4. Notes

1. Detailed information on the resistance if the used BAC/PAC clone can be found on the supplier info sheet.
2. The exciter and emitter ranges of the filter should cover the absorption and emission maximums of the used fluorophore respectively. For spectrum green the absorption maximum is 496 nm and the emission maximum is 520 nm; for spectrum orange the absorption maximum is 552 nm and the emission maximum is 576 nm.
3. The BACs/PACs can be conveniently chosen using one of the genome browsers UCSC Genome Browser—BAC end pairs or Ensembl Human—Tilepath).

   The design of probes differs depending on the intended application.

   (a) To identify copy number changes of the region of interest (for example a gene) select clones spanning the region (**spanning probe**). Besides, you will need a reference probe against which the number of fluorescent signals will be evaluated. Commercially available chromosome enumeration probes (CEP) for the same chromosome as the analyzed region can be used although the disadvantage of such combination is that commercial probes usually have a higher hybridisation efficiency resulting in stronger signal intensities than locus-specific probes. This may lead to false

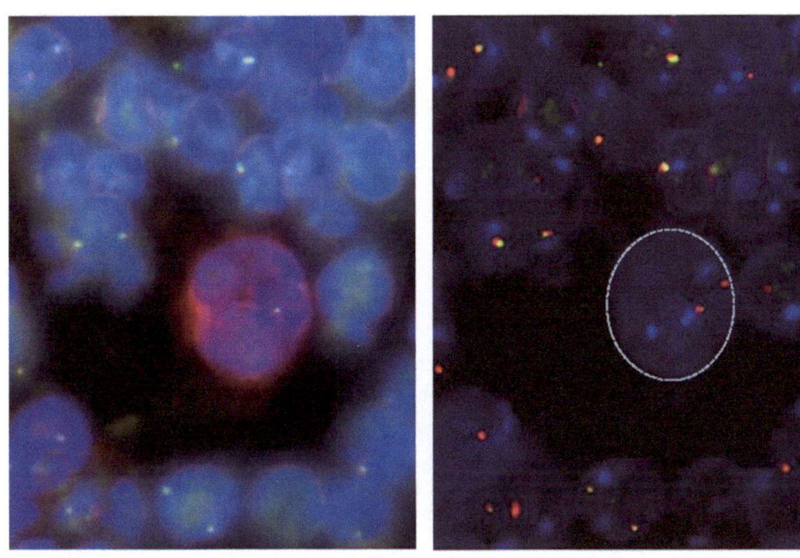

Fig. 2. Copy number analysis of Hodgkin and Reed-Sternberg cells in a classical Hodgkin lymphoma with tetraploid chromosome number. The FICTION technique was used combining immunofluorescent staining against the surface marker CD30 (left image, *red*) and fluorescent probes and counterstained nuclei (*blue*, DAPI). Loss of *CHD2* locus (co-localized *green* and *red* signals) compared to a CEP10 control (*blue* signals) and the ploidy level (right image). Not all signals in focus plane; false color display.

positive results. The locus-specific clone and the reference probe must be labelled with different fluorophores for easy identification of the signals (Fig. 2).

(b) To identify translocation breakpoints select two clones flanking the potential breakpoint (**break-apart probe BAP**). For example the clones may be located outside the 5′ and 3′ regions of a putative translocation involved gene. Label the two clones with different fluorophores. The genomic distance between clones shall normally be less than 1 Mb so they appear as a co-localized signal. We recommend choosing green and red appearing fluorophores that usually merge into yellow signal indicating no break in the investigated region. Using this probe a clearly distinct green and red signal indicates a breakpoint within the flanked region (Fig. 3a).

(c) To identify fusion genes select two clones each spanning one gene putatively involved in the fusion (**dual-fusion probe**). Label the two clones with different fluorophores. Distinct green and red signals will indicate no fusion of the spanned genes. Any merge signal (that may appear in

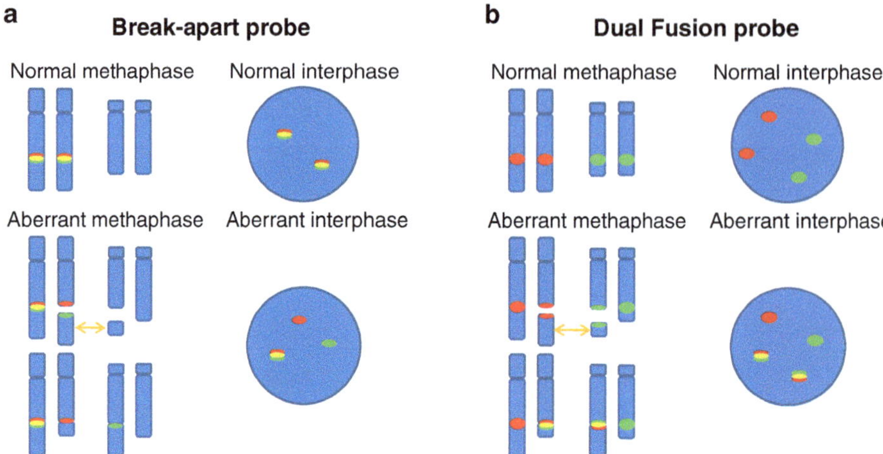

Fig. 3. (**a**) Signal pattern expected for a break-apart probe in normal and aberrant metaphase or nucleus. (**b**) Signal pattern expected for a dual fusion probe in normal and aberrant metaphase or nucleus (14).

yellow) in a significant proportion of cells indicates gene fusion (Fig. 3b).

4. We recommend using vented culture tubes. Alternatively use standard 15 ml tubes but ensure not to close the cap tightly to allow gas exchange.
5. The given volumes may be scaled down to a mini- or micro-scale preparation according to the instructions of the PhasePrep™ BAC DNA Kit supplier and depending on the quantity of BAC DNA that is needed for FISH/FICTION experiments. In our experience a mini-scale preparation yields at least 20 μg of DNA. In case of using mini-scale preparation all subsequent centrifugation steps can be performed in 50 ml tubes. For micro-scale preparations 15 ml tubes can be used.
6. In contrast to the supplier protocol we recommend dissolving the final DNA precipitate in 40–100 μl of distilled water.
7. You can alternatively digest the slides in 99 ml distilled water with 5 mg pepsin and 1 ml 1 M HCl warmed to 37°C for 5 min.
8. The washing protocol can be automated using the VP 2000 Processor (Abbott GmbH & Co. KG, Wiesbaden-Delkenheim, Germany) or an alternative.
9. Use Isis-Workstation Metasystems software or an alternative for documentation.
10. As the preparations of sections may truncate cells resulting in false positive results the threshold may be up to 15% especially for complex probes. In research projects aimed at detecting copy number losses in tumor sections, to avoid false positive results, the authors sometimes set the threshold arbitrary to 30%.

11. For the analysis of classical Hodgkin lymphoma to stain the HRS cells we use the anti-CD30 (*BER-H2*) monoclonal antibody (dilution 1:20–1:50 the exact dilution must be determined experimentally) as the primary antibody. As secondary antibody we use the Alexa 594-conjugated rabbit anti-mouse antibody (Invitrogen, GmbH, Darmstadt, Germany) (dilution 1:50). The antibodies are diluted with PNM buffer.
12. Further antibodies may be used if necessary. These should not cross-react with other antibodies used. Always wash the slide with 100 μl of the PN buffer prior the incubation with a new antibody.
13. Avoid any type of digestion after the deparaffinization as it destroys the cells' phenotype and thus no immunophenotyping will be possible.

Acknowledgments

The work of the authors on FISH/FICTION has been supported by the Deutsche Krebshilfe, the Wilhelm Sander Stiftung, the Kinderkrebsinitiative Buchholz/Holm-Seppensen, the Jose Carreras foundation, the BMBF, and the Polish Ministry of Science and Higher Education (grant N301 238736). MG was supported by the FEBS long-term fellowship and the Support for International Mobility of Scientists fellowships of the Polish Ministry of Science and Higher Education. We thank Reina Zühlke-Jenisch and Ursula Schnaidt for excellent technical assistance.

References

1. Cheung SW, Tishler PV, Atkins L, Sengupta SK, Modest EJ, Forget BG (1977) Gene mapping by fluorescent in situ hybridization. Cell Biol Int Rep 1:255–262
2. Rudkin GT, Stollar BD (1977) High resolution detection of DNA-RNA hybrids in situ by indirect immunofluorescence. Nature 3: 472–473
3. Bauman JG, Wiegant J, Borst P, van Duijn P (1980) A new method for fluorescence microscopical localization of specific DNA sequences by in situ hybridization of fluorochromelabelled RNA. Exp Cell Res 128:485–490
4. Langer-Safer PR, Levine M, Ward DC (1982) Immunological method for mapping genes on Drosophila polytene chromosomes. Proc Natl Acad Sci USA 79:4381–4385
5. Van Prooijen-Knegt AC, Van Prooijen-Knegt AC, Van Hoek JF, Bauman JG, van Duijn P, Wool IG, van der Ploeg M (1982) In situ hybridization of DNA sequences in human metaphase chromosomes visualized by an indirect fluorescent immunocytochemical procedure. Exp Cell Res 141:397–407
6. Trask BJ (2002) Human cytogenetics: 46 chromosomes, 46 years and counting. Nat Rev Genet 3:769–778
7. Levsky JM, Singer RH (2003) Fluorescence in situ hybridization: past, present and future. J Cell Sci 116:2833–2838
8. Siebert R, Weber-Matthiesen K (1997) Fluorescence in situ hybridization as a diagnostic tool in malignant lymphomas. Histochem Cell Biol 108:391–402
9. Schröck E, Padilla-Nash H (2000) Spectral karyotyping and multicolor fluorescence in situ hybridization reveal new tumor-specific chromosomal aberrations. Semin Hematol 37:334–347

10. Oscier DG, Gardiner AC (2001) Lymphoid neoplasms. Best Pract Res Clin Haematol 14: 609–630
11. Weber-Matthiesen K, Winkemann M, Müller-Hermelink A, Schlegelberger B, Grote W (1992) Simultaneous fluorescence immunophenotyping and interphase cytogenetics: a contribution to the characterization of tumor cells. J Histochem Cytochem 40:171–175
12. Weber-Matthiesen K, Deerberg J, Müller-Hermelink A, Schlegelberger B, Grote W (1993) Rapid immunophenotypic characterization of chromosomally aberrant cells by the new FICTION method. Cytogenet Cell Genet 63:123–125
13. Weber-Matthiesen K, Pressl S, Schlegelberger B, Grote W (1993) Combined immunophenotyping and interphase cytogenetics on cryostat sections by the new FICTION method. Leukemia 7:646–649
14. Ventura RA, Martin-Subero JI, Jones M, McParland J, Gesk S, Mason DY et al (2006) FISH analysis for the detection of lymphoma-associated chromosomal abnormalities in routine paraffin-embedded tissue. J Mol Diagn 8: 41–51
15. Martín-Subero JI (2001) Development of the multicolor FICTION technique. Applications to the diagnosis and research of malignant lymphomas. *Doctoral thesis.*

Chapter 14

Identification of Pathogenetically Relevant Genes in Lymphomagenesis by shRNA Library Screens

Vu N. Ngo

Abstract

RNA interference (RNAi) is a conserved posttranscriptional gene silencing mechanism that has recently emerged as a breakthrough genetic tool in functional genomics and drug target discovery. An increasing number of studies applying RNAi in high-throughput screens have begun to unravel complex signaling networks underlying diverse cellular processes. This chapter describes an approach to construct a conditional small-hairpin (sh)RNA library and its application in human lymphoma cell lines. A library cloning procedure outlines the incorporation of shRNA sequences and random 60-mer "bar code" oligonucleotides, enabling rapid identification of the hairpin by microarrays. Lymphoma cell lines are optimized for efficient retroviral transduction and tetracycline inducibility. The shRNA library is suitable for identifying molecular targets in cancer, but also versatile for various screening strategies.

Key words: RNA interference, RNAi, Conditional shRNA library, High-throughput screen, Bar codes, Functional genomics, Cancer, Lymphoma

1. Introduction

The principle of RNA interference (RNAi) involves double-stranded (ds)RNA-mediated degradation of messenger (m)RNA molecules, thereby preventing protein synthesis (1). When introduced into cells, long dsRNAs are cut into small fragments approximately 20 base pairs (b.p.) in length by an endoribonuclease Dicer. These short interfering (si)RNA fragments are processed by an RNA-induced silencing complex (RISC), which unwinds the RNA duplexes and selects appropriate strands to target corresponding mRNAs for degradation (2). Originally observed in a flowering plant, RNAi principles are widespread in eukaryotic cells (1, 3–5). The combination of speed, efficiency, and exquisite specificity has made RNAi a valuable approach that is quickly adopted for the study of gene function in model organisms such as yeast, worms,

Ralf Küppers (ed.), *Lymphoma: Methods and Protocols*, Methods in Molecular Biology, vol. 971,
DOI 10.1007/978-1-62703-269-8_14, © Springer Science+Business Media, LLC 2013

and fruit flies. In mammalian cells, long dsRNAs (over 30 b.p.) elicit strong interferon responses, resulting in general shutdown of protein translation (6). Researchers have circumvented such adverse effects by using shorter siRNA fragments (~21–25 b.p.), which are equally efficient in achieving specific gene silencing (7).

With the available sequence of the human genome, RNAi can be developed into a powerful high-throughput screen for genetic determinants of various biological processes including cancer. A loss-of-function RNAi screen can identify genes that are essential for cancer's growth and survival and therefore have potential as drug targets. Typical cancer genes harbor mutations and alterations; however, RNAi screen can reveal targets that are not necessarily mutated but remain essential for cancer phenotype. In this regard, RNAi-based screen is a powerful tool complementary to the arsenal of existing functional genomics methods such as high-throughput DNA sequencing, single-nucleotide polymorphism genotyping, array-based comparative genomic hybridization, gene expression profiling, and various proteomics techniques. In our studies, we used shRNA library screening to identify essential genes in an NF-κB-dependent, activated B-cell-like (ABC) subgroup of diffuse large B cell lymphoma (DLBCL). We have uncovered molecular determinants upstream of NF-κB including CARD11, MALT1, CSNK1A1, BTK, CD79A/B, and MYD88, which are important for survival of ABC DLBCL (8–10). These findings and screening results from other libraries (reviewed in ref. (11)) have demonstrated potential values of the RNAi-based high-throughput approach in revealing new functional molecules in normal and pathological conditions.

In contrast to single-gene knockdown experiments, large-scale RNAi screens require further consideration on the format, readout, and library complexity. Two frequently used approaches are arrayed and pooled library screens. In the arrayed format, cells are transiently transfected with individual siRNA duplexes or shRNA vectors stored separately in 96- or 384-well plates. This approach allows image-based and flow cytometry screening for many morphological and biochemical parameters of target cells. By contrast, the simplified pooled library format combines shRNA vectors into pools and is more suitable for screening growth phenotype that requires long-term cell culture. Depending on the phenotype to be screened, readout assays of arrayed screens are direct for individual RNAi sequences and often laborious and time-consuming, thus necessitating automation. Pooled library screens, on the other hand, require multiplex technologies such as microarrays or high-throughput DNA sequencing. The speed of screening also depends on parameters that determine the library size such as complexity and coverage. Most established libraries have typical 3–6 shRNA constructs per gene, but vary in coverage from a few hundred genes to the entire genome. These features of existing shRNA libraries have been discussed in several excellent reviews (11–13); however, detailed practical information on development and usage of this important technology still remains in need.

We have developed an inducible, bar coded shRNA library and established a protocol to screen for essential genes in human lymphoma cell lines. The shRNA library method consists of four major procedures: (1) shRNA library construction; (2) generation of doxycycline-inducible cell lines; (3) performing library screen for growth phenotype; and (4) detection of candidate shRNAs. shRNA library construction begins with selection of appropriate RNAi expression vectors. Retroviral vectors are mainly used for actively dividing cells whereas lentiviral vectors can efficiently transduce nondividing cells. These vectors integrate their DNA into the host cell genome allowing stable RNAi expression. Adenoviral vectors provide an alternative system with likely higher, but transient, shRNA expression levels. Our library vector was developed from the pSUPER retroviral backbone, which uses an RNA polymerase III-mediated H1 promoter to drive shRNA expression (14). We further modified the vector by cloning a tetracycline-inducible element and a random 60-mer oligonucleotide sequence. The inducible RNAi retroviral vector ensures that shRNA expression is regulatable. The unique 60-mer bar code cloned adjacent to each shRNA facilitates detection of the hairpin abundance in a pooled RNAi screen using a microarray. Successful RNAi screens also depend on efficient introduction of shRNA constructs into target cells. We exploit the high transduction efficiency of the ecotropic retroviral system by expressing a murine ecotropic receptor in human lymphoma cell lines and pseudotyping the RNAi library vector with an ecotropic envelope protein. These cell lines are subsequently engineered to express the bacterial tetracycline repressor (TR) protein, which conditionally regulates shRNA expression. We describe an approach to perform a growth phenotype-based screen in a pooled library fashion and identify potential hits using microarray hybridization. The protocol details procedures used for human lymphoma cell lines but can be readily applied to various cell systems.

2. Materials

2.1. shRNA Library Construction

2.1.1. Cloning 60-Mer Bar Code Oligonucleotides

1. Random 60-mer oligonucleotide templates (Table 1). Make a 1.5 μM working dilution.
2. Xho1-F-primer and Mfe1-R-primer (Table 1). Make a 10 μM working dilution.
3. dNTPs (10 mM).
4. Taq DNA polymerase (5 U/μl).

Table 1
Random 60-mer oligonucleotide template and bar code primers

Name	Sequence
Random 60-mer oligonucleotide template	5′-GATGCGCTCGGTATGCTACTCGAG *N(60)* CAATTGGCT TGGATCTGCGACTAG[a]-3′
XhoI-F-primer	5′-GATGCGCTCGGTATGCTACTC-3′
MfeI-R-primer	5′-CTAGTCGCAGATCCAAGCCAA-3′

[a]Underlined sequences indicate XhoI and MfeI restriction sites. N(60) indicates 60 random nucleotides

5. A thermal cycler.
6. Agarose.
7. Ethidium bromide.
8. Ultraviolet (UV) light box.
9. QIAEX gel extraction kit (Qiagen, Valencia, CA, USA).
10. XhoI and MfeI restriction enzymes (New England Biolabs, Ipswich, MA, USA).
11. Precast 20% polyacrylamide gel (Life Technology, Grand Island, NY, USA).
12. Retroviral vector pRSMX (see Note 1).
13. T4 ligase, 400,000 U/ml (New England Biolabs).
14. Competent DH5α bacteria (Life Technologies).
15. Luria Bertani (LB) agar (with ampicillin) plates.
16. QIAGEN plasmid Maxi kit (Qiagen).

2.1.2. Cloning shRNA Inserts

1. Commercially synthesized shRNA oligonucleotides at 100 μM (see Note 2).
2. 10× annealing buffer: 100 mM Tris–HCl, pH 7.5–8.0, 500 mM NaCl, 10 mM EDTA.
3. Beckman Multimek Liquid Handler (Beckman Coulter, Indianapolis, IN, USA).
4. Multichannel pipet.
5. BglII and HinDIII restriction enzymes (New England Biolabs).
6. Competent DH5α bacteria, 15 μl aliquot per well in 96-well PCR microplates, stored at −80°C.
7. 12-lane trays (Scinomix, Earth City, MO, USA).
8. TrakMates Tubes, 1.4 ml 2D bar coded 96 tubes in one latch rack (Fisher Scientific, Pittsburgh, PA, USA).
9. AirPore Tape Sheets (Qiagen).

Table 2
PCR primers for bar code and shRNA insert

Name	Sequence
pMSCV-5′	5′-CCCTTGAACCTCCTCGTTCGACC-3′
pMSCV-3′	5′-GAGACGTGCTACTTCCATTTGTC-3′

10. pMSCV-5′ and pMSCV-3′ primers at 100 pmol/μl working dilution (Table 2).
11. Autoclaved 80% glycerol.
12. QuickStep 2 96-well PCR Purification Kit (EdgeBio, Gaithersburg, MD, USA).
13. Deep well 96-well blocks (Corning, Lowell, MA, USA).
14. DNA sequencing services.
15. Bioinformatic services.

2.1.3. Generation of Bar Code Microarrays

1. Commercially synthesized 60-mer bar code oligonucleotides at 100 μM in 96-well plates (see Note 3).
2. 20× Saline Sodium Citrate (SSC) buffer: 3 M NaCl, 300 mM sodium citrate.
3. Polypropylene 384-well microplates (Nunc, Rochester, NY, USA).
4. UltraGAPS slides (Corning).
5. OmniGrid 100 microarrayer (GeneMachines, San Carlos, CA, USA).
6. Slide storage box.

2.2. Generation of Ecotropic Receptor- and Tetracycline Repressor-Expressing Cell Lines

1. FLYRD18/mCAT-IRES-Bleo cell line (see Note 4).
2. HEK 293 T cell line.
3. TR-expressing retroviral vector (see Note 5).
4. pCMV-TO/eGFP retroviral plasmid.
5. 0.45 μm syringe-driven filter unit (Millipore, Billerica, MA, USA).
6. Polybrene (hexadimethrine bromide; Sigma, St. Louis, MO, USA).
7. Sorvall RT7 Plus centrifuge with microplate carriers (or equivalent).
8. Doxycycline (Sigma).
9. Fetal bovine serum (FBS) without tetracycline.

Table 3
PCR primers for genomic DNA

Name	Sequence
Primer 1	5′-CTAATACGACTCACTATAGGGAGATGTCGCTATGTGTTCTGGGAAATCAC[a]
Primer 2	5′-GGTTAAGATCAAGGTCTTTTCACCTGGC

[a]The underlined indicates the T7 promoter sequence

2.3. shRNA Library Screen for Growth Phenotype

1. shRNA library DNA in pools of 96 constructs (see Subheading 3.1.2, step 9).
2. Mutant ecotropic envelope-expressing helper plasmid pHIT/EA6x3* (see Note 6).
3. Gag/pol-expressing helper plasmid pHIT60 (kindly provided by Stephen Goff) (15).
4. Lipofectamine 2000 (Life Technologies) (or equivalent).
5. OptiMEMI (Life Technologies).
6. Puromycin (Sigma).

2.4. Detection of Candidate shRNAs

2.4.1. Recovery of shRNA-Associated Bar Code Sequences

1. QIAamp DNA mini kit (Qiagen).
2. RNAse (4 mg/ml; Sigma).
3. 95–100% ethanol.
4. 10 mM Tris–HCl buffer pH 7.5.
5. Primers (Table 3). Make a 10 μM working dilution.
6. Sephadex G-50 columns (Amersham Biosciences).

2.4.2. Generation of Fluorescently Labeled RNA Probes

1. MAXIscript T7 kit (Ambion).
2. Sephadex G-50 columns (Amersham Biosciences).
3. ULS aRNA Labeling Kit (Kreatech Diagnostics, The Netherlands).
4. Microcon centrifugal filter YM-30 (Millipore).

2.4.3. Microarray Hybridization

1. Bar code microarrays (see Subheading 3.1.3).
2. UV Crosslinker (Stratagene).
3. Prehybridization solution: 5× SSC, 0.1% SDS, 0.1% BSA.
4. Microscope slide-staining troughs.
5. Microscope slide-staining racks.
6. Isopropanol (Sigma).
7. Hybridization solution: 40% formamide, 5× SSC, 0.2% SDS, 0.1 μg/μl COT-1 DNA (Life Technologies).
8. Corning hybridization chambers (Sigma).
9. Water bath at 55°C.

2.4.4. Post-hybridization and Analysis

1. Wash buffer 1: 2× SSC, 0.1% SDS.
2. Wash buffer 2: 0.1× SSC, 0.1% SDS.
3. Wash buffer 3: 0.1× SSC.
4. Double-distilled water.
5. GenePix 4000 microarray scanner (Molecular Devices, Sunnyvale, CA, USA).
6. GenePix Pro software (Molecular Devices).

3. Methods

3.1. shRNA Library Construction (Fig. 1)

3.1.1. Cloning 60-Mer Bar Code Oligonucleotides (See Note 7)

1. Prepare a PCR mixture of 1 µl of random 60-mer oligonucleotide templates, 1 µl of Xho1-F-primer, 1 µl of Mfe1-R-primer, 1 µl of dNTPs, 2.5 µl of 10× Taq DNA polymerase buffer, 1 µl of Taq DNA polymerase, and 17.5 µl of H_2O.
2. Run PCR program: 95°C for 1 min, 40 cycles of (95°C for 30 s, 60°C for 30 s, 72°C for 10 s), 72°C for 5 min, 4°C, and pause.

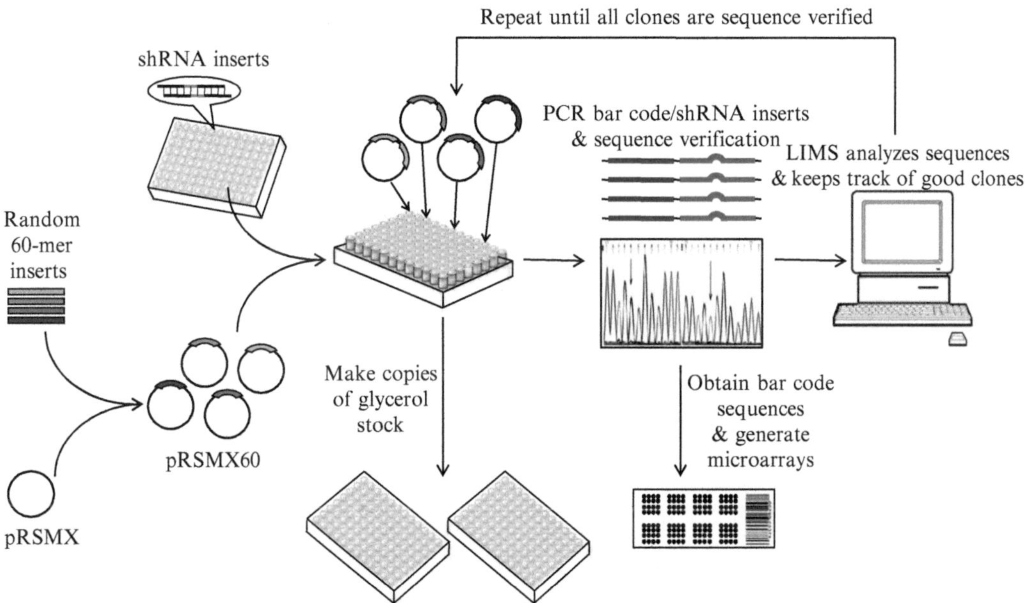

Fig. 1. shRNA library construction. Random 60-mer bar code nucleotides are synthesized, PCR amplified, and cloned into the inducible RNAi retroviral vector pRSMX. Approximately 30,000 colonies, each carrying a unique bar coded vector (pRSMX60), are combined and plasmid DNAs are generated. About 10,000 individual shRNA inserts in 96-well plates are prepared by annealing forward and reverse shRNA oligonucleotides and cloned into the bar coded pRSMX60 vector. Individual clones are sequenced to verify the shRNA inserts and identify the associated bar code sequence for each hairpin. Cloning procedures are facilitated by an LIMS, which is developed using in-house bioinformatics resource. Final constructs are replicated in copies of bacterial glycerol stocks. Bar code sequences of the completely sequence-verified library are used to generate microarrays for detection of shRNA hits.

3. PCR products (108 b.p.) are analyzed in 2% agarose gel containing ethidium bromide and visualized under UV light.
4. The PCR product is cleaned with a QIAEX gel extraction kit following the manufacturer's instructions.
5. Digest PCR products with XhoI and MfeI for 2 h at 37°C.
6. Separate XhoI/MfeI digested fragments using 20% polyacrylamide gel and isolate the bar codes (~66 b.p.) using QIAEX gel extraction kit.
7. Digest the retroviral vector pRSMX with XhoI and MfeI for 2 h at 37°C. Separate and purify the digested vector using 1% agarose gel and QIAEX gel extraction kit, respectively.
8. Prepare a ligation mixture of 3 μl of random 60-mer bar code DNA fragments prepared in step 6, 1 μl of vector pRSMX prepared in step 7, 0.5 μl of 10× ligase buffer, and 0.5 μl of T4 ligase. Perform ligation at 16°C for 16 h.
9. Transform 5 μl of ligation reaction into 50 μl of chemically competent DH5α bacteria by heat shock as recommended by the manufacturer. Transformed bacteria are plated on LB agar (with ampicillin) plates and inoculated at 37°C for 16–18 h.
10. Count bacterial colonies and repeat the ligation–transformation steps as needed until a desired number of colonies is obtained. This step will generate a library of vectors carrying random 60-mer bar code sequences. To ensure that each vector receives a unique bar code, we empirically generate threefold as many bacterial colonies as the number of shRNA constructs in a library. For example, we generate approximately 30,000 transformed bacterial colonies—representing 30,000 unique bar codes—to cover a library of 10,000 shRNA constructs.
11. Plasmid DNAs are recovered from fresh transformed bacterial colonies by flooding the agar surface with liquid LB-ampicillin medium (for example, use 50 ml for a 150 × 15 mm petri dish) and scraping off the colonies. Combine all bacterial suspensions and prepare plasmid DNA using a Qiagen Maxiprep kit. Name the new vector library pRSMX60.

3.1.2. Cloning shRNA Inserts

The following cloning procedure will be performed in 96-well plates (standard U-bottom or PCR microplate format when indicated) using a robotic liquid handler and multichannel pipet. Pipeting liquid volumes are suggested for individual wells in a 96-well plate. Cloning procedures are summarized in Fig. 1.

1. shRNA oligonucleotides are designed and obtained as described in Note 2. Forward and reverse 62-mer shRNA oligonucleotide strands are annealed by mixing 5 μl of each oligo strand (100 μM) with 10 μl of 10× annealing buffer and 80 μl of H_2O. Using a thermal cycler, the mixture is heated up at 95°C

for 2 min, ramp cooled to 25°C for 1 h, and kept at 4°C until ready for use.

2. Annealed oligos are approximately at 200 ng/μl and further diluted to 0.05 ng/μl by mixing 3 μl of annealed oligos (200 ng/μl) with 187 μl of H_2O and repeat the dilution one more time.

3. Digest a library of bar coded vectors pRSMX60 (Subheading 3.1.1, step 11) with BglII and HindIII. Separate and purify the vector using 1% agarose gel and QIAEX gel extraction kit. Adjust the final DNA concentration to 10 ng/μl. This step should be scaled up to prepare sufficient vector for subsequent cloning. We estimate that 300 μg of pRSMX60 vectors are needed for cloning approximate 30,000 shRNA constructs.

4. Ligate annealed oligos from step 2 above into BglII/HindIII digested pRSMX60 vectors by mixing a 3:1 molar ratio of insert:vector. Briefly, prepare a mixture of 3 μl of diluted annealed oligos (0.15 ng), 0.5 μl of pRSMX60 vector (5 ng), 0.5 μl of 10× ligase buffer, 0.3 μl of T4 ligase, and 0.7 μl of H_2O (see Note 8). If using a robotic liquid handler (recommended), first aliquot 3 μl of diluted annealed oligos to a PCR microplate. Then use a multichannel pipet to add 2 μl of the ligation master mix (0.5 μl of pRSMX60 vector, 0.5 μl of 10× ligase buffer, 0.3 μl of T4 ligase, and 0.7 μl of H_2O) into each well. Spin the plate quickly using a tabletop centrifuge to bring the content down to the bottom of each well. Perform ligation at 16°C for 16–18 h.

5. Transform 2 μl of ligation reaction to 15 μl of chemically competent DH5α bacterial aliquot in 96-well PCR plates: Using a multichannel pipet, add 2 μl of ligated products from step 4 above to competent cells previously prepared in 96-well plates and incubate on ice for 30 min. Heat shock at 42°C for 30 s and then incubate on ice for 2–3 min. Add 100 μl of LB medium and incubate at 37°C for 1 h. Transfer 25–30 μl of bacteria into LB/agar/ampicillin in 12-lane trays and allow the bacterial droplet to cover the agar surface by turning the tray with a tilting motion. Incubate the trays at 37°C for a maximum of 14 h.

6. Inoculate one bacterial colony from each lane into a corresponding TrakMates tube secured in a 96-position rack and containing 400 μl of LB/ampicillin medium. To minimize errors, use sterile pipet tips to pick the colonies, and drop and leave the tips in the TrakMates tubes. When colony transfer for one 96-tube rack is completed, use a multichannel pipet to remove the tips. Cover the TrakMates tubes with a porous paper seal (AirPore Tape Sheet) and incubate at 37°C in an orbital shaker (250 rpm) for 16 h.

7. PCR amplification of shRNA and 60-mer bar code inserts: Prepare a PCR mixture of 2 μl of bacterial culture from step 6, 0.04 μl of pMSCV-5′ primer, 0.04 μl of pMSCV-3′ primer, 0.25 μl of dNTPs, 2.7 μl of 10× Taq DNA polymerase buffer, 0.25 μl of Taq DNA polymerase, and 21.72 μl of H_2O. Robotic liquid handling of multiple 96-well plates is recommended at this step. Carry out PCR program: 95°C for 10 min, 30 cycles of (94°C for 20 s, 58°C for 20 s, 72°C for 30 s), 72°C for 5 min, 4°C, and pause. Immediately freeze the remaining bacterial cultures by adding 100 μl of autoclaved 80% glycerol and storing at −80°C.
8. Verification of shRNA sequences and identification of 60-mer bar code sequences: PCR products in 96-well plates from **step** 7 are cleaned using a QuickStep 2 96-well PCR Purification Kit and sequenced using pMSCV 5′ primer (Table 2). A laboratory information management system (LIMS) is essential to facilitate this cloning step. LIMS will process sequencing results, confirm or reject incorrect shRNA sequences, and identify 60-mer bar code sequences (see Note 9). Incorrect clones from the original removable TrakMates tubes in the 96-tube rack are discarded and replaced with correct ones to preserve clone positions in original 96-well plates. Repeat step 6 as needed until all clones are sequence verified. On average, each round of sequence verification yields about 50% of clones that harbor correct shRNA sequences and acceptable 60-mer bar codes (having 30–70% G/C content and being unique among previously identified bar code sequences).
9. Completely sequence-verified shRNA library constructs are stored as bacterial glycerol stocks in replicate copies at −80°C. Plasmid DNA is prepared by thawing and inoculating 10 μl of bacterial glycerol stocks in 300 μl of LB/ampicillin medium prefilled in a deep well 96-well block for 6 h with shaking (250 rpm). Combine bacterial cultures from all 96-wells in the block with 1 l of LB/ampicillin medium and continue inoculation at 37°C for 16 h. Use QIAGEN plasmid maxi kit to prepare for DNA from overnight bacterial culture.

3.1.3. Generation of Bar Code Microarrays (See Note 10)

1. 60-mer bar code sequences are obtained from sequence analysis in Subheading 3.1.2, step 8, commercially synthesized in 96-well plates and diluted to 100 μM in H_2O.
2. Mix 25 μl of 100 μM bar code oligonucleotides with 75 μl of 4× SSC to make an oligo solution at a final concentration of 25 μM and 3× SSC.
3. Transfer 40 μl of oligos from 96-well plates to 384-well plates using the Multimex liquid handler and store oligos at −20°C.

4. Generate bar code microarrays by printing oligos from 384-well plates onto UltraGAPS slides using the OmniGrid 100 microarrayer according to manufacturer's instructions.
5. Store printed slides in a dry storage and use within 6 months.

3.2. Generation of Ecotropic Receptor- and Tetracycline Repressor-Expressing Cell Lines

1. Use approximately 2 ml of 0.45 μm filtered viral supernatant from a confluent monolayer of FLYRD18/mCAT-IRES-Bleo cells to transduce 1×10^6 lymphoma cells in the presence of 4 μg/ml polybrene in 6-well plates by centrifugation using a Sorvall RT7 Plus centrifuge with microplate carriers (1,260 × *g*, 1.5 h, room temperature). Repeated spin infection on the following day may improve transduction efficiency. Transduced cells are selected by bleomycin (Zeocin) 2 days post spin infection. Appropriate bleomycin concentration should be determined for specific cell lines. For human lymphoma lines, we typically use at 25–250 μg/ml.
2. To establish tetracycline-inducible cell lines, ecotropic receptor-expressing cells are subsequently transduced by TR-expressing retroviruses, using a viral packaging and transduction method as detailed in Subheading 3.3.1. Single-cell clones by limiting dilution are isolated and screened for those that have tight TR regulation. We have developed a functional screening assay using a retroviral eGFP reporter vector, pCMV-TO/eGFP, driven by a tetracycline-responsive CMV promoter. Once single-cell clones are expanded, they are transduced with a retrovirus generated from pCMV-TO/eGFP vector. Clones that express eGFP only when doxycycline is added will be selected.

3.3. shRNA Library Screen for Growth Phenotype (Fig. 2)

3.3.1. Packaging and Transduction of Retroviral Library

Depending on the organism and cell type used for screening experiments, several transfection methods and packaging cell systems are available. Experimentations with each method are recommended to achieve optimal transduction efficiency in target cells. The following protocol describes an approach using Lipofectamine 2000 transfection reagent and HEK 293 T packaging cell line with some modifications.

1. Plate HEK 293T cells in 10 ml of DMEM–10% FBS medium in 10-cm tissue culture plates so that the cells reach 60–70% confluence on the following transfection day. On transfection day, replace 5 ml of existing medium with 5 ml of fresh DMEM–10% FBS medium and grow cells for 3 h. Immediately before transfection, carefully wash without dislodging cells with 10 ml of OptiMemI medium once and place the cells back into a CO_2 incubator in 5 ml of OptiMemI medium.
2. Combine ten pools of shRNA constructs by mixing 5 μg of DNA from each pool of 96 shRNA constructs (see Note 11). Mix this 50 μg of library DNA with 25 μg of pHIT/EA6x3* and 25 μg

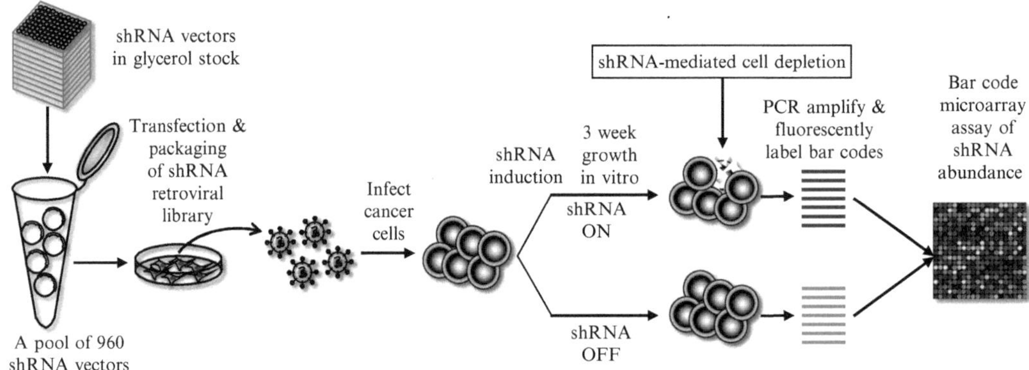

Fig. 2. Pooled shRNA library screen for genes essential in cell growth and proliferation. Retroviral library is generated from batches of 960 shRNA constructs and transduced into human lymphoma cell lines. After puromycin selection for transduced cells, shRNA expression is induced in one population by doxycycline and kept uninduced in a parallel reference population. Cells are grown in culture for 3 weeks at which genomic DNA is collected and bar code sequences are recovered by PCR. Fluorescently labeled bar codes from each population are combined and hybridized to a microarray containing complementary bar code sequences. The screen is repeated four times to enable statistical analysis of the results.

of pHIT60 in 1.5 ml of OptiMemI medium. Incubate the mixture at room temperature for 5 min.

3. Mix 30 μl of Lipofectamine 2000 in 1.5 ml of OptiMemI medium (see Note 12). Incubate the mixture at room temperature for 5 min.
4. Combine diluted DNA and Lipofectamine 2000 (total 3 ml), mix by briefly vortexing and centrifugation, and incubate the mixture at room temperature for 30 min.
5. Carefully add 3 ml of DNA/Lipofectamine 2000 mixture to HEK 293T cells prepared in step 1 without causing cell dislodgement by slowly pipeting the mixture to the wall of the plate. Similarly add 0.5 ml of FBS to the culture. Mix by gently swirling the culture plate, place cells back to a CO_2 incubator, and incubate for 48–72 h.
6. Retroviral supernatants from 48 h post transfection are filtered through a 0.45 μm filter and used to transduce target cells by a centrifugation method in the presence of 4 μg/ml polybrene. We use 2 ml of viral supernatant and transduce 1.5×10^6 target cells in each well of a 6-well plate. Spin infection is carried out at $1,200 \times g$ for 1.5 h at room temperature using a Sorvall RT7 Plus centrifuge. After centrifugation, add 4–6 ml of fresh growth medium into each well. We routinely repeat the spin infection for the same target cells on the following day using additional retroviral supernatants from 72 h post transfection to achieve higher transduction efficiency. The total amount of viral supernatant produced in one 10-cm plate is sufficient for four independent spin infections for one target cell line.

3.3.2. Selection of Transduced Cells and shRNA Induction

1. Transduced cells are selected by puromycin starting 48 h after the second spin infection and completing over a course of 3–6 days. Optimal puromycin concentrations are predetermined for each cell line. We use 1–2 μg/ml of puromycin for selection of transduced DLBCL cell lines.
2. Once puromycin selection is completed, cells are pelleted by centrifugation at 290×*g* for 5 min and resuspended in fresh growth medium. A minimum of 1×10^6 cells are seeded into each of duplicate wells in a 6-well plate. shRNA expression is induced by adding 20 ng/ml of doxycycline into one well and keeping the other well as an uninduced control. Samples from this "day 0" time of shRNA induction should contain most of shRNA constructs represented in the library and could be used as a reference sample. It is therefore recommended that genomic DNA be harvested from this time point, particularly for those cell lines that have "leaky" TR regulation (see Subheading 3.4.1 for genomic DNA preparation).
3. Uninduced and doxycycline-induced cultures are maintained over a period of 21 days by splitting the cultures at 1:1 ratio (that is, removing 3 ml of culture and adding back 3 ml of fresh medium) and replenishing the doxycycline at a relatively constant concentration of 20 ng/ml every 2 or 3 days as doxycycline half-life is about 48 h (see Note 13).

3.4. Detection of Candidate shRNAs

3.4.1. Recovery of shRNA-Associated Bar Code Sequences

1. The bar code sequences are recovered from genomic DNA harvested from cell populations. Typically, pellets of $2–4\times10^6$ cells (approximately 2–4 ml of culture) from each culture are collected and subject to genomic DNA extraction using the QIAamp DNA mini kit with the following modifications. We include 1 μl of 4 mg/ml RNAse to the sample lysis buffer and incubate the samples at 37°C for 30 min. The additional RNA digestion ensures that genomic DNA is free of RNA. Immediately after RNA digest, proteinase K solution (provided from the kit) is added and samples are incubated at 56°C for 1 h. Longer proteinase K incubation would help remove trace of RNAse and minimize degradation of subsequent RNA probes. The last modification includes the use of Tris–HCl buffer (10 mM, pH 7.5) to elute genomic DNA.
2. After determining concentrations of genomic DNA by a UV spectrophotometer, set up PCR reactions by preparing a PCR mixture of 2 μl (~200 ng) of gDNA templates, 0.8 μl of primer 1 (Table 3), 0.8 μl of primer 2 (Table 3), 1 μl of dNTPs, 4 μl of 10× Taq DNA polymerase buffer, 0.5 μl of Taq DNA polymerase, and 31 μl of H_2O. Run PCR program: 95°C for 2 min, 35 cycles of (95°C for 45 s, 60°C for 45 s, 72°C for 45 s), 72°C for 5 min, 4°C, and pause. After confirming a product band of 566 b.p. in 1% agarose gel containing ethidium bromide

by visualizing under UV light, PCR products are cleaned by spinning through Sephadex G-50 columns at 800 × *g* for 2 min (see Note 14).

3.4.2. Generation of Fluorescently Labeled RNA Probes

Preparation of fluorescently labeled RNA probes is carried out by in vitro transcription of PCR products generated in Subheading 3.4.1, step 2, using the Ambion's MAXIscript T7 Kit and labeling of RNA products using the Kreatech's Universal Linkage System. It is recommended that the bar code microarrays generated in Subheading 3.1.3 be prepared and ready as described in Subheading 3.4.3, step 1, *before* labeled RNA probes are produced.

1. Following instructions from the Ambion's MAXIscript T7 Kit, prepare a mixture of 10 μl of PCR products from Subheading 3.4.1, step 2, 4 μl of rNTPs, 2 μl of 10× buffer, 2 μl of T7 polymerase, and 2 μl of H_2O. Incubate the mixture at 37°C for 2 h.
2. Add 1 μl of DNAse and incubate at 37°C for 15 min. Add 1 μl of 0.5 M EDTA to stop the reaction.
3. RNA products are confirmed by running 1 μl in 1% agarose gel containing ethidium bromide and visualized under UV light. RNAs are cleaned by spinning through Sephadex G-50 columns at 800 × *g* for 2 min and concentrations are determined by a UV spectrophotometer.
4. Label 1–2 μg of RNA from uninduced or induced samples with ULS Cy3 or Cy5 fluorescent dyes according to the supplier's instructions.
5. Combine Cy3- and Cy5-labeled samples and remove unincorporated dyes by spinning the samples through Sephadex G-50 columns at 800 × *g* for 2 min.
6. Load flow-through from the Sephadex G-50 columns into YM30 columns. Wash the labeled probes in YM30 columns by adding 0.5 ml of RNAse-free water and spinning at 12,900 × *g* for approximately 8 min in a microfuge. Repeat the wash once. Concentrate the probes by carefully monitoring the centrifugation time without drying out the samples. Typically, spinning the samples in YM30 columns in a microfuge at 12,900 × *g* for an additional 2 min after the second wash is sufficient to concentrate the probes to a preferred volume of approximately 7–9 μl.
7. Recover the concentrated probes by inverted spin of the YM30 columns into a collection Eppendorf tube.

3.4.3. Microarray Hybridization

1. Bar code microarrays generated in Subheading 3.1.3 are UV cross-linked at 600 mJ and pre-hybridized in a prehybridization solution at 50°C for 30 min. The arrays are washed several times in double-distilled water at room temperature to remove traces of SDS and rinsed once in isopropanol. Immediately dry

the arrays by placing them in a slide staining rack and spinning at 61 ×*g* for 2 min in a Sorvall RT7 Plus centrifuge.

2. Two micrograms of each labeled probe from induced and uninduced samples were combined and hybridized to DNA microarrays in 40% formamide, 5× SSC, 0.2% SDS, and 0.1 μg/μl COT-1 DNA at 55°C for 16 h.
3. Make a fresh hybridization solution by mixing 3 μl of 1 μg/μl COT-1 DNA, 0.6 μl of 10% SDS, 12 μl of formamide, and 7.5 μl of 20× SSC.
4. Add 23 μl of the hybridization mixture to 7 μl of combined Cy3/Cy5-labeled RNA probes prepared above and denature the probes by heating at 95°C for 5 min. Briefly spin to collect the samples at the bottom of an Eppendorf tube and add the probes onto the arrays. Place the arrays in hybridization chambers. Remember to add water into reservoirs at both ends of the hybridization chamber to keep a humidified condition inside the chamber.
5. Incubate hybridization chambers at 55°C for 16 h.

3.4.4. Post-hybridization and Analysis

1. After overnight hybridization, arrays are washed twice in Wash buffer 1 at 68–70°C for 5 min each, twice in Wash buffer 2 at 48–50°C for 5 min each, and once in Wash buffer 3 at room temperature for 1 min. Arrays may be rinsed with double-distilled water several times at room temperature to remove traces of SDS. Spin to dry arrays at 61 ×*g* for 2 min.
2. Quickly scan the arrays using the GenePix 4000 scanner. Fluorescent images are analyzed by GenePix software. Each bar code experiment is performed in quadruplicate, and the microarray results for each bar code are averaged (see Note 15).

4. Notes

1. The vector pRSMX is derived from the self-inactivating retroviral backbone pSUPER (16), which is modified to become doxycycline inducible using a strategy previously described by van de Wetering et al. (17). This vector contains a puromycin selectable marker for selection of infected cells (Fig. 3).
2. Sequence templates for 62-mer forward and reverse shRNA oligonucleotides are shown in Fig. 3. Annealed forward and reverse oligos yield shRNA inserts with 5′ overhang ends compatible for ligation into the pRSMX vector at BglII and HindIII cloning sites. shRNA sequences are selected by following published guidelines (18), using an in-house generated bioinformatics software. In brief, selected sequences satisfy many of the following criteria: (1) A/T content is more at the 3′ end than

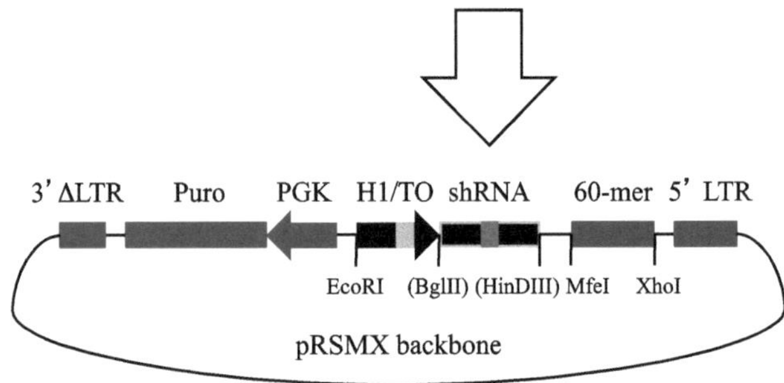

Fig. 3. shRNA vector construction. Shown are templates for 21-mer shRNA forward and reverse oligonucleotides in an annealed configuration with 5′ overhang ends compatible with BglII and HindIII restriction sites (*top*). The pRSMX backbone is a self-inactivating retroviral construct due to a deletion in the 3′ LTR (*bottom*). The chimeric H1 promoter contains a tet operator element (TO) enabling the H1 promoter regulatable by binding of a tetracycline repressor protein. Bar code sequences (60-mer) are cloned into the vector at Mfe1 and Xho1 sites. BglII and HindIII sites are used to clone shRNA inserts and will be destroyed after ligation. The vector contains a puromycin selectable marker for selection of transduced cells.

that at the 5′ end of the sense strand; (2) overall G/C content is low, ranging from 30 to 52%; (3) there is no long stretch of over 6 Gs or Cs or over 3 As or Ts; (4) there is no overlap with the cloning sites EcoRI and Xho1; and (5) there is no overlap of 15 or greater nucleotides with other genes in the genome. Forward and reverse oligos are commercially synthesized and normalized at 100 μM in H_2O.

3. Bar code sequences are obtained from sequence analysis in Subheading 3.1.2, step 8, commercially synthesized, normalized at 100 μM in H_2O, and delivered in 96-well plates.

4. To exploit high transduction efficiency of ecotropic retroviruses, human lymphoma cell lines are engineered to express a murine ecotropic receptor. We use a previously established human fibrosarcoma cell line FLYRD18/mCAT-IRES-Bleo that constitutively produces feline endogenous viruses in the culture supernatant. These viruses carry a murine ecotropic receptor (mCAT)-expressing cassette and can transduce most human hematopoietic cells (19). Transduced cells that express the mCAT are selected by the antibiotic selectable marker bleomycin. FLYRD18/mCAT-IRES-Bleo cells are grown in DMEM medium supplemented with 10% FBS.

5. Tetracycline repressor (TR)-expressing vector is derived from the retroviral backbone pBMN-IRES (kindly provided by Gary Nolan). TR was amplified by PCR from the pcDNA6/TR plasmid template (Life Technologies) and cloned into the pBMN-IRES vector before the IRES sequence. A drug selectable marker for hygromycin or neomycin is cloned downstream of the IRES sequence, enabling selection of TR-expressing clones.

6. The pHIT/EA6x3* helper plasmid expresses the EA6 chimeric ecotropic envelope protein that harbors three mutations: G541R, N261I, and E311V. The G541R mutation does not decrease viral titer, but minimizes cell–cell fusion (as compared to that of wild-type envelope), thereby reducing syncytium-induced cytopathogenicity during retroviral transduction (20).
7. The bar code approach helps expedite identification of shRNAs from complex library pools. Several groups have applied different barcoding strategies, including the use of the hairpin, half-hairpin, and a separate bar code sequence (21–23). We have compared these different bar code approaches and found that external 60-mer bar code sequences yield stronger microarray signals (data not shown), in agreement with earlier report by Paddison et al. (13).
8. The amounts of insert and vector corresponding to the molar ratio of 3:1 are calculated using the following formula:

$$\text{ng of Insert} = \frac{(\text{ng of Vector} \times \text{Insert size} \times \text{Insert} : \text{Vector molar ratio})}{\text{Vector size}}.$$

$$\text{Insert size} = 0.062\ \text{kb}; \text{Insert} : \text{Vector molar ratio} = 3;$$
$$\text{Vector size} = 6.3\ \text{kb}.$$

9. Commercially synthesized DNA oligonucleotides contain errors such as point mutations and sequence deletion that compromise shRNA effectiveness. Sequence verification of all individual constructs in the library is therefore necessary. We develop an LIMS using in-house bioinformatics capability to manage this labor-intensive and time-consuming procedure. LIMS is a software suit consisting of many computer programs, each of which handles a specific task such as selecting and formatting shRNA sequences ready for synthesis, verifying shRNA sequences from bacterial clones, identifying unique 60-mer bar code sequences, and maintaining positional accuracy of all constructs throughout the cloning procedure.
10. Production of bar code microarrays is a time-consuming, labor-intensive process that requires customized automation equipment. Recent improvements in in situ oligonucleotide microarray synthesis technology have provided an attractive commercial alternative to the current bar code microarray self-production. Monitoring of shRNA abundance in a pooled library screen can also be accomplished by directly sequencing the hairpin using high-throughput sequencing technology, as demonstrated by Bassik et al. (24).
11. Optimal number of shRNA constructs that can be efficiently transduced into target cells and produce detectable microarray signals must be determined. We have performed optimization

screens comparing microarray signal intensities from test pools of approximately 500, 1,000, and 2,000 shRNA constructs. Strong, reproducible signals are readily detectable for pools of 500 and 1,000 whereas signals from pools of 2,000 constructs are significantly decreased (data not shown). Pools of about 1,000 constructs are therefore selected to minimize screening samples in subsequent experiments.

12. The minimal volume of Lipofectamine 2000 required for the indicated amount of library DNA is empirically determined without adversely affecting the viability of the packaging cell line HEK 293T and subsequent retroviral titers.
13. For growth phenotype screening, we collected samples at days 7, 14, and 21 after doxycycline induction. We discovered that many shRNAs that are toxic to cells at day 21 are unable to cause toxicity at earlier time points. We therefore select day 21 as the experiment endpoint to capture shRNAs that are more slowly toxic to cells.
14. PCR product size of 566 b.p. is chosen as microarray probes from earlier optimization experiments because they produce higher microarray hybridization signals than other tested shorter fragments at 427 or 243 b.p. (data not shown).
15. To enable statistical analysis, we repeat RNAi library transductions four times and perform paired Student's *t*-tests to calculate a two-tailed *P* value. shRNA hits are validated by cloning individual shRNAs and confirming their toxicity in lymphoma lines. Knockdown specificity is assessed by identifying at least two independent shRNA sequences that target the same gene and yield phenotypic similarities. To rule out off-target effects that often plague RNAi screens, more rigorous confirmation is needed, including performing complementary experiments in which exogenous cDNA can rescue the toxicity of an shRNA.

Acknowledgments

I would like to thank Xin Yu, Hong Zhao, Laurence Lamy, Haihua Chu, Jenny Zhang, Cailin Collins, Weihong Xu, and Yandan Yang for the construction of the library. I thank Liming Yang, Wenming Xiao, John Powell, and George Wright for bioinformatics and statistical supports and R. Eric Davis for the FLYRD18/mCAT-IRES-Bleo cell line and packaging vectors. I am grateful to the generous support and helpful advice from Louis M. Staudt. This work was supported in part by the Damon Runyon Cancer Research Foundation to V.N.N. and the Intramural Research Program of the National Institute of Health, National Cancer Institute, to L.M.S.

References

1. Fire A, Xu S, Montgomery MK et al (1998) Potent and specific genetic interference by double-stranded RNA in Caenorhabditis elegans. Nature 391:806–811
2. Hannon GJ (2002) RNA interference. Nature 418:244–251
3. Kennerdell JR, Carthew RW (1998) Use of dsRNA-mediated genetic interference to demonstrate that frizzled and frizzled 2 act in the wingless pathway. Cell 95:1017–1026
4. Ngo H, Tschudi C, Gull K et al (1998) Double-stranded RNA induces mRNA degradation in Trypanosoma brucei. Proc Natl Acad Sci U S A 95:14687–14692
5. Napoli C, Lemieux C, Jorgensen R (1990) Introduction of a chimeric chalcone synthase gene into petunia results in reversible co-suppression of homologous genes in trans. Plant Cell 2:279–289
6. Manche L, Green SR, Schmedt C et al (1992) Interactions between double-stranded RNA regulators and the protein kinase DAI. Mol Cell Biol 12:5238–5248
7. Elbashir SM, Harborth J, Lendeckel W et al (2001) Duplexes of 21-nucleotide RNAs mediate RNA interference in cultured mammalian cells. Nature 411:494–498
8. Ngo VN, Young RM, Schmitz R et al (2011) Oncogenically active MYD88 mutations in human lymphoma. Nature 470:115–119
9. Bidere N, Ngo VN, Lee J et al (2009) Casein kinase 1alpha governs antigen-receptor-induced NF-kappaB activation and human lymphoma cell survival. Nature 458:92–96
10. Ngo VN, Davis RE, Lamy L et al (2006) A loss-of-function RNA interference screen for molecular targets in cancer. Nature 441: 106–110
11. Mullenders J, Bernards R (2009) Loss-of-function genetic screens as a tool to improve the diagnosis and treatment of cancer. Oncogene 28:4409–4420
12. Downward J (2004) Use of RNA interference libraries to investigate oncogenic signalling in mammalian cells. Oncogene 23:8376–8383
13. Paddison PJ, Silva JM, Conklin DS et al (2004) A resource for large-scale RNA-interference-based screens in mammals. Nature 428: 427–431
14. Brummelkamp TR, Bernards R, Agami R (2002) Stable suppression of tumorigenicity by virus-mediated RNA interference. Cancer Cell 2:243–247
15. Markowitz D, Goff S, Bank A (1988) A safe packaging line for gene transfer: separating viral genes on two different plasmids. J Virol 62: 1120–1124
16. Brummelkamp TR, Bernards R, Agami R (2002) A system for stable expression of short interfering RNAs in mammalian cells. Science 296:550–553
17. van de Wetering M, Oving I, Muncan V et al (2003) Specific inhibition of gene expression using a stably integrated, inducible small-interfering-RNA vector. EMBO Rep 4:609–615
18. Reynolds A, Leake D, Boese Q et al (2004) Rational siRNA design for RNA interference. Nat Biotechnol 22:326–330
19. Cosset FL, Takeuchi Y, Battini JL et al (1995) High-titer packaging cells producing recombinant retroviruses resistant to human serum. J Virol 69:7430–7436
20. O'Reilly L, Roth MJ (2003) G541R within the 4070A TM protein regulates fusion in murine leukemia viruses. J Virol 77:12011–12021
21. Berns K, Hijmans EM, Mullenders J et al (2004) A large-scale RNAi screen in human cells identifies new components of the p53 pathway. Nature 428:431–437
22. Silva JM, Li MZ, Chang K et al (2005) Second-generation shRNA libraries covering the mouse and human genomes. Nat Genet 37: 1281–1288
23. Schlabach MR, Luo J, Solimini NL et al (2008) Cancer proliferation gene discovery through functional genomics. Science 319:620–624
24. Bassik MC, Lebbink RJ, Churchman LS et al (2009) Rapid creation and quantitative monitoring of high coverage shRNA libraries. Nat Methods 6:443–445

Chapter 15

Studying MicroRNAs in Lymphoma

Joost Kluiver, Izabella Slezak-Prochazka, and Anke van den Berg

Abstract

MicroRNAs (miRNAs) play important roles in development, differentiation, homeostasis, and also in diseases such as lymphoma. This chapter describes methods to study the role of miRNAs in lymphoma. First, we describe a multiplex RT reaction followed by qPCR that can be used to determine differential expression of candidate miRNAs. Second, we provide a protocol for stable overexpression of miRNAs using lentiviral-based ectopic expression systems. Third, we describe a straightforward assay to determine whether the candidate miRNA may function as an oncogene or a tumor suppressor gene in lymphomagenesis.

Key words: B-cell lymphoma, microRNA, qRT-PCR, miRNA cloning, growth competition assay, lentivirus

1. Introduction

MicroRNAs (miRNAs) belong to a large and diverse family of noncoding (nc) RNAs. Mature miRNAs are formed from longer primary transcripts called pri-miRNAs via two processing steps. The first step, which occurs in the nucleus, is carried out by the Drosha–DGCR8 complex and results in the formation of a ~65 nt hairpin precursor (pre-) miRNA. This pre-miRNA is transported to the cytoplasm and further processed to a ~22 nt miRNA/miRNA* duplex by a protein complex containing Dicer and TRBP. The more abundant strand of the duplex is defined as the mature miRNA and the less abundant partner is referred to as the miRNA* or star strand. The mature strand is incorporated into the RNA-induced silencing complex (RISC) that consists of an Argonaute protein, DICER, and TRBP (1). The mature miRNA guides the RISC to regions in target gene transcripts with sequence complementarity to the miRNA and this results in inhibition of protein synthesis. Nucleotide 2–8 of the miRNA, the seed sequence, is

Ralf Küppers (ed.), *Lymphoma: Methods and Protocols*, Methods in Molecular Biology, vol. 971, DOI 10.1007/978-1-62703-269-8_15, very important for the recognition of miRNA targets. It is thought that 30–50% of all genes can be regulated by miRNAs indicating the widespread involvement of miRNAs in the regulation of gene expression (2).

MiRNAs have been shown to be involved in almost all known biological processes including cell cycle, apoptosis and other major signaling pathways. Direct links between miRNA levels and certain diseases indicate that these molecules play critical roles in pathogenetic processes (3). The first oncogenic miRNA reported to play a role in lymphoma is miR-155, which is processed from the pri-miR-155/BIC transcript. This miRNA is highly expressed in Hodgkin lymphoma, primary mediastinal B-cell lymphoma and diffuse large B-cell lymphoma (4, 5). Eμ-miR-155 transgenic mice develop polyclonal B-cell proliferations that can progress into B-cell leukemia or lymphoma (6). The miR-17~2 cluster that consists of six miRNAs has also been widely implicated in lymphomagenesis. The region containing this miRNA cluster is often amplified in lymphomas, including diffuse large B-cell lymphoma, mantle cell lymphoma and Burkitt lymphoma (7–9). MiR-17~92 knockout mice have an impaired B-cell development and conversely (10), miR-17~92 transgenic mice develop lymphoproliferative disease (11). In the Eμ-MYC model, miR-17~92 was shown to accelerate B-cell lymphomagenesis with miR-19a and miR-19b being the most oncogenic miRNAs (12–14). MiR-21 is highly expressed in lymphomas and many other types of cancer. Conditional overexpression of miR-21 in a mouse model resulted in pre-B cell malignancies that were addicted to miR-21 expression (15). Upon downmodulation of miR-21 tumors regressed completely, indicating that miRNAs can function as genuine oncogenes.

MiRNAs can also have tumor suppressor activity. The miR-15a/16-1 containing chromosomal region is frequently deleted in chronic lymphocytic leukemia (CLL) (16) and mice with a deletion of the corresponding chromosomal region developed CLL-like disease (17). Interestingly, miR-155 may also function as a tumor suppressor by regulating the expression of AID and thereby controlling the formation of Myc-Igh translocations in activated B cells (18). In line with this finding, miR-155 is expressed at very low levels in Burkitt lymphomas which are characterized by Myc translocations (19). MiR-150 is downregulated in Hodgkin lymphoma, NK/T cell lymphoma, and mantle cell lymphoma (20, 21). MiR-150 transgenic mice have a block in early B cell development (22). Overexpression of miR-150 in NK/T cell lines resulted in reduced proliferation and increased apoptosis (23).

A large variety of techniques can be used to generate genome-wide miRNA expression profiles. The most commonly used methods are microarray, quantitative reverse transcription PCR (qRT-PCR), and next generation sequencing (24–27). MiRNAs that are differentially expressed in a certain type of lymphoma in comparison to their normal counterparts or other subtypes of lymphoma may be selected for further study. The role of the candidate miRNAs in specific cells can be determined by gain-of-function or loss-of-function studies. Increase in miRNA levels can be achieved using miRNA mimics (short-term) and miRNA expression vectors (long-term). For miRNA loss-of-function studies, miRNA antisense oligos (short-term) or so-called miRNA sponges (long-term) can be used (28). To determine the mechanism by which a specific miRNA exerts its function, the target genes that are regulated by the miRNA need to be identified. To this end, several methods to identify miRNA targets have been developed, e.g., RIP-CHIP, HITS-CLIP, SILAC, and PAR-CLIP (29–32). Further validation of identified targets is often performed by luciferase-based assays and/or western blotting.

In this chapter, we discuss how to follow up from miRNA expression profiles to functional studies. First, we describe how selected candidate miRNAs can be validated by a multiplex cDNA synthesis qPCR. Secondly, we show a miRNA cloning strategy that is useful for miRNA gain-of-function studies. Lastly, we describe a straightforward assay that can be used to functionally screen candidate miRNAs for oncogenic or tumor suppressor activity in lymphoma cells.

2. Materials

2.1. Quantification of miRNA Levels by qRT-PCR

2.1.1. Multiplex miRNA Reverse Transcription Reaction

1. Trizol (Invitrogen, Carlsbad, CA).
2. TaqMan MicroRNA Reverse Transcription (RT) Kit (Applied Biosystems, Carlsbad, CA, USA): 100 nM dNTPs (with dTTP), 50 U/μl MultiScribe Reverse Transcriptase, 10× Reverse Transcription buffer, 20 U/μl RNase inhibitor, nuclease-free water.
3. 5× miRNA-specific RT primers, purchased as a kit with qPCR primers and probe (Applied Biosystems).
4. 5× mature miRNA control RT primers, for RNU24 Assay (Assay ID: 001001), RNU44 (Assay ID: 001094), RNU48 (Assay ID: 001006), RNU49 (Assay ID: 001005) (all Applied Biosystems).
5. Thermocycler.

2.1.2. qPCR

1. 2× qPCR MasterMix Plus (RT-QP2X-03-50+, Eurogentec, Liege, Belgium): dNTP/dUTP, HotGoldStar DNA polymerase, Uracil-*N*-glycosylase, $MgCl_2$ (5 mM final concentration), stabilizers and ROX passive reference.
2. 20× miRNA-specific TaqMan qPCR miRNAs primers purchased as a set with RT-primers (Applied Biosystems).
3. 20× mature miRNA control TaqMan qPCR primers and probe, purchased as a kit with RT primers (Applied Biosystems).

4. Nuclease-free water.
5. MicroAmp® Optical 384-well reaction plate with barcode (Applied Biosystems).
6. Real-time PCR thermocycler ABI7900HT (Applied Biosystems).
7. Sequence detection system software (SDS, version 2.1; Applied Biosystems).

2.2. miRNA Overexpression

2.2.1. miRNA Cloning

1. DNA clean and concentrator-5 PCR purification kit (Zymo Research, Irvine, CA, USA).
2. Zymoclean gel DNA recovery kit (Zymo Research).
3. T4 DNA ligase 5 U/μl (Invitrogen).
4. Top ten chemically competent *E. coli* cells (Invitrogen).
5. Primers (IDT, Leuven, Belgium).
6. Lentiviral expression vector (e.g. ,vectors from SBI, Mountain View, CA, USA).

2.2.2. Generation of Viral Particles and Viral Infection

1. 2× HBS: dissolve 8.2 g NaCl, 5.95 g HEPES acid (tissue culture grade), and 0.105 g Na_2HPO_4 (anhydrous) in approx. 400 ml of water, adjust pH to 7.06 with 10 N NaOH and add water to 500 ml. Filter-sterilize with a 0.20 μM filter.
2. 2.5 M $CaCl_2$: dissolve 36,75 g $CaCl_2{\cdot}2H_2O$ in approx. 80 ml water and add water to 100 ml. Filter-sterilize with a 0.20 μM filter.
3. 4 μg/μl Polybrene (Sigma, Seelze, Germany, Hexadimethrine bromide): dissolve 0.4 g polybrene in approx. 80 ml water and add water to 100 ml. Filter-sterilize with an 0.2 μM filter.
4. 293T cells (DSMZ, Braunschweig, Germany ACC-635).
5. 293T culture medium: DMEM supplemented with 2 mM ultra-glutamine, 100 U/ml penicillin/streptomycin, and 10% fetal bovine serum (Cambrex Biosciences, Walkersville, USA).
6. 0.45 μM PVDF Luer-Lock filter (Millipore, Amsterdam, the Netherlands, #SLHV013SL).
7. 5 ml Luer-Lock syringes (BD Biosciences, Erembodegem-Aalst, Belgium).
8. 0.2 μM Luer-Lock filter (Corning, Amsterdam, the Netherlands).

2.3. Growth Competition Assay

1. Lymphoma cell lines.
2. Retro- or lentiviral particles.
3. Flowcytometer with appropriate filter settings to measure GFP fluorescence.

3. Methods

3.1. Quantification of miRNA Levels by qRT-PCR

3.1.1. Multiplex miRNA Reverse Transcription Reaction

1. Isolate total RNA using Trizol according to the manufacturer's protocol (see Note 1).
2. Prepare a master mix containing 0.15 μl of dNTPs (100 nM), 1 μl of 50 U/μl MultiScribe reverse transcriptase, 1.5 μl of 10× reverse transcription buffer, 0.19 μl of 20 U/μl RNase inhibitor, 1 μl of 1–6 different 5× miRNA-specific RT primers, and 1 μl of 5× miRNA endogenous control RT primer (see Note 2) and nuclease-free water to a total volume of 10 μl per reaction. Before performing an experiment, determine which small ncRNA is most suitable as endogenous housekeeping control (Fig. 1, see Note 3).

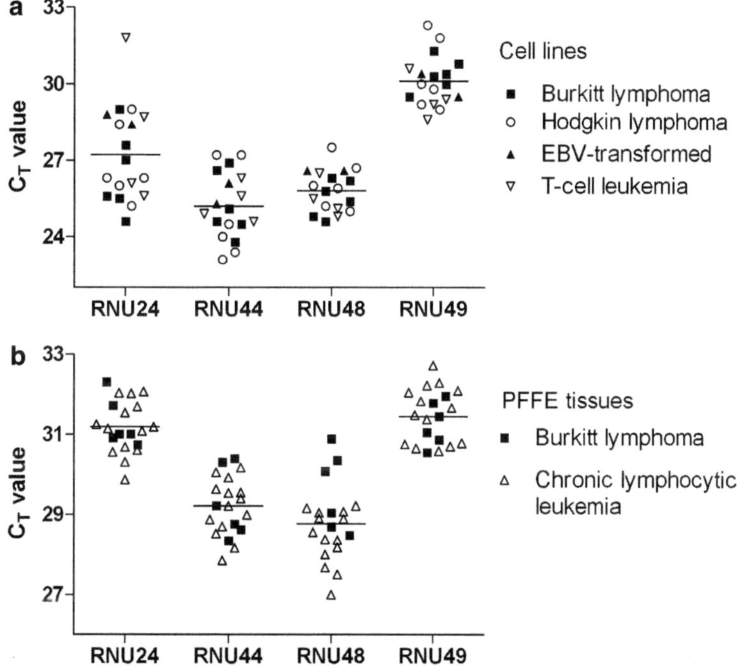

Fig. 1. Comparison of endogenous control genes for miRNA normalization. (**a**) Levels of four endogenous miRNA controls (RNU24, RNU44, RNU48, and RNU49) were compared in Burkitt lymphoma (Ramos, ST486, CA46, BJAB, DG75, BL65), Hodgkin lymphoma (DEV, KMH2, L540, SUPHD1, L1236, L428), EBV-transformed (1H2, 15E6), and T-cell leukemia cell lines (Molt4, Molt16, ATL16T, Jurkat). Separate RT reactions were performed for each snoRNA and cDNA was diluted 4× for the qPCR. (**b**) Levels of the same endogenous miRNA controls as in (**a**) in 6 Burkitt lymphoma and 14 chronic lymphocytic leukemia cases. RNA was isolated from paraffin-embedded tissues, multiplex RT reaction was performed with pooled primers for all snoRNAs and cDNA was diluted 20× for the subsequent qPCR.

3. Add 5 µl total RNA (1–10ng) to 10 µl of master mix, mix gently by pipetting.
4. Incubate the tubes on ice for 10 min and keep on ice until needed.
5. Run the RT reactions in a thermocycler. The RT reaction program is as follows: 16°C for 30 min, 42°C for 30 min, 85°C for 5 min, 4°C.
6. Dilute cDNA at least 4× (see Note 4).

3.1.2. qPCR

1. Prepare qPCR master mix containing 5 µl of 2× qPCR MasterMix Plus, 0.5 µl of 20× TaqMan microRNA Assay, 2 µl of nuclease-free water per one reaction.
2. Add 7.5 µl of master mix to 2.5 µl of diluted cDNA (see Note 5). Run reactions in triplicate on a 384-well plate.
3. The qPCR program is as follows: 50°C for 2 min, 95°C for 10 min (AmpliTaq Gold enzyme activation), followed by 40 cycles of 15 s at 95°C and 1 min at 60°C.
4. Mean cycle threshold (C_T) values for each miRNA can be quantified with the sequence detection system software (SDS, version 2.1; Applied Biosystems).
5. To correct for potential differences in cDNA input, the difference between the C_T value of each miRNA and endogenous control (ΔC_T) is calculated. The relative level of miRNA is calculated as $2^{-\Delta C_T}$.

3.2. miRNA Overexpression

3.2.1. miRNA Cloning

1. Design primers on the genomic sequence around the pre-miRNA hairpin including approximately 100–250 nt flanking sequence on each site of the hairpin. Location of the pre-miRNA on the genome can be found using the miRbase records (http://www.mirbase.org/). Add restriction enzyme sites to the 5′ ends of the forward and reverse primers to enable directional cloning to a (viral) expression vector of choice. Ensure that the restriction sites that will be used for cloning are not present in the amplicon. Add four random nucleotides at the 5′ end of the forward and reverse primer to allow efficient digestion of the amplified DNA by the restriction enzymes (see example below).

	Random	restriction site	miRNA specific
Forward (*Xho*I):	5′ ***nnnn***	-CTCGAG	-nnnnnnnnnnnnnnnnnnn 3′
Reverse (*Eco*RI):	5′ ***nnnn***-	GAATTC	-nnnnnnnnnnnnnnnnnnn 3′

2. Perform a standard PCR with genomic DNA of a healthy control as a template to amplify the miRNA gene of interest using the following PCR mix: 3 µl of forward and 3 µl of reverse

primer (each 10 μM), 3 μl dNTP (2 mM), 3 μl 10× PCR buffer, 0.2 μl Taq DNA polymerase (5 U/μl), 100 ng genomic DNA and add water to a total volume of 30 μl.

3. Run the following PCR program: 5 min 94°C, 30 cycles of 30 s at 94°C, 45 s at 55°C, and 45 s at 72°C, and as a final step 7 min at 72°C. The annealing temperature should be adjusted to the melting temperature of the primers (melting temperature calculated based on the miRNA gene-specific sequence part of the primers). Generally, an annealing temperature that is 5°C lower than the melting temperature of the primers is optimal for PCR.
4. Visualize a small portion of the PCR product by agarose gel electrophoresis to confirm the size of the amplicon.
5. Purify the PCR product using a PCR purification kit.
6. Digest the purified PCR product using the appropriate restriction enzymes. For example prepare a mix of: 10 μl purified PCR product, 6 μl H_2O, 2 μl *Eco*RI buffer, 1 μl 20 U/μl *Xho*I, and 1 μl 20 U/μl *Eco*RI. Incubate for 2 h at 37°C.
7. Run the digest on an agarose gel, cut out the desired product and purify from gel.
8. Digest the expression vector with the same (or compatible) restriction enzymes as the enzymes used to digest the PCR product and purify.
9. Set up a ligation reaction with your predigested viral vector of choice and the purified digested PCR product. Use 50 ng of vector in a 1:3 ratio with PCR product in a reaction volume of 10 μl.
10. Transform 1 μl of the ligation reaction to *E. coli* and analyze clones for inserts the next day. Generally, sequence verification of three clones will suffice to be able to pick a clone with the correct insert.

3.2.2. Generation of Viral Particles and Viral Infection

For the generation of viral particles retroviral and lentiviral systems can be used. Lentiviral systems often give higher infection rates for most lymphoma cell lines making this system more attractive as compared to retroviral systems. To generate lentiviral particles, we use a third generation lentiviral system with 293T cells as packaging cells (33). Below is the description for the generation of lentiviral particles using $CaPO_4$-based transfection in a six-well plate format. It can be easily scaled to for instance 10 cm dish if larger quantities are desired.

1. Plate 7×10^5 293T cells in 3 ml culture medium per well (six-well plate) approximately 16 h before transfection.
2. The next day (day of transfection) cells should be 70–80% confluent.

3. Prepare a FACS tube with 150 μl 2× HBS. In a separate tube prepare the following plasmid mix:

   Optional: 0.5 μg pSuper-DGCR8 (see Note 6).

   – 15 μl 2.5 M $CaCl_2$
   – 1 μg pMD2.G
   – 1 μg pRSV.REV
   – 1 μg pMDL-g/pRRE
   – 2 μg lentiviral vector (third generation compatible)
   – Add H_2O to a total volume of 150 μl
4. Add the plasmid mix to the FACS tube containing 2× HBS dropwise while bubbling air through the 2× HBS solution with a 2 ml pipet (tap big drops on the wall of the tube gently to the bottom).
5. Leave the tube at RT for 2 min with the cap on (the mixture should become slightly cloudy/milky).
6. Mix once by pipetting up and down and add the 300 μl mix dropwise to the cells (still in 3 ml culture medium) and gently rock/tap the plate to spread the precipitate (after a couple of minutes calcium phosphate crystals should be visible under a microscope).
7. Incubate cells with the precipitate overnight.
8. Gently take off the media and replace it with 2 ml of culture medium (or 1 ml to get a higher virus titer).
9. Harvest the supernatant containing the virus 48 h after transfection and filter using a 0.45 μM low protein binding filter. The virus can be used directly or stored at −80°C (freezing will decrease the titer with approximately 50%).
10. Mix 1 ml of target cell suspension with 1 ml of virus supernatant, add 2 μl polybrene (4 μg/ml final concentration), mix and incubate overnight (less virus can be used in case the infection rates are too high).
11. Wash cells once with PBS after 24 h and culture in their normal culture media.
12. Good expression of the miRNA is normally achieved within 2–3 days after infection.
13. Overexpression of the miRNA can be confirmed by sorting cells based on fluorescent markers such as GFP or by selection using for instance puromycin (depending on the vector) and performing qRT-PCR as described above (see Note 7).

3.3. Growth Competition Assay

For the growth competition assay (or GFP competition assay) cells are infected with viral constructs that contain the miRNA and a GFP marker (other fluorescent markers may also be used).

This assay can be used to determine whether the overexpressed miRNA has tumor suppressor or oncogenic activity. The power of the assay is that within a cell culture the growth of infected and noninfected cells are directly compared under the same circumstances (see Note 8).

1. Aim at reaching an infection percentage of 20–50% by varying the amount of virus and cells used for infection. This will allow for the detection of both increases and decreases in the percentage of GFP positive cells.
2. Start the growth competition assay 3 or 4 days after infection by determining the GFP percentage using a flow cytometer.
3. Determine the GFP percentage every 2–3 days for 3 weeks. As a control an empty vector or the miRNA with a scrambled seed sequence can be used.
4. Normalize the GFP percentages to the first day of measurement (day 3 or 4 after infection) to calculate the fold changes in GFP-positive cells over time. An increase in GFP-positive cells indicates that the cells that overexpress the miRNA have gained a growth advantage and that the miRNA thus has oncogenic potential. Vice versa, a drop in GFP-positive cells indicates that the miRNA has tumor suppressor activity as cells grow slower or die upon overexpression of the miRNA.

4. Notes

1. As an alternative, RNA can be isolated with the miRNeasy kit (Qiagen, Venlo, the Netherlands) or other commercially available kits that does not cause depletion of small RNAs. For good-quality RNA isolated from cell lines or fresh-frozen tissues, DNase treatment is not necessary. However, for poor-quality RNA from paraffin-embedded tissues, DNAse treatment is essential. RNA integrity can be monitored using 1% agarose gel or using Experion chip analysis (Bio-Rad, Veenendaal, the Netherlands).
2. For the relative quantification of miRNA levels using endogenous controls, qRT-PCR assays for various small noncoding RNA (ncRNA) species have been developed. These include assays for transfer RNA (tRNA), small nuclear RNA (snRNA), and small nucleolar RNA (snoRNA) (Applied Biosystems). From diverse small RNA-seq and cloning studies it is known that the miRNA sequence can vary, especially at the 3' end. Therefore, it is important to verify that the primers and probe are designed against the most abundant miRNA variant. Ideally, this should be checked using an RNA-seq experiment in the

cell type of interest. Alternatively, the miRbase records (http://www.mirbase.org/) can be used to identify the most abundant miRNA variant as determined in multiple small RNA-seq experiments performed on many different tissue types. The design of qRT-PCR assays for small ncRNAs is similar to that for miRNAs allowing for a multiplex RT reaction with RT primers for small ncRNAs and miRNAs. Pooling RT primers in a multiplex RT reaction has several advantages over separate RT reactions for the miRNA and the control. It increases the robustness of the relative miRNA quantification as potential differences in RNA input may be introduced when separate RT reactions are performed for the control and the miRNA. Moreover, less RNA is needed when only one RT reaction is performed and it is more cost- and time-effective. Using 1 μl instead of 3 μl of each 5× RT primer does not influence the quantification of miRNA or small ncRNA levels. Using our protocol, up to 7 RT reactions can be pooled, including at least one RT-reaction for an endogenous control.

3. Depending on the quality, origin and treatment of the tissues/cells, different small ncRNAs may be optimal to serve as endogenous control. This should be tested for each type of experiment by analyzing the expression of 3–4 different small ncRNAs in a representative set of samples and identifying the small ncRNA that is abundant and most consistent. For instance, we determined that of four snoRNAs that were tested (Fig. 1), RNU48 (also known as SNORD48) is the best control for a comparison of miRNA levels between Burkitt lymphoma, Hodgkin lymphoma, and Epstein-Barr virus-transformed cell lines. In contrast, RNU49 is superior when comparing Burkitt lymphoma and chronic lymphocytic leukemia paraffin-embedded tissues. Moreover, RNU24, RNU44, or RNU48 can serve as a control for anti-IgM or PMA/ionomycin treatment, at least for Ramos and CA46 cells. RNU24 or RNU44 are abundant and invariant upon PBMC stimulation with anti-CD3 and anti-CD28 antibodies.
4. A 4× dilution of cDNA is recommended for single RT reaction. For multiplex cDNA synthesis (seven pooled RT primers), we recommend at least 16× dilution to achieve the same qPCR efficiency as for single RT reaction.
5. According to the manufacturer's protocol, the reaction volume for qPCR in a 384-well plate is 20 μl. However, a total volume of 10 μl also results in robust miRNA detection as we observe similar efficiencies for qRT-PCR performed in 10 and 20 μl.
6. The viral RNA transcripts that are produced in the 293T cells are packaged into viral particles. However, these transcripts contain miRNA hairpins that can be recognized by the miRNA processing machinery of 293T cells. This may result in

inefficient packaging of full-length viral RNA transcripts and thus lower viral titers. We have observed that virus titers can be two- to tenfold increased by using a shRNA against DGCR8 such as the pSuper-DGCR8 (shRNA against DROSHA decreased viral titers). Higher viral titers were observed especially when viral particles were generated with miRNA cluster containing transcripts.

7. The fold overexpression of a candidate miRNA is dependent on several factors. For miRNAs expressed at low endogenous levels >1,000-fold increase can be obtained, whereas for miRNAs with higher endogenous levels substantially lower folds of induction are observed. The level of overexpression also depends on the promoter, popular strong promoters in hematopoietic cells include PGK, MSCV, and EF1-α.

8. MiRNAs may also regulate the expression of secreted factors. These factors may affect the growth of both the infected and noninfected cells. This could, in theory, result in missing effects on cell growth when the GFP competition assay is used. Proliferation assays on sorted GFP positive cells may be used to circumvent these effects.

Acknowledgment

This work was supported by the Dutch Cancer Society (RUG 2009-4279 to A.v.d.B.).

References

1. Chendrimada TP, Gregory RI, Kumaraswamy E et al (2005) TRBP recruits the Dicer complex to Ago2 for microRNA processing and gene silencing. Nature 436:740–744
2. Lewis BP, Burge CB, Bartel DP (2005) Conserved seed pairing, often flanked by adenosines, indicates that thousands of human genes are microRNA targets. Cell 120:15–20
3. Sandhu SK, Croce CM, Garzon R (2011) Micro-RNA expression and function in lymphomas. Adv Hematol 2011:347137
4. Eis PS, Tam W, Sun L et al (2005) Accumulation of miR-155 and BIC RNA in human B cell lymphomas. Proc Natl Acad Sci USA 102:3627–3632
5. Kluiver J, Poppema S, de Jong D et al (2005) BIC and miR-155 are highly expressed in Hodgkin, primary mediastinal and diffuse large B cell lymphomas. J Pathol 207:243–249
6. Costinean S, Zanesi N, Pekarsky Y et al (2006) Pre-B cell proliferation and lymphoblastic leukemia/high-grade lymphoma in E(mu)-miR155 transgenic mice. Proc Natl Acad Sci USA 103:7024–7029
7. Robertus JL, Kluiver J, Weggemans C et al (2010) MiRNA profiling in B non-Hodgkin lymphoma: a MYC-related miRNA profile characterizes Burkitt lymphoma. Br J Haematol 149:896–899
8. Ota A, Tagawa H, Karnan S et al (2004) Identification and characterization of a novel gene, C13orf25, as a target for 13q31-q32 amplification in malignant lymphoma. Cancer Res 64:3087–3095
9. Scholtysik R, Kreuz M, Klapper W et al (2010) Detection of genomic aberrations in molecularly defined Burkitt's lymphoma by array-based, high resolution, single nucleotide

polymorphism analysis. Haematologica 95: 2047–2055

10. Ventura A, Young AG, Winslow MM et al (2008) Targeted deletion reveals essential and overlapping functions of the miR-17 through 92 family of miRNA clusters. Cell 132:875–886
11. Xiao C, Srinivasan L, Calado DP et al (2008) Lymphoproliferative disease and autoimmunity in mice with increased miR-17-92 expression in lymphocytes. Nat Immunol 9:405–414
12. He L, Thomson J.M., Hemann M.T., *et al.* A microRNA polycistron as a potential human oncogene. *Nature* 2005; 435: 828–33.
13. Mu P, Han YC, Betel D et al (2009) Genetic dissection of the miR-17 92 cluster of microRNAs in Myc-induced B-cell lymphomas. Genes Dev 23:2806–2811
14. Olive V, Bennett MJ, Walker JC et al (2009) miR-19 is a key oncogenic component of mir-17-92. Genes Dev 23:2839–2849
15. Medina PP, Nolde M, Slack FJ (2010) OncomiR addiction in an in vivo model of microRNA-21-induced pre-B-cell lymphoma. Nature 467:86–90
16. Calin GA, Dumitru CD, Shimizu M et al (2002) Frequent deletions and down-regulation of micro- RNA genes miR15 and miR16 at 13q14 in chronic lymphocytic leukemia. Proc Natl Acad Sci USA 99:15524–15529
17. Klein U, Lia M, Crespo M et al (2010) The DLEU2/miR-15a/16-1 cluster controls B cell proliferation and its deletion leads to chronic lymphocytic leukemia. Cancer Cell 17:28–40
18. Dorsett Y, McBride KM, Jankovic M et al (2008) MicroRNA-155 suppresses activation-induced cytidine deaminase-mediated Myc-Igh translocation. Immunity 28:630–638
19. Kluiver J, Haralambieva E, de Jong D et al (2006) Lack of BIC and microRNA miR-155 expression in primary cases of Burkitt lymphoma. Genes Chromosomes Cancer 45:147–153
20. Gibcus JH, Tan LP, Harms G et al (2009) Hodgkin lymphoma cell lines are characterized by a specific miRNA expression profile. Neoplasia 11:167–176
21. Zhao JJ, Lin J, Lwin T et al (2010) microRNA expression profile and identification of miR-29 as a prognostic marker and pathogenetic factor by targeting CDK6 in mantle cell lymphoma. Blood 115:2630–2639
22. Zhou B, Wang S, Mayr C, Bartel DP, Lodish HF (2007) miR-150, a microRNA expressed in mature B and T cells, blocks early B cell development when expressed prematurely. Proc Natl Acad Sci USA 104:7080–7085
23. Xiao C, Calado DP, Galler G et al (2007) MiR-150 controls B cell differentiation by targeting the transcription factor c-Myb. Cell 131:146–159
24. Chen C, Ridzon DA, Broomer AJ et al (2005) Real-time quantification of microRNAs by stem-loop RT-PCR. Nucleic Acids Res 33:e179
25. Wang H, Ach RA, Curry B (2007) Direct and sensitive miRNA profiling from low-input total RNA. RNA 13:151–159
26. Landgraf P, Rusu M, Sheridan R et al (2007) A mammalian microRNA expression atlas based on small RNA library sequencing. Cell 129:1401–1414
27. Basso K, Sumazin P, Morozov P et al (2009) Identification of the human mature B cell miR-Nome. Immunity 30:744–752
28. Ebert MS, Neilson JR, Sharp PA (2007) MicroRNA sponges: competitive inhibitors of small RNAs in mammalian cells. Nat Methods 4:721–726
29. Vinther J, Hedegaard MM, Gardner PP, Andersen JS, Arctander P (2006) Identification of miRNA targets with stable isotope labeling by amino acids in cell culture. Nucleic Acids Res 34:e107
30. Tan LP, Seinen E, Duns G et al (2009) A high throughput experimental approach to identify miRNA targets in human cells. Nucleic Acids Res 37:e137
31. Hafner M, Landthaler M, Burger L et al (2010) Transcriptome-wide identification of RNA-binding protein and microRNA target sites by PAR-CLIP. Cell 141:129–141
32. Chi SW, Zang JB, Mele A, Darnell RB (2009) Argonaute HITS-CLIP decodes microRNA-mRNA interaction maps. Nature 460: 479–486
33. Dull T, Zufferey R, Kelly M et al (1998) A third-generation lentivirus vector with a conditional packaging system. J Virol 72:8463–8471

Chapter 16

Molecular Methods of Virus Detection in Lymphoma

Ruth F. Jarrett, Alice Gallagher, and Derek Gatherer

Abstract

The herpesviruses Epstein-Barr virus (EBV) and human herpesvirus 8 and the retrovirus human T-cell leukemia virus type 1 are directly implicated in the pathogenesis of lymphoma and leukemia in man. EBV is associated with an expanding spectrum of lymphomas and it would appear likely that additional, possibly novel, viruses will be implicated in lymphoma pathogenesis in the future. This chapter describes techniques that may be useful in the analysis of viruses and lymphoma including a standard EBV EBER in situ hybridization assay and a degenerate PCR assay for detection of novel herpesviruses. Lastly, a method for analysis of next-generation sequences in the quest for novel viruses is described.

Key words: Virus, Herpes virus, Epstein-Barr virus, EBER, PCR, Virus discovery, Degenerate PCR, Next-generation sequencing, Digital transcriptome subtraction

1. Introduction

Viruses are associated with a significant minority of lymphomas and it is probable that further associations will be uncovered in the future. The herpesvirus Epstein-Barr virus (EBV) was initially isolated from cultures of Burkitt's lymphoma and was the first human tumor virus to be discovered (1). It is now associated with a range of lymphoid and other disorders and, as more laboratories test for the presence of this virus in histopathological sections, it is likely that this spectrum will expand still further (2, 3). The human T-cell leukemia virus type 1 (HTLV-1) was the first human retrovirus to be identified and is associated with adult T-cell leukemia, an aggressive malignancy of mature T-cells (4). HTLV-2 is a related virus that shows a less certain relationship with T-cell malignancy (3, 5). Like EBV, both HTLV-1 and 2 were discovered by culturing tumor cells in vitro and stimulating production of viral particles (4, 5). In 1994, human herpesvirus 8 (HHV-8 or KSHV) was identified in Kaposi's sarcoma biopsies using the molecular

Ralf Küppers (ed.), *Lymphoma: Methods and Protocols*, Methods in Molecular Biology, vol. 971,
DOI 10.1007/978-1-62703-269-8_16, © Springer Science+Business Media, LLC 2013

technique, representational difference analysis (6). This virus was subsequently linked to primary effusion lymphoma and multicentric Castleman's disease (3). The above viruses are all directly associated with leukemia/lymphoma; an increased incidence of lymphoma is also seen in the context of human immunodeficiency virus (HIV) and hepatitis C virus infection but here the mechanism is indirect (3).

EBV has a worldwide distribution and almost all healthy adults are infected with the virus (7). In contrast, HTLV-1 and to a lesser extent HHV-8 have a more restricted distribution and therefore demonstration of infection, using techniques such as serology, is a first step in making the diagnosis of virally associated disease. Following primary infection, EBV persists in the host for life and establishes a reservoir in the memory B-cell compartment. Since most adults, therefore, will have evidence of EBV infection, serology and PCR cannot be used to make a diagnosis of EBV-associated lymphoma; it is essential to use assays which combine EBV detection with cellular localization of the virus. The EBV genome encodes a large number of viral proteins and RNAs but in EBV-associated malignancies, which are associated with latent infection, only a small group of proteins and RNAs is expressed (7). The EBER RNAs, two small noncoding RNAs, are abundantly expressed in all cells latently infected by EBV and EBER in situ hybridization is the method of choice for detection of EBV in biopsy specimens. This method is described below. Immunohistochemistry can also be used to detect EBV latent proteins; however, caution must be exercised since the pattern of expression of EBV proteins varies in different lymphomas. EBV LMP-1 immunohistochemistry is widely used and is useful in the evaluation of Hodgkin lymphoma and post-transplant lymphoproliferative disease but LMP-1 is not expressed in all EBV-associated tumors, notably Burkitt's lymphoma (3).

As mentioned above, it is likely that additional viruses will be associated with lymphoma pathogenesis in the future. In the past two decades molecular methods have been used increasingly in virus discovery projects and the advent of next-generation sequencing (NGS) has revolutionized this research area. Several techniques are available to preferentially expand viral or exogenous sequences from biological specimens and these can be coupled to NGS to identify novel viruses. In the case of lymphoma, it is likely that any virus present in the lymphoma cells will be in a latent state, i.e., very few or no viral gene products will be expressed and virions will not be released from infected cells. Techniques that rely on extraction of nucleic acid from viral particles present in body builds are therefore unlikely to be useful in the analysis of lymphoma although have been used successfully in other situations (8–10). These methods could, however, be applied to the analysis of tissue culture supernatants from short term cultures of lymphomas. The method of choice for virus detection will depend on the particular condition and whether there is a suspicion that a particular virus family is involved.

Degenerate PCR assays, which rely on sequences that are well conserved among members of a virus family or subfamily, are particularly useful. These assays identify nucleotide sequences encoding proteins related to known proteins and have many applications in biology. Our protocol for detection of herpesviruses is described below but this methodology can be applied to other virus families. We have also described a degenerate PCR for detection of polyomaviruses (11); this assay is less sensitive than the herpesvirus assay and does not detect the Merkel cell polyomavirus although does detect KI and WU polyomaviruses (all discovered after the assay was designed) (8, 9, 12). Similar assays have been designed for detection of novel retroviruses but these have the drawback of amplifying endogenous retroviral sequences if applied to genomic DNA.

NGS can be applied to many different nucleic acid samples including: viral preparations amplified using nonspecific methods such as SISPA (sequence-independent single primer amplification) (13, 14); degenerate PCR products; rolling circle amplification products (15); transcriptomes (12); and complete genomes. The depth of coverage required will clearly depend on the complexity of the starting sample and the precise analysis plan will depend on both the starting sample and the index of suspicion that a virus is involved. This is a rapidly evolving field and as sequencing techniques improve and costs decrease it is likely that genomic sequencing will be used increasingly in virus discovery. In this chapter we describe our protocol for analyzing transcriptomes of tumor cells but this method can be adapted to the analysis of other samples.

2. Materials

2.1. EBV EBER In Situ Hybridization

1. Sections of formalin-fixed, paraffin-embedded lymphoma biopsy, 4 μm thickness on coated slides (see Note 1).
2. Positive control: section of EBV-positive lymphoma biopsy (Fig. 1) (see Note 2).
3. Humid chamber: commercially available chambers are available but we use Perspex boxes, manufactured in-house, that hold the slides in a horizontal position above moistened paper wipes.
4. Glass staining jars and slide holders (Fisher Scientific, Loughborough, UK).
5. Incubator or hybridization oven, set at 55°C.
6. Water bath adjusted to 55°C.
7. Light microscope.
8. Dako Pen (Code S2002).
9. Coverslips.

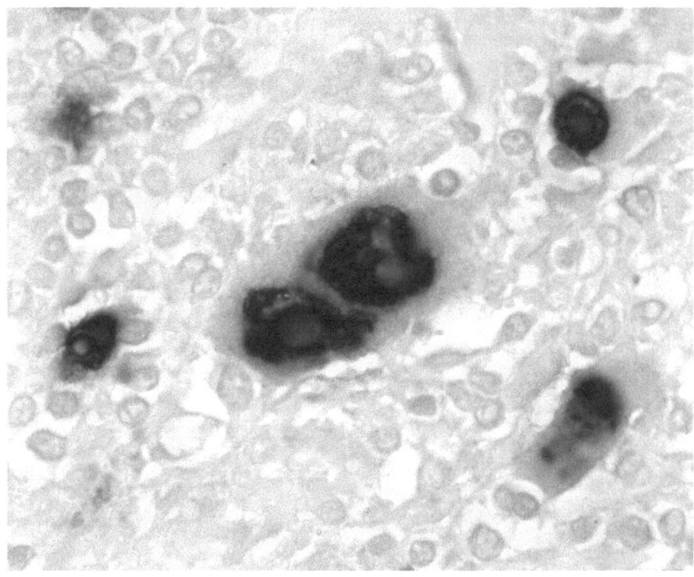

Fig. 1. Typical appearance of positive EBV EBER staining in a case of EBV-associated Hodgkin lymphoma (×1,000).

10. Water: Milli-Q or distilled.
11. Histoclear® (National Diagnostics, Hessle, UK).
12. Ethanol (Fisher Scientific), 99% and diluted to 70% in water.
13. Pronase (Roche Diagnostics, West Sussex, UK) and buffer (see Note 3): prepare a 10 mg/ml stock solution of pronase in water, aliquot, and store at –20°C. Prepare a 1 mg/ml solution of pronase by adding 500 μl of pronase to 4.5 ml of pronase buffer and mix by inversion.
14. Pronase buffer: add 25 ml of 1 M Tris pH 7.5 and 5 ml of 0.5 M EDTA and make up to 500 ml with distilled water.
15. Glycine (Sigma-Aldrich Company Ltd., Dorset, UK) solution: add 37.5 ml of 1 M NaCl, 25 ml of Tris pH7.5, 0.5 g of glycine and make up to 250 ml.
16. Tris buffered saline (TBS): add 6.06 g Tris–HCl, 1.39 g Tris base and 8.40 g NaCl and make up to 1 L with distilled water.
17. Dako Epstein-Barr Virus (EBER) PNA Probe/Fluorescein (Code Y5200).
18. Dako PNA ISH Detection Kit (Code K5201). Kit includes: control probes (see Note 2); TBS; Stringent Wash Solution; alkaline phosphatase-conjugated rabbit anti-FITC F(Ab') fragment; chromogenic substrate with inhibitor of endogenous alkaline phosphatase (BCIP/NBT/levamisole); and proteinase K (see Note 3).

19. Aqueous mountant—Aqua-mount (Fisher Scientific), Faramount (Dako) or Glycergel (Dako).
20. DPX mountant (Sigma-Aldrich Company Ltd.), optional.

2.2. Degenerate PCR for Detection of Herpesviruses

1. High molecular weight DNA from sample of interest extracted using standard procedures, e.g., QIAamp® DNA Blood Mini Kits (Qiagen, Crawley, UK) or, for larger samples, proteinase K digestion and organic solvent extraction. Amplifiability of test DNA samples is checked prior to analysis using non-labelled β-globin primers and conventional gel electrophoresis (see Note 4).
2. NanoDrop® ND-1000 spectrophotometer (Labtech International, East Sussex, UK).
3. Vortex mixer, e.g., Whirlimixer (Fisher Scientific).
4. GeneAmp PCR System 2400 (Applied Biosystems, Warrington, UK).
5. ABI PRISM® 3130xl Genetic Analyzer or equivalent.
6. GeneScan® (Applied Biosystems) or Peak Scanner™ Software v1.0 (available free to download from Applied Biosystems Web site).
7. Plugged pipette tips (Anachem, Bedfordshire, UK).
8. 0.2 ml PCR tubes (Thermo Fisher Scientific, Epsom, UK).
9. Water: distilled and aliquoted.
10. Primers (see Note 5):

    Outer primers:

    5′ Primer 1A 5′ *GACTTTCCAAGTTTC*TAYCCNAGYATHAT 3′

    5′ Primer 1B 5′ *GACTTTCCAAGTTTC*TAYCCNTCNATHAT 3′

    3′ Primer 2 5′ **ACAAACATACAGTCC*GTRTCNCCRTADAT 3′

    Nested primers:

    5′ Primer 3A 5′ #*GTTTGATGCCCGACCT*TAYGGNTTYACNGG 3′

    5′ Primer 3B 5′ #*GTTTGATGCCGACCT*TAYGGNGTNACNGG 3′

    N=A+G+C+T; Y=C+T; H=A+C+T; R=A+G and D=A+G+T. *=FAM label, #=HEX label. Clamp sequences are shown in italics. Dilute primers to 80 μM in water (see Note 6) and store at −20°C.
11. dNTPs (Amersham Pharmacia Biotech, Buckinghamshire, UK): combine and dilute to 2 mM (each nucleotide) in water and store at −20°C.

12. HotStarTaq DNA Polymerase (Qiagen).
13. 10× PCR buffer containing 15 mM $MgCl_2$ (Qiagen) (see Note 7).
14. 5× Q-solution (Qiagen) (see Note 8).
15. Positive controls: The following templates are used as controls for three of the four primer combinations; there is no known herpesvirus control for the fourth primer combination. In order to assess assay sensitivity, we normally use positive controls at two different concentrations.

    Plasmid containing HCMV *pol* gene sequences (500 and 5,000 copies).

    Plasmid containing EBV *pol* gene sequences (500 and 5,000 copies).

    Plasmid containing HHV-6 *pol* gene sequences (500 and 5,000 copies).

    DNA from Raji or Namalwa cell line, EBV positive.

    Plasmid controls are generally spiked into herpesvirus-negative high molecular weight DNA (1 μg per reaction).
16. Negative controls:

    Water (no template control).

    Herpesvirus-negative DNA from same species as test samples (species control).
17. QuickStep™ 2 PCR purification kit (VH Bio Ltd., Gateshead, UK).
18. Hi-Di Formamide (Life Technologies, Paisley, UK).
19. GS 350 and GS 1000 size marker (Applied Biosystems).

2.3. Analysis of Output of Next-Generation Sequencing for the Identification of Novel Viruses

Software

Newbler (Roche).

Tablet 1.11.05.03 (16).

Codonwibbler (D. Gatherer, unpublished, available on request).

ehmmpfam (17) on Pfam (18) as implemented in EMBOSS (19).

3. Methods

3.1. EBV EBER In Situ Hybridization

EBV EBER in situ hybridization is an extremely robust method for detection of EBV latent infection in biopsy specimens. A variety of EBER probes are available including riboprobes, DNA oligonucleotides, and peptide nucleic acids (PNAs). We currently use the

Dako PNA probe but have used oligonucleotide probes successfully in the past (20).

All wash steps are performed by placing slides in a rack and immersing in the appropriate buffer in a glass staining jar. Incubations are performed with the slides in a level, horizontal position in a humid chamber. Wear gloves to avoid RNase contamination and do not allow sections to dry out.

1. Preparation of sections: To dewax sections, immerse slides in Histoclear® for 15 min. To prepare for pronase digestion, immerse slides in 99% ethanol for 2 min and then in 70% ethanol for 2 min. Dry around the section with soft tissue and allow to air dry. Encircle the specimen using a Dako pen. Prepare pronase solution as described above (see Note 3). Add ~100 μl of pronase solution to section and incubate for 5 min at room temperature in a humid chamber. Prepare glycine solution. Immerse the slides in glycine solution for 2 min. Immerse slides in 70% ethanol for 2 min and then in 99% ethanol for 2 min. Air dry the slides.
2. Hybridization. Add 1–2 drops (~35 μl) of PNA-FITC-conjugated probe to the section, ensuring the sample is covered. Apply coverslip of appropriate size, place in a humid chamber and incubate at 55°C for 1.5 h.
3. Stringent wash: Dilute "Stringent Wash Solution" 1:60 and preheat to 55°C. Remove slides from the humid chamber and immerse in TBS, with gentle agitation, until the coverslips come off easily. Do not force the coverslips off as this will damage the sections. Immerse in fresh TBS for 2 min. Immerse slides in warmed Stringent Wash Solution for 25 min with intermittent shaking. Immerse slides in fresh TBS for 5 min. Remove staining rack from TBS and remove slides from staining rack individually. Carefully wipe excess fluid from around the sections.
4. Detection: Apply sufficient antibody conjugate (alkaline phosphatase-conjugated rabbit anti-FITC F(Ab') fragment) to completely cover the section. Incubate for 30 min at room temperature in a humid chamber. Tap off conjugate from each slide and immerse in fresh TBS for 5 min. Remove slides from TBS and carefully wipe excess fluid from around the sections. Cover sections with BCIP/NBT/levamisole substrate and incubate in a humid chamber in the dark, e.g., cover chamber in foil and/or place in drawer, at room temperature for 1 h. Check color development on the positive control section. Tap off substrate from the slides and immerse in distilled water for 5 min.
5. Mounting slides: Add a small drop of aqueous mountant and then slowly lower a coverslip on to the section from an angle of 45°C to avoid bubbles. Leave overnight at room temperature to dry. Alternatively, add aqueous mountant and leave to dry

overnight and then, for longer term storage, dip into Histoclear®, drain off excess moisture, cover with DPX mountant followed by the coverslip and leave to dry on a hotplate at 70°C overnight.

6. Visualization: Visualize under a light microscope. Positive staining is blue/black, nuclear and has a distinctive appearance (Fig. 1). In many situations it is important to determine the cell type that is staining; for instance, in Hodgkin lymphoma it is crucial to determine whether any positive staining is in the Hodgkin and Reed-Sternberg cells.

3.2. Degenerate PCR for Detection of Herpesviruses

Degenerate PCR assays are used to identify new members of a virus family or subfamily. We have successfully used the assay described below to detect known and previously unknown herpesvirus sequences (21). The sequences of the primers are derived from well-conserved sequences in herpesvirus polymerase proteins (21). As more sequences have been deposited in GenBank, it is apparent that the pentapeptide sequence used to derive 3′ Primer 2 (described below) is not perfectly conserved across all mammalian herpesviruses; therefore, we have designed modified versions of this primer (see Note 4). The assay is run in semi-nested format and the products of both first and second round PCRs are analyzed. The forward primer in both PCRs is split into two syntheses in order to limit degeneracy and increase sensitivity; thus, six PCRs are performed and analyzed for each sample. In semi-nested format the assay can detect <100 genomes in a background of 1 μg high molecular weight DNA.

1. To avoid contamination, prepare and store all reagents, including water, in single-use aliquots. Use plugged pipette tips at all stages prior to sample analysis. Set up first and second round PCR reactions and prepare samples for capillary electrophoresis in three separate locations. Do not handle positive controls or second round PCR products in PCR set-up areas. Include frequent "no template controls"—we include a negative control after every two test samples.
2. First round PCR: The first round includes two separate PCRs for each sample—primer combinations 1A plus 2 and 1B plus 2. Prepare mastermix for 14 × 50 μl reactions as follows:

   3.5 μl HotStarTaq DNA polymerase

   70 μl 10× PCR buffer

   70 μl 2 mM mix of nucleotides

   70 μl primers (35 μl of each primer)

   140 μl 5× Q-solution

   206.5 μl sterile water

   Vortex briefly.

3. Aliquot 40 μl of the mastermix into each 0.2 ml PCR tube. Add 1 μg of template in a 10 μl volume.
4. Perform thermal cycling using the following cycling parameters: 95°C for 15 min followed by 5 cycles of 94°C for 60 s; 44°C for 2 min; 72°C for 3 min, followed by 35 cycles as above but with an annealing temperature of 55°C, followed by a final extension at 72°C for 7 min.
5. Second round PCR: The second round includes four PCRs for each sample—primer combinations 3A plus 2 and 3B plus 2 for each of the two first round reactions. Prepare mastermix for 14×50 μl reactions as above but add 332.5 μl sterile water. Vortex briefly.
6. Aliquot 49 μl of the mastermix into each 0.2 ml PCR tube. Add 1 μl of first round product.
7. Perform thermal cycling as described for the first round PCR.
8. Capillary electrophoresis (see Note 9): Purify 30 μl of each reaction (both first and second round products) using the QuickStep™ 2 PCR purification kit, according to the manufacturer's instructions—this step removes dye-labelled primers. Mix 1 μl of purified PCR product with 12 μl of Hi-Di formamide and 0.25 μl of GS 350 or GS 1000 size marker as appropriate. Heat at 95°C for 5 min and cool on ice before subjecting to capillary electrophoresis on an ABI PRISM® 3130xl Genetic Analyzer, or equivalent.
9. Analyze results using GeneScan® or Peak Scanner™ software. It is important to compare results from the species control with test samples in order to correctly identify nonspecific amplification products. Novel herpesviruses are expected to give rise to amplicons of similar size to known viruses, e.g., first round product sizes are: EBV 532 base pairs; human cytomegalovirus 598 base pairs; HHV-8 520 base pairs; and second round products sizes are: EBV 235 base pairs; human cytomegalovirus 310 base pairs; and HHV-8 235 base pairs.
10. Clone and sequence or directly sequence PCR products in the anticipated size range for herpesvirus amplicons. Determine whether sequence is known or novel herpesvirus (or artifact) by comparison with known herpesvirus sequences.

3.3. Analysis of Output of Next-Generation Sequencing for the Identification of Novel Viruses

NGS can be performed using an increasing number of platforms. Important considerations in choice of platform are depth of coverage and length of sequence read; the latter is particularly important in the identification of novel viral sequences. The technology is rapidly evolving and routine read lengths from the Illumina HiSeq, which offers excellent coverage, have increased from 35 bases in summer 2008 to over 100 bases in spring 2012. At the present time, most laboratories send their DNA or cDNA samples of interest to genomic facilities for library preparation and sequencing and so the protocol described below is restricted to analysis of the resulting sequences. However, library preparation techniques can bias the content of the sequence libraries (22), and in-house preparation potentially avoids the problems of cross-contamination from other sequencing projects; there are, therefore, advantages in performing this step at the host laboratory. The precise analysis method will depend on the starting sample (body fluid, transcriptome, genome) and the sequencing platform.

The following is our current protocol for analysis of transcriptomes in a project investigating viral involvement in Hodgkin lymphoma. We have used two sequencing platforms: the Illumina HiSeq, which gives good depth of coverage, and the Roche 454 GS-FLX System (454), which gives greater sequence read length. Illumina HiSeq reads are of equal length and in open text FASTQ format. 454 reads are of variable length and in binary SFF format. Both SFF and FASTQ are processed using the Newbler package v.2.6 (Roche) on a SuSe Linux 11.4 system with 145 Gb of RAM. Read length and quality can be assessed using QUASR (http://sourceforge.net/projects/quasr/). Subheading 3.3.1 describes a search for known viruses. Subheading 3.3.2 describes the search for hitherto unknown viral sequences when no known viruses are identified using the method described in Subheading 3.3.1.

3.3.1. Known Viruses

1. To search for known viruses, the gsMapper application within Newbler is used. Download template sequences from GenBank and align both the FASTQ and SFF reads against them. To make the process more computationally tractable, analyze the following template sets separately for each sample:

   Complete viral genomes, of which there were 3933 in summer 2011.

   The complete GenBank fractions for the following sets of viruses, obtained using the NCBI Taxonomy Browser:

   (a) Double-stranded DNA
   (b) Single-stranded DNA
   (c) Double-stranded RNA
   (d) Single-stranded RNA
   (e) Retroviruses
   (f) Unclassified viruses

2. View output, specified in ACE format, in Tablet (The James Hutton Institute, http://bioinf.scri.ac.uk/tablet/) (16). In addition, from the Tablet window, generate spreadsheets tabulating the sequences to which the reads align and documenting the number of sequences for each. View each hit carefully in Tablet to confirm that the alignment quality is good (see Note 10).

3. Where the quality of alignment is satisfactory, extract the aligned region of the virus from GenBank and search against human genome and human protein sequences (or other host organism if applicable) using BLAST (see Note 11). Hits which pass the visual alignment quality checks and do not demonstrate any similarity to human genome sequences, are candidates for known viruses present within the sample.

3.3.2. Novel Viral Sequences

1. Use the Golden Path (University of California Santa Cruz, http://hgdownload.cse.ucsc.edu/downloads.html#human) (23) version of the human genome sequence as an alignment template in gsMapper, in exactly the same way as described for the viral templates above. The objective of this alignment to the human genome is not to visualize output (which is in any case unwieldy), but to clean out reads matching the human genome.
2. Using ReadStatus.txt file from the mapping directory of the gsMapper output and some basic Unix commands, obtain a list of the status of each read in the alignment. Reads are classified in the following categories:
   (a) Full—these are reads mapped along their entire length to the template
   (b) Partial—reads only mapped over part of their length
   (c) Chimeric—reads with area mapping to noncontiguous parts of the template
   (d) Too Short—self-explanatory, but also includes longer reads with low quality
   (e) Repeat—usually low complexity sequences
   (f) Unmapped—reads not passing any of the criteria for mapping

   Once lists of the reads falling into each category have been obtained, the full sample of reads can be subsetted into its individual categories. For SFF reads use sfffile, a command line tool from the Newbler package. For FASTQ reads a custom Perl script has been developed in-house and is available on request. The Unmapped fraction is of most interest as this represents reads which have passed the quality scoring thresholds but which do not align to the human Golden Path genome.
3. Perform a de novo assembly of the Unmapped reads using the gsAssembler application from the Newbler package. This attempts to extend the reads by detecting overlaps, producing output as FASTA-formatted sequence contigs.
4. View output in ACE format using Tablet. Sort contigs according to size and perform BLAST analysis of the largest contigs by hand to make a quick assessment of the main non-host content

of the read sample (see Note 12). Contigs from viruses related to known viruses and novel viruses will have non-exact BLAST hits to known viruses and no BLAST hits, respectively.

5. Perform automated BLAST searching of the SwissProt protein database (www.uniprot.org/downloads) using the blastall application (24), and assess all contigs down to around 100 base pairs. With SFF format reads, some individual reads are long enough to be considered contigs in their own right. With FASTQ format reads, a read needs to overlap with at least one other read before sufficient length is obtained for BLAST searching to be considered. At this stage, exclude de novo contigs with perfect matches to known viruses or identifiable contaminants.

6. In our experience, the number of contigs with no BLAST hits can be large. Two approaches are used to assist with prioritization of these sequences for further analysis (see Notes 13 and 14).

   (a) Codon bias analysis using codonwibbler (D.Gatherer, unpublished, available on request).

   (b) Hidden Markov Model analysis using ehmmpfam (17) on Pfam (18) as implemented in EMBOSS (19).

   Priority is given to sequences with a strong hit to Pfam using ehmmpfam (thus suggesting a protein-coding function), or sequences with higher levels of codon bias in one of their three reading frames (thus suggesting that they include an open reading frame).

7. If potentially novel viral sequences are identified, laboratory-based validation and replication studies are required to determine whether the sequence/contig is present in the starting sample and other similar samples, respectively.

   It should be emphasized that all the software used in the processes described above operates algorithmically to align, assemble and annotate sequences. It does not give an intelligent assessment of whether or not a novel virus has been discovered. At all stages in the process it is important that the output is examined carefully by the eyes of a trained virologist. Determining whether a sequence is a novel virus or a contaminant or artifact requires a background knowledge of the whole project including the origin of the samples, how they were processed in the laboratory, how the libraries were prepared for sequencing, what other sequencing projects were being carried out, what other organisms are cultured in the laboratory and so on. It also requires an experienced eye for looking at read alignments in Tablet, for looking at BLAST output, and last but not least a capacity to know when one has crossed the boundary into wishful thinking.

4. Notes

1. We use formalin-fixed specimens but other fixatives have been used with success.
2. Controls: An EBV-associated lymphoma specimen is an excellent control since both positive and negative cells will be present in the section. Cells should only be scored as positive if the typical nuclear staining pattern is observed. Nonspecific staining of plasma cells and eosinophils is sometimes present and should not be confused with real staining. We generally hybridize each sample with a negative control probe or omit the probe; however, since EBV-associated lymphoma sections contain both positive and negative cells this negative control is probably unnecessary. Many lymph node sections also contain scattered positive lymphocytes providing an internal positive control. The use of control probes, such as U6 or oligo-dT, to assess the RNA quality in the section has also been advocated (25); however, in our experience, EBER staining can sometimes be observed when the control staining has not worked.
3. The Dako ISH kit includes reagents for a proteinase K digestion step but we, in part for historical reasons, prefer to use a pronase digestion and have therefore described this method here. Other minor modifications from the protocol provided with the kit are described. This method also works with oligonucleotide probes although the hybridization temperature and stringency of the wash steps may have to be altered.
4. This β-globin PCR gives an amplification product of 514 base pairs and is used to check the amplifiability of test samples.

   5′ primer: 5′ AGCCACCTACATTTGCTTCTGACAC 3′

   3′ primer: 5′ CCCAGACTCACCCTGAAGTTCTCA 3′

   For each 25 μl reaction, add 12.5 μl HotStarTaq Mastermix Plus, 1 μM primers and 100 ng DNA and perform thermal cycling using the following parameters: 95°C for 5 min followed by 40 cycles of 94°C for 30 s, 55°C for 1 min and 72°C for 2 min, followed by a final extension step at 72°C for 7 min.
5. Amino acid sequences of polymerase proteins from mammalian and avian herpesviruses were aligned and conserved blocks of sequence identified. Conserved pentapeptide motifs were used to design the degenerate portion of each primer (14 bases) and the 5′ clamp sequences were derived from the consensus nucleotide sequence at the relevant position. Primers 1A and B are based on the conserved amino acid sequence YPSII/M and differ in the codon usage for serine (S). Primers 3A and 3B are based on the sequence YGF/VTG, with 3A

covering the sequence YGFTG and 3B covering YGVTG. Primer 2 is based on the sequence IYGDT; however this sequence is not perfectly conserved with some viruses having the sequence VYGDT. To cover this variation, we have synthesized two variants of primer 2:

3′ primer 2A: 5′ *GATTAATACGGAGTCA*GTRTCNCCRT ANA 3′

3′ primer 2B: 5′ *TTGATTAAGACGGAG*TCNGTRTCNC CRTA 3′

Primer 2A lacks the extreme 3′ nucleotide of the original primer and has N in place of D at the 3′ end. Primer 2B is based on the perfectly conserved YGDTD motif. Primers 2A and 2B have 64-fold degeneracy whereas the original primer 2 has 48-fold degeneracy.

6. Final primer concentrations in the range 4–8 μM give optimal sensitivity but inhibition is apparent at concentrations ≥16 μM.
7. The final concentration of $MgCl_2$ using the buffer supplied by Qiagen is 1.5 mM; however, optimization experiments suggested that a $MgCl_2$ concentration of 2.0 mM was marginally more sensitive.
8. Addition of a co-solvent is necessary for the amplification of some templates. We routinely use Q solution but glycerol (10% final concentration) or 0.5 M betaine may be substituted.
9. PCR products can be analyzed using conventional slab gel electrophoresis; however, capillary electrophoresis gives greater sensitivity and superior size resolution of fragments. The latter enables more accurate discrimination of viral sequences from nonspecific amplification products, which are an inherent feature of these assays and will be present in test samples and the species control. We label the 5′ and 3′ primers in the second round, semi-nested PCR with different dyes and therefore a genuine herpesvirus product will be labelled with both dyes. This improves confidence in the interpretation of results as singly labelled products can be excluded.
10. Artifactual hits may be caused by reads containing repetitive sequences aligning to similar low complexity regions inside viral sequences.
11. This allows identification of hits caused by the second artifactual source, human genome sequences with close homology to viral sequences, often found in endogenous retroviruses. Hits which pass the visual alignment quality checks and do not demonstrate any similarity to human genome sequences, are candidates for known viruses present within the sample. Even at this stage, artifacts may still remain, caused by contamination in sample preparation or during the sequencing process.

This possibility should be particularly borne in mind if the number of aligned sequences is low. One regular artifact is alignment of reads to bacteriophage sequences. This is caused by contamination of the sample at some point with cloning vectors containing parts of phage genomes.

12. Contigs obtained from the Unmapped reads include:

    (a) Human cDNAs—where reads span splice junctions, they may not match any contiguous area in the Golden Path human genome template. Newbler may assess some of these reads as Chimeric, but others may slip though as Unmapped, and then be de novo assembled into contigs corresponding to parts of host cDNAs.

    (b) Known viruses—these are often the same viruses identified in the previous alignments to viral template sets.

    (c) Bacterial, fungal or mycoplasmal contaminants—these may have arisen as laboratory contaminants or may be cross-contamination from other sequencing projects.

    (d) Other identifiable contaminants—for example if a mouse feeder cell line has been used at some point, there may be murine genomic sequences. Murine retroviruses, present in reagents used in sample preparation, may also be identified.

    (e) Contigs with non-exact BLAST hits to known viruses.

    (f) Contigs with no BLAST hits.

    The latter two categories are obviously of the most interest, as they may represent novel viruses that are relatives of known viruses or completely novel viruses.

13. Codonwibbler is an in-house Perl script that implements a variety of codon bias metrics on all three reading frames of the contigs. The rationale for this process is that coding sequences frequently exhibit codon bias and contigs with noticeable differences in codon bias between one of the reading frames and the other two are candidates for coding sequences. Codon bias analysis is preferable to a search of open reading frames (ORFs) since even small errors in sequencing or assembly can produce nonsense mutations that remove ORFs. Codon bias is less sensitive to errors of this sort.

14. Hidden Markov Modelling uses statistical models (HMMs) of sequence variation in protein families to identify distant members of these families. It is more sensitive than BLAST. Ehmmpfam is a tool from the EMBOSS suite that compares a DNA sequence with the Pfam library of HMMs. Contigs that have a HMM hit are candidates for coding sequences. Unlike codon bias analysis, which merely indicates the presence of a potential ORF, HMM analysis gives some indication of the type of protein that may be encoded by that ORF.

Acknowledgments

Work in our laboratory is supported by Leukaemia Lymphoma Research, the Kay Kendall Leukaemia Fund, and the Medical Research Council. I am grateful to Arjan Diepstra for helpful discussion.

References

1. Epstein MA, Achong BG, Barr YM (1964) Virus particles in cultured lymphoblasts from Burkitt's lymphoma. Lancet 1:702–703
2. Dojcinov SD, Venkataraman G, Pittaluga S et al (2011) Age-related EBV-associated lymphoproliferative disorders in the Western population: a spectrum of reactive lymphoid hyperplasia and lymphoma. Blood 117: 4726–4735
3. Jarrett RF (2006) Viruses and lymphoma/leukaemia. J Pathol 208:176–186
4. Poiesz BJ, Ruscetti FW, Gazdar AF, Bunn PA, Minna JD, Gallo RC (1980) Detection and isolation of type C retrovirus particles from fresh and cultured lymphocytes of a patient with cutaneous T-cell lymphoma. Proc Natl Acad Sci USA 77:7415–7419
5. Kalyanaraman VS, Sarngadharan MG, Robert-Guroff M, Miyoshi I, Golde D, Gallo RC (1982) A new subtype of human T-cell leukemia virus (HTLV-II) associated with a T-cell variant of hairy cell leukemia. Science 218:571–573
6. Chang Y, Cesarman E, Pessin MS et al (1994) Identification of herpesvirus-like DNA sequences in AIDS-associated Kaposi's sarcoma. Science 266:1865–1869
7. Rickinson AB, Kieff E (2007) In: Knipe DM, Howley PM (eds.), Epstein-Barr virus. Fields virology. Lippincott Williams & Wilkins: Philadelphia, pp. 2655–2700
8. Allander T, Andreasson K, Gupta S et al (2007) Identification of a third human polyomavirus. J Virol 81:4130–4136
9. Gaynor AM, Nissen MD, Whiley DM et al (2007) Identification of a novel polyomavirus from patients with acute respiratory tract infections. PLoS Pathog 3:e64
10. Palacios G, Druce J, Du L et al (2008) A new arenavirus in a cluster of fatal transplant-associated diseases. N Engl J Med 358:991–998
11. Wilson KS, Gallagher A, Freeland JM, Shield LA, Jarrett RF (2006) Viruses and Hodgkin lymphoma: No evidence of polyomavirus genomes in tumor biopsies. Leuk Lymphoma 47:1315–1321
12. Feng H, Shuda M, Chang Y, Moore PS (2008) Clonal integration of a polyomavirus in human Merkel cell carcinoma. Science 319: 1096–1100
13. Reyes GR, Kim JP (1991) Sequence-independent, single-primer amplification (SISPA) of complex DNA populations. Mol Cell Probes 5:473–481
14. Allander T, Emerson SU, Engle RE, Purcell RH, Bukh J (2001) A virus discovery method incorporating DNase treatment and its application to the identification of two bovine parvovirus species. Proc Natl Acad Sci USA 98:11609–11614
15. Jarrett RF, Johnson D, Wilson KS, Gallagher A (2006) Molecular methods for virus discovery. Dev Biol (Basel) 123:77–88
16. Milne I, Bayer M, Cardle L et al (2010) Tablet–next generation sequence assembly visualization. Bioinformatics 26:401–402
17. Eddy SR (1998) Profile hidden Markov models. Bioinformatics 14:755–763
18. Punta M, Coggill PC, Eberhardt RY et al (2012) The Pfam protein families database. Nucleic Acids Res 40:D290–D301
19. Rice P, Longden I, Bleasby A (2000) EMBOSS: the European Molecular Biology Open Software Suite. Trends Genet 16:276–277
20. Armstrong AA, Weiss LM, Gallagher A et al (1992) Criteria for the definition of Epstein-Barr virus association in Hodgkin's disease. Leukemia 6:869–874
21. Gallagher A, Perry J, Shield L, Freeland J, MacKenzie J, Jarrett RF (2002) Viruses and Hodgkin disease: no evidence of novel herpesviruses in non-EBV-associated lesions. Int J Cancer 101:259–264
22. Oyola SO, Otto TD, Gu Y et al (2012) Optimizing Illumina Next-Generation Sequencing library preparation for extremely AT-biased genomes. BMC Genomics 13:1
23. Kent WJ, Sugnet CW, Furey TS et al (2002) The human genome browser at UCSC. Genome Res 12:996–1006

24. Altschul SF, Gish W, Miller W, Myers EW, Lipman DJ (1990) Basic local alignment search tool. J Mol Biol 215:403–410

25. Gulley ML (2001) Molecular diagnosis of Epstein-Barr virus-related diseases. J Mol Diagn 3:1–10

Chapter 17

High-Throughput RNA Sequencing in B-Cell Lymphomas

Wenming Xiao, Bao Tran, Louis M. Staudt, and Roland Schmitz

Abstract

High-throughput mRNA sequencing (RNA-seq) uses massively parallel sequencing to allow an unbiased analysis of both genome-wide transcription levels and mutation status of a tumor. In the RNA-seq method, complementary DNA (cDNA) is used to generate short sequence reads by immobilizing millions of amplified DNA fragments onto a solid surface and performing the sequence reaction. The resulting sequences are aligned to a reference genome or transcript database to create a comprehensive description of the analyzed transcriptome. This chapter describes a protocol to perform RNA-seq using the Illumina sequencing platform, presents sequencing data quality metrics and outlines a bioinformatic pipeline for sequence alignment, digital gene expression, and mutation discovery.

Key words: Next-generation sequencing, Transcriptome, Gene expression, B cell, B-cell lymphoma, Immunoglobulin genes, Mutation

1. Introduction

Advances in high-throughput sequencing have opened new horizons in cancer research. The recent application of this method has led to immense progress in our understanding of biology and pathogenesis of B-cell lymphomas by identifying gene mutations and unbalanced expression of transcript isoforms (1–6). These studies included analyses of whole genomes, exomes (the coding sequences of the genome) or transcriptomes using genomic DNA, enriched exomic DNA or mRNA, respectively. High-throughput sequencing of mRNA represents an elegant way to study both gene mutations as well as the type and quantity of the transcripts in a cell. Compared with alternative high-throughput gene expression technologies, such as microarrays, RNA-seq achieves a base pair-level resolution, a higher dynamic range of expression levels and lower background level (7). Owing to the fact that RNA-Seq does not

Ralf Küppers (ed.), *Lymphoma: Methods and Protocols*, Methods in Molecular Biology, vol. 971,
DOI 10.1007/978-1-62703-269-8_17, © Springer Science+Business Media, LLC 2013

rely upon a predetermined set of probe sequences it also provides information on alternative spliced isoforms and novel transcripts. RNA-seq can detect rare transcripts with an abundance of 1–10 RNA molecules per cell (8) facilitating the detection of mutations in oncogenes and tumor suppressor genes even though these transcripts might be present at low abundance (9).

This chapter contains the relevant protocols for performing RNA-seq on Illumina Genome Analyzer or Illumina Hiseq2000 Systems using kits and reagents from commercial sources. Following RNA sample quantification and quality control, a double stranded cDNA library is generated, controlled for size range and quantified. In the cluster station, the denatured cDNA library is annealed to oligonucleotides that are covalently bound to the surface of the flow-cell. These oligonucleotides prime the synthesis of a complementary DNA strand creating a double stranded DNA template of which one strand is immobilized on the flow cell. Following denaturation, the free ends of these covalently attached DNA molecules are bound by neighboring oligonucleotides on the surface of the flow cell, which in turn prime the synthesis of a complementary DNA strand that is consequently also covalently bound to the flow cell. This process is repeated to create local "clusters" of identical DNA molecules. The sequencing reaction in the Genome Analyzer or HiSeq2000 is performed using a sequencing primer that is complementary to the adapter sequence attached to each template strand. The four different bases (A, G, T, and C) are labeled with different colored fluorophores. Following incorporation of the labeled bases, the Genome Analyzer's camera records the colors for each cluster and Illumina software modules (Firecrest and Bustard) converts the fluorophore information into DNA sequence data with associated intensity and quality scores. The subsequent bioinformatic analysis of RNA-seq is done in several stages. In the first step the sequencing reads are mapped. The correct alignment of short RNA sequences is a crucial, albeit complex, task. Depending on the experimental setup, RNA-seq generates tens of millions of sequence reads corresponding to complex transcriptomes. Several short-read assembly programs have been developed to tackle these problems applying different mathematical algorithms and scoring schemes (10). The mapping strategy outlined in this chapter uses sequential alignment steps using the mapping software BWA (11) and proved to be robust yet computationally inexpensive. Once the data has been mapped to a reference, the aligned reads are parsed to assign single nucleotide variants (SNV), which are validated by additional alignments using Bowtie (12) and Novoalign (http://www.novocraft.com). In a final step, an expression score for every mapped read is computed.

2. Materials

2.1. Sample Quality Control

1. Agilent RNA 6000 Nano Kit (Agilent Technologies).
2. Rat Brain Total RNA (Life Technologies).
3. Agilent 2100 Bioanalyzer (Agilent Technologies).
4. RNaseZAP (Ambion).
5. DNAse/RNAse-free water (DEPC-treated).
6. Bioanlayzer Chip vortexer (IKA MS 3).

2.2. Library Preparation

1. TruSeq RNA Sample Preparation Kit-Set A (Illumina).
2. 0.2 mL clear thin wall PCR strip tubes.
3. DynaMag-96 Side (Invitrogen).
4. SuperScript II Mix (Invitrogen).
5. Agencourt AMPure XP 60 mL (Beckman Coulter).
6. Ethanol absolute.
7. Tris-Cl 10 mM, pH 8.5 with 0.1% Tween 20.

2.3. Library Quality Control

1. Agilent DNA 1000 Kit (Agilent Technologies).

2.4. qPCR

1. Illumina Eco Real-Time PCR System (Illumina).
2. SYBR FAST Master Mix (KAPA Biosystems).
3. Illumina GA Primer Premix (KAPA Biosystems).
4. 6 x Illumina GA DNA Standards (KAPA Biosystems).
5. DNAse/RNAse-free Water.
6. 10 mM Tris-Cl + 0.05% Tween-20 (DNase/RNase-free).
7. Eco Plates (Illumina).
8. Eco Adhesive Seals (Illumina).

2.5. Cluster Generation and Sequencing

Illumina Cluster Station

Illumina HiSeq2000 or GAIIx Systems

2.6. RNA-Seq Data Quality Control and Assessment

Software:

Illumina Software: http://www.illumina.com/support/sequencing/sequencing_software.ilmn

FastQC: http://www.bioinformatics.bbsrc.ac.uk/projects/fastqc

FASTX_toolkit: http://hannonlab.cshl.edu/fastx_toolkit

2.7. Alignment of Sequences to Reference

Software:

BWA: http://bio-bwa.sourceforge.net/

IGV: http://www.broadinstitute.org/igv/home

Download reference sequences:

RefSeq database: ftp://ftp.ncbi.nlm.nih.gov/refseq/release/vertebrate_mammalian

Ensembl database: http://useast.ensembl.org/info/data/ftp/index.html

Genome: ftp://ftp.ncbi.nlm.nih.gov/genomes/

2.8. Extraction of Putative Single Nucleotide Variants

Software:

GATK: http://www.broadinstitute.org/gsa/wiki/index.php/Home_Page

PERL V5.8.8 or later

Download transcriptome mapping files from UCSC Human Genome Browser Project: http://genome.ucsc.edu/cgi-bin/hgTables?command=start

2.9. Verification of Putative Sequence Variants with Novoalign and Bowtie

Software:

Bowtie: http://bowtie-bio.sourceforge.net/index.shtml

Novoalign: http://www.novocraft.com/main/index.php

2.10. Annotation of Sequence Variants

Software:

PERL V5.8.8 or later

2.11. Digital Gene Expression

Software:

PERL V5.8.8 or later

2.12. Immunoglobulin Gene Assembly

Software:

BLAST: http://blast.ncbi.nlm.nih.gov

SSAKE: http://www.bcgsc.ca/platform/bioinfo/software/ssake

Formatdb: ftp://ftp.ncbi.nih.gov/blast/documents/formatdb.html

Download reference sequences:Immunoglobulin gene segments: http://www.imgt.org/IMGTdownloads.html

3. Methods

3.1. Sample Quality Control

RNA quality is very important for successful library preparation. Degraded RNA can result in low yield, over representation of the 5′ ends of RNA, or failure of library preparation. As a proxy for

RNA quality, the Agilent Bioanalyzer measures the fluorescence intensity of 18S and 28S ribosomal RNA and calculates from that the RNA integrity number (RIN).

1. Prepare the gel by pipetting 550 μL of RNA-6000 nano gel matrix into a spin filter. Centrifuge at 1,500 × *g* for 10 min at room temperature. Aliquot 65 μL into RNase-free microfuge tubes (see Note 1).
2. Vortex the RNA 6000 Nano dye concentration for 10 s, spin down and add 1 μL of dye into a 65 μL aliquot of filtered gel. Vortex and spin at 13,000 × *g* for 10 min (see Note 2).
3. Heat denature RNA ladder for 2 min at 70°C and cool on ice for 1 min (see Note 3).
4. Put an RNA 6000 Nano chip on the chip priming station. Pipette 9.0 μL of gel–dye mix in the well marked G. Position the plunger at 1 mL and close the chip priming station. Press plunger until it is held by the clip. Wait for exactly 30 s and release clip. Wait for additional 5 s. Pull plunger slowly back to 1 mL position. Open chip priming station and pipette 9.0 μL of gel–dye mix in the wells marked G.
5. Pipette 5 μL of RNA 6000 Nano marker in all 12 sample wells and in the well marked with the ladder symbol.
6. Pipette 1 μL of prepared ladder in well marked with the ladder symbol. Pipette 1 μL of sample in each of the 11 sample wells and Rat Brain Total RNA control in well 12 (see Note 4). Pipette 1 μL of RNA 6000 Nano Marker in each unused sample well. Put the chip in the adapter of the chip vortexer and vortex for 1 min at 2,400 rpm. Run the chip in the Agilent 2100 Bioanalyzer within 5 min using the program RNA-Eukaryote Total RNA Nano Series II.xsy.

3.2. Library Preparation

The library preparation is outlined in Fig. 1.

1. 1 μg of total RNA is required for cDNA library construction (see Note 5). Dilute the total RNA with DNAse/RNAse-free water to final volume of 50 μL.

Purification and Fragmentation of mRNA

2. Add 50 μL of RNA Purification Beads to each well of the RBP plate using a multichannel pipette to bind mRNA to oligo-dT magnetic beads (see Note 6). Mix by gently pipetting up and down for six times.
3. Seal the plate and denature RNA in a thermal cycler (65°C for 5 min, 4°C hold). Remove from thermal cycler.
4. Incubate at room temperature for 5 min to allow binding to beads.

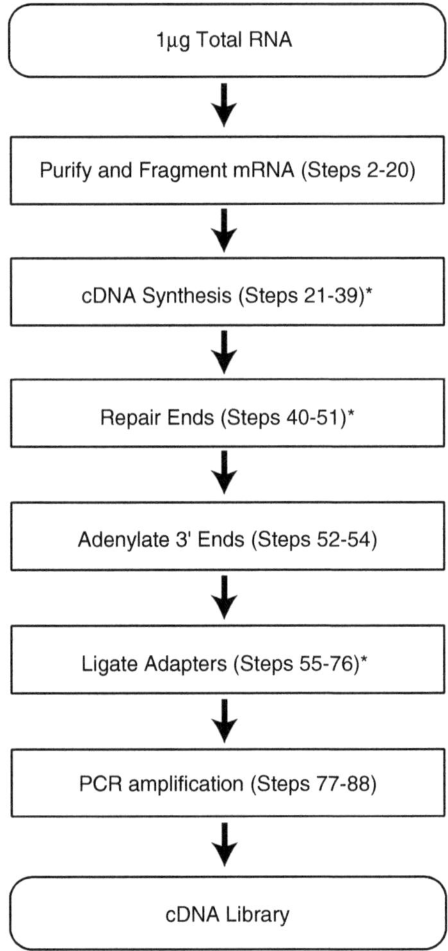

Fig. 1. Workflow for cDNA library preparation. *Asterisks* Indicates steps after which library preparation can be interrupted.

5. Place the plate on the magnetic stand at room temperature for 5 min to separate beads from the solution (see Note 7).
6. Carefully remove and discard supernatant using a multichannel pipette. Remove the plate from the magnetic stand.
7. Wash beads by adding 200 μL of Bead Washing Buffer to remove unbound RNA (see Note 8).
8. Place the plate on the magnetic stand at room temperature for 5 min.
9. Carefully remove and discard supernatant. Remove the plate from the magnetic stand.
10. Add 50 μL of Elution Buffer. Mix gently by pipetting.
11. Seal the plate and elute mRNA in a thermal cycler (80°C for 2 min, 25°C hold).

12. Remove the plate from thermal cycler and add 50 μL of Bead Binding Buffer and incubate at room temperature for 5 min.
13. Place the plate on the magnetic stand at room temperature for 5 min (see Note 7).
14. Carefully remove and discard entire supernatant.
15. Remove the plate from the magnetic stand and wash beads by adding 200 μL of bead wash buffer to remove unbound RNA.
16. Place the plate on the magnetic stand at room temperature for 5 min.
17. Carefully remove and discard supernatant. Remove the plate from the magnetic stand.
18. Add 19.5 μL of Elute, Fragment Mix. Mix gently by pipetting (see Note 9).
19. Place the sealed plate in a thermal cycler and elute fragment, and prime RNA using the program: 94°C for 8 min, 4°C hold.
20. Remove from thermal cycler and spin briefly. Proceed immediately to Synthesize First Strand cDNA.

Synthesize First Strand cDNA

21. Place the plate on the magnetic stand at room temperature for 5 min.
22. Transfer 17 μL of supernatant (fragmented and primed mRNA) to a new plate.
23. Add 8 μL of First Strand Master Mix and SuperScript II mix. Mix by pipetting.
24. Incubate the plate in a thermal cycler: 25°C for 10 min, 42°C for 50 min, 70°C for 15 min, 4°C hold.
25. Remove from thermal cycler and proceed immediately to Synthesize Second Strand cDNA.

Synthesize Second Strand cDNA

26. Add 25 μL of second strand master mix. Mix by pipetting.
27. Incubate in a thermal cycler, at 16°C for 2.5 h.
28. Remove the plate from thermal cycler. Bring the reaction mixture to room temperature.

Ampure XP Clean Up

29. Vortex the Ampure XP beads until they are well dispersed and add 90 μL Ampure XP beads to reaction mixture. Mix by vortexing.
30. Incubate at room temperature for 15 min.
31. Place the plate on the magnetic stand at room temperature for 5 min (see Note 7).
32. Carefully remove and discard 135 μL of supernatant. Some liquid may remain in the wells (see Note 10).

33. With the plate remaining on the magnetic stand, carefully add 200 μL of freshly prepared 80% EtOH.
34. Incubate at room temperature for 30 s and carefully remove and discard entire supernatant. Repeat steps 33 and 34 once for a total of two 80% EtOH washes.
35. Dry beads at room temperature for 15 min and remove the plate from the magnetic stand.
36. Add 52.5 μL Resuspension Buffer. Mix by pipetting.
37. Incubate at room temperature for 2 min.
38. Place the plate on the magnetic stand at room temperature for 5 min (see Note 7).
39. Transfer 50 μL of supernatant (double stranded cDNA) from to a new PCR plate. Library preparation can be interrupted at this step. cDNA can be stored up to 7 days at −20°C.

End Repair

40. Add 10 μL of Resuspension Buffer.
41. Add 40 μL of End Repair Mix. Mix by pipetting.
42. Incubate in a thermal cycler at 30°C for 30 min. Library preparation can be interrupted at this step. cDNA can be stored up to 7 days at −20°C.

Ampure XP Clean Up

43. Add 160 μL of mixed Ampure XP beads. Mix by pipetting.
44. Incubate at room temperature for 15 min.
45. Place the plate on the magnetic stand at room temperature for 5 min (see Note 7).
46. Carefully remove and discard 127.5 μL of supernatant. Some liquid may remain in the wells (see Note 10).
47. Wash as in steps 33–35.
48. Resuspend the dried pellet with 52.5 μL Resuspension Buffer. Mix well by pipetting.
49. Incubate at room temperature for 2 min.
50. Place the plate on the magnetic stand at room temperature for at least 5 min (see Note 7).
51. Transfer 17.5 μL of supernatant from to a new PCR plate. Library preparation can be interrupted at this step. cDNA can be stored up to 7 days at −20°C.

Adenylate 3′ Ends

52. Add 12.5 μL of A-Tailing Mix. Mix by pipetting.
53. Incubate at 37°C for 30 min.
54. Proceed immediately to Ligate Adapters.

Ligate Adapters

55. Add 2.5 μL of DNA Ligase Mix, 2.5 μL of Resuspension Buffer, and 2.5 μL of each RNA Adapter Index. Mix by pipetting.
56. Incubate in a thermal cycler at 30°C for 10 min.
57. Remove the plate from thermal cycler.
58. Inactivate ligation by adding 5 μL of Stop Ligase Mix. Mix by pipetting.

Ampure XP Clean Up

59. Add 42 μL of mixed Ampure XP beads. Mix by pipetting.
60. Incubate at room temperature for 15 min.
61. Place the plate on the magnetic stand at room temperature for at least 5 min (see Note 7).
62. Carefully remove and discard supernatant (see Note 10).
63. Wash as in steps 33–35.
64. Resuspend the dried pellet with 52.5 μL Resuspension Buffer. Mix by pipetting.
65. Incubate at room temperature for 2 min.
66. Place the plate on the magnetic stand at room temperature for at least 5 min (see Note 7).
67. Transfer 50 μL of supernatant into new plate.
68. Add 50 μL of mixed Ampure XP beads. Mix by pipetting.
69. Incubate at room temperature for 15 min.
70. Place the plate on the magnetic stand at room temperature for at least 5 min (see Note 7).
71. Carefully remove and discard 95 μL of supernatant. Some liquid may remain in each well (see Note 10).
72. Wash as in steps 33–35.
73. Add 22.5 μL Resuspension Buffer. Mix by pipetting.
74. Incubate at room temperature for 2 min.
75. Place the plate on the magnetic stand at room temperature for at least 5 min (see Note 7).
76. Transfer 20 μL of supernatant to corresponding wells of the plate of step 67. Library preparation can be interrupted at this step. cDNA can be stored up to 7 days at −20°C.

Enrich DNA Fragments

77. Add 5 μL of PCR Primer Cocktail.
78. Add 25 μL of PCR Master Mix. Mix by pipetting.
79. Amplify cDNA library using the PCR program:

98°C for 30 s, 15×(98°C for 10 s, 60°C for 30 s, 72°C for 30 s), 72°C for 5 min, hold at 4°C.

80. Add 50 μL of mixed Ampure XP beads. Mix by pipetting.
81. Incubate at room temperature for 15 min.
82. Place the plate on the magnetic stand at room temperature for at least 5 min (see Note 7).
83. Carefully remove and discard 95 μL of the supernatant. Some liquid may remain in the wells (see Note 10).
84. Wash as in steps 33–35.
85. Resuspend dried pellet with 15 μL Resuspension Buffer. Mix by pipetting.
86. Incubate at room temperature for 2 min.
87. Place the plate on the magnetic stand at room temperature for at least 5 min (see Note 7).
88. Carefully transfer 15 μL of supernatant to a new plate.

3.3. Library Quality Control

The cDNA library is analyzed using an Agilent DNA 1000 Bioanalzyer chip to validate quality, size range and quantity of the library. The anticipated yield of the cDNA library preparation is at least 10 nM. The expected size of most of the cDNA molecules is typically 300–350 bp. In case of high quantities of primer-dimers that are typically around 50 bp in size an additional Ampure XP Clean Up step is recommended.

1. Bring DNA dye concentrate and DNA gel matrix to room temperature. Vortex the DNA dye concentrate for 10 s and spin down (see Note 11).
2. Pipette 25 μL of the dye concentrate into DNA gel matrix vial.
3. Vortex for 10 s. Check proper mixing of gel and dye.
4. Transfer the gel–dye mix to the top receptacle of a spin filter. Place the spin filter in a microcentrifuge and spin for 15 min at room temperature at 2,240×*g*.
5. Place DNA chip on the chip priming station.
6. Pipette 9.0 μL of the gel–dye mix at the bottom of the well marked G (see Note 12). Position plunger at 1 mL and close the chip priming station.
7. Press the plunger of the syringe down until it is held by the clip. Wait for exactly 60 s and then release the plunger with the clip release mechanism.
8. Wait for 5 s, then slowly pull back the plunger to the 1 mL position. Open the chip priming station.
9. Pipette 9.0 μL of the gel–dye mix in each of the wells marked.
10. Pipette 5 μL of DNA marker into the well marked with the ladder symbol and into each of the 12 sample wells (see Note 13).

Table 1
qPCR reaction mix

(μL)	Reagent
5	2× SYBR Master Mix
1	10× Illumina GA Primer Premix
2	Water
2	DNA template, GA Standard, or water

11. Pipette 1 μL of the DNA ladder in the well marked with the ladder symbol. In each of the 12 sample wells pipette 1 μL of sample for used wells or 1 μL of deionized water for unused wells.
12. Place the chip horizontally in the adapter of the chip vortexer. Vortex for 60 s at 2,400 rpm. Run within 5 min using the program dsDNA-DNA 1000 Series II.xsy.

3.4. qPCR

Accurate quantification of the cDNA library is critical for successful cluster generation. Low concentration of cDNA libraries result in lower cluster density and reduced sequencing yield, whereas too high concentrations can cause too dense clusters leading to poor cluster resolution during the sequencing.

1. Dilute cDNA Libraries 1:10,000 (see Note 14).
2. Pipette qPCR reaction mix (Table 1):
3. Place a 48-well qPCR plate on the dock assembly and adjust the angle of the dock as desired.
4. Pipette 8 μL of the qPCR reaction mix into in the sample plate.
5. Add 2 μL of either DNA, Illumina DNA Standards (1–6) or water into sample wells.
6. Seal the 48-well microplate with adhesive film.
7. Vortex and spin plate. No air bubbles should be visible.
8. Run the sample plate in the Eco qPCR System using the PCR program: 95°C 5 min, 35× (95°C for 30 s, 60°C 45 s) (see Note 15).

3.5. Cluster Generation and Sequencing

The cDNA samples are clustered on the Flowcell using the protocol: cBot_PE_Amplification_Linearization_Blocking_PrymHyb_v8.0.xml. Sequencing is performed on an Illumina GAIIx or HiSeq2000 instrument.

3.6. RNA-Seq Data Quality Control and Assessment

There is a variety of software tools, including both published open-source algorithms and programs developed in-house, for RNA-Seq data QC and analysis. This section focuses on data quality control and assessment based on Illumina software tools.

3.6.1. Instrument Run Monitoring and Data Evaluation

During the Illumina instrument run for both HiSeq2000 and GAIIx platforms, the data quality control can be performed based on the output from instrument control software and off instrument (remote) Sequencing Analysis Viewer software. Table 2 shows software tools and metrics for data evaluation.

Table 3 lists acceptable run performance metrics based on control sample (e.g., Phix). Performance may vary for a given mRNA sample based on sample quality, cluster density and other experimental factors.

Table 2
Instrument run monitoring and data evaluation

Platform	Quality metrics	Software	Software function
HiSeq2000 GAIIx	Raw Cluster Numbers and %PassFilter Clusters allow the assessment of cluster density and performance	HCS/SCS (Instrument Control Software)	Instrument Control Software provides a graphical interface which allows user view run status and statistics while running the instrument
	Intensity and Phasing/ Prephasing values allow assessment of the sequencing chemistry performance	RTA (Real-Time Analysis)	RTA is a component within the Instrument Control Software that monitors the run's progress, optimizes run conditions and provides run-time quality statistics
	Basecall Quality Scores allow assessment of base call quality cycle by cycles	SAV (Sequencing Analysis Viewer)	SAV is an application that allows user, in real time, to view run quality metrics generated by RTA from an off-instrument remote location
	Focus quality can alert any focus issues		

Table 3
Run performance metrics

Matrix	Description	Normal value range
Cluster Density (1,000/mm^2)	Density of clusters detected by image analysis	200–820 K/mm^2
Cluster PF (%)	Percentage of clusters passing intensity filter criteria	60–90%
Phasing/Prephasing	Value used by RTA for the percentage of molecules in a cluster for which sequencing falls behind (phasing) or jumps ahead (prephasing) the current cycle within a read	Phasing < Prephasing Phasing <0.5 Prephasing <0.7
Q30+ (%)	Percentage of bases with a Phred score of 30 or higher	≥80%
Mismatch Error Rate	The calculated error rate, as determined by the alignment of control (e.g., Phix)	<2%

Table 4
Quality assessment using FastQ

Matrix	Fail QC value range
Per base quality score	If the lower quartile for any base is less than 5 or if the median for any base is less than 20
Per sequence quality score	If the most frequently observed mean quality for sequences is below 20
Per base sequence content	If the difference between A and T, or G and C is greater than 20% in any position
Sequence GC content	If the sum of the deviations from the normal distribution represents more than 30% of the reads
PCR duplicates	If non-unique sequences make up more than 50% of the total
Overrepresented sequences	If any sequence is found to represent more than 1% of the total

3.6.2. Post Sequencing Data Processing

1. Convert sequencing data to FASTQ files using Illumina CASAVA or OLB software. There is an option to automatically filter low quality reads that have failed Illumina specific thresholds.
2. Generate quality reports using the FastQC software (Table 4).
3. Perform adapter clipping and low quality trimming using FASTX_toolkit software.
4. Specify the adapter and contamination files such as ribosomal DNA or other contaminant sequences in the CASAVA contaminant file path, using CASAVA eland_rna module for alignment and automatically filtering the contaminants during mapping.
5. Align reads to reference genome and splice junctions by using CASAVA pipeline eland_paired and eland_rna modules for sequence quality evaluation.

3.7. Alignment of Sequences to Reference

In essence, there are two assembly strategies for RNA-seq data. The sequence reads can be directly aligned to a reference such as the genome or transcriptome, or assembled de novo and subsequently aligned. Whilst the use of de-novo assembly provides the opportunity to construct a whole genome transcriptome map including the discovery of previously unknown transcripts this method requires greater computational resources and is less sensitive in mapping transcripts at low abundance. Thus, except for immunoglobulin gene assembly (Subheading 3.12), this chapter will focus on reference-based mapping strategies. Alignment to the genome allows mapping of reads from unannotated loci. However, this method alone produces a high risk of false positive SNVs due to mismapped reads (e.g., spanning one or more intron-exon junctions). To address these problems, we outline a sequential alignment

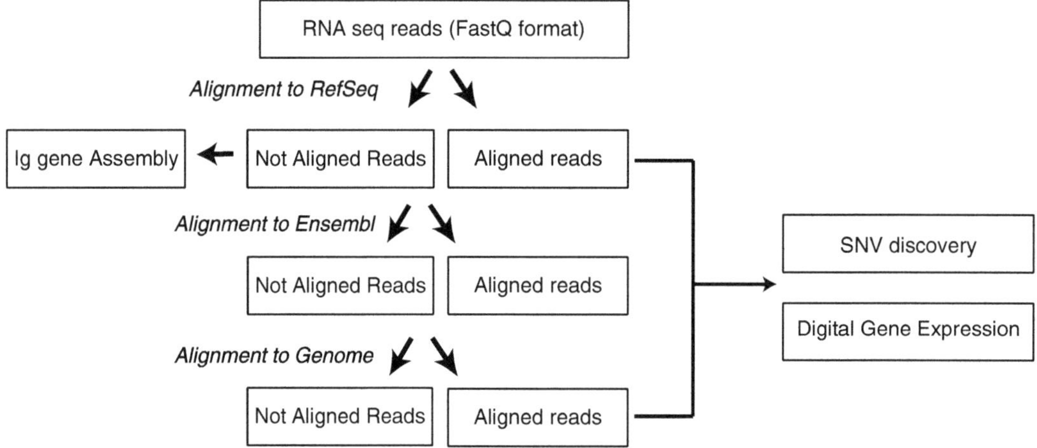

Fig. 2. Flowchart of sequential alignment process using BWA. In the first step of the mapping process, RNA-Seq reads are aligned to the RefSeq database. Aligned and not aligned reads are stored in separate SAM format files. Not aligned reads are mapped and assembled for Ig transcripts, or aligned to the Ensembl database. Reads that fail to align this step are mapped to the genome.

strategy using the BWA alignment tool (11) (Fig. 2). Following download and indexing, RNA-seq reads (in fastq format) are aligned to the RefSeq database (see Note 16). The aligned and unaligned reads are stored in separate files in SAM format. Unaligned reads from this step will be used immunoglobulin gene assembly in Subheading 3.12. In the next alignment step, reads that failed to map to RefSeq are mapped to the Ensembl database, which includes additional transcripts and pseudogenes (13). Reads that do not align via Ensembl and then aligned to the Genome sequence. This sequential mapping strategy usually results in the following mapping efficiencies: RefSeq: 70–80%, Ensembl: 5% and Genome: 5–10%. The three final alignment files are merged and converted into BAM format for visualization of aligned target sequences in the Integrated Genome Viewer (IGV) (14). The converted BAM file (the binary version of a SAM file) is sorted and indexed before being imported into IGV (see Note 17).

3.8. Extraction of Putative Single Nucleotide Variants

The resulting alignments are utilized to generate single nucleotide variants (SNV) calls. To reduce the impact of method-specific artifacts, the following considerations should be taken into account: Reads that map to identical starting position (redundant reads) are discarded. Only reads that have a Phred quality score of 20 or more will be used to create SNV calls. To ensure both a high sensitivity and a reduction of false positive results, SNVs can be included, if these are observed in more than three reads and the ratio of mutant reads vs. total coverage is greater than 20%. A wide variety of software is available to make variant calls, such as mpileup in SAMtools combining with BCFtools or Variant Discovery Tools in the Genome Analysis Toolkit (GATK). Alternatively, individual PERL scripts can be developed to "look up" mismatches in alignments.

Based on mapping of RefSeq and Ensembl transcripts to the genome, which is available from UCSC Human Genome Browser Project, the identified SNVs are subsequently assigned to their corresponding genomic coordinates.

3.9. Verification of Putative Sequence Variants with Novoalign and Bowtie

A major challenge of variant detection with RNA-seq is the reduction of false positive SNVs. High sequencing coverage of an observed variant can minimize incorrect variant calls. However, false positive SNVs can also result from mismapping of sequencing reads to highly homologous sequences (e.g., pseudogenes) present in the genome or transcriptome database. To tackle this problem, it is helpful to utilize multiple alignment algorithms with different mathematical scoring metrics. To this end, reads containing SNVs from the first alignment using BWA (Subheading 3.7) are extracted from the SAM file (e.g., using PERL) and are split based on exon structure using RefSeq or Ensembl transcriptome definitions as the reference. Reads less than 16 bp long are discarded. These reads are aligned to the human genome using Bowtie and Novolalign mapping tools. Since both programs can generate alignment results in SAM format, the subsequent analyses can be performed with the same tools as used for BWA alignment. A variant is declared confirmed if it is called by BWA and either the Bowtie or Novoalign algorithms. As determined by Sanger resequencing, the combination of BWA with Bowtie and Novoalign yields a true positive rate of up to 95% (6).

3.10. Annotation of Sequence Variants

To understand the biological effects of a SNV, the annotation of the observed variant is paramount. While the majority of SNVs discovered by RNA-Seq are single nucleotide polymorphisms (SNP), the main focus in cancer biology lies on the identification of non-SNP mutations with the potential to activate or inactivate gene functions. In the first step of the annotation process, SNVs that corresponded to known SNPs are identified based on the genomic coordinates. If an SNV occurs within the coding region of a transcript, a PERL script can be used to determine the resulting amino acid substitution. For this conversion, the relative position of a given SNV within the coding sequences is used to determine the resulting codon change. Using the genetic code, this codon change is translated into an amino acid substitution. The interpretation of non-synonymous SNVs can be facilitated using the SIFT (Sorting Intolerant from Tolerant) algorithm that can predict the effect of amino acid substitutions (15). SIFT presumes that functionally important amino acids will be conserved within protein families. Thus, amino acid substitutions at conserved positions tend to be predicted as deleterious. SIFT also considers the type of amino acid change. Based on the amino acids appearing at each position in the alignment, SIFT calculates the probability that an amino acid at a position is tolerated conditional on the most frequent amino acid being tolerated.

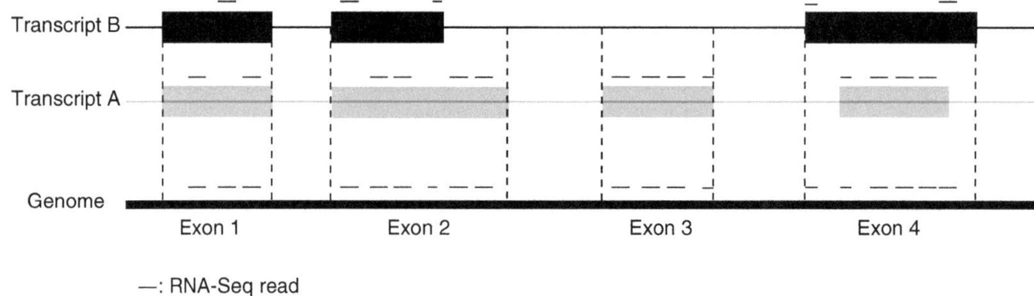

Fig. 3. Digital gene expression. RNA-Seq reads from BWA alignment are assigned to genes based on their genomic coordinates using RefSeq and Ensembl mapping results from UCSC Human Genome Browser Project. The total length of a gene is the sum of all exon covered by RNA-seq reads.

3.11. Digital Gene Expression

Digital gene expression can be derived from the initial transcriptome alignment by BWA. RNA-Seq reads are assigned to genes based on their genomic coordinates assigned in Subheading 3.8 (Fig. 3). The total extend of a gene is considered to be the union of all exons that match with RNA-seq reads. RPKM (reads per kilobase per million) is calculated based on the total reads that hit on the gene, the length of this gene and total number of reads from the test sample.

$$\text{RPKM} = \text{Number of reads for a gene} / (\text{gene length[kb]} * (\text{Number of reads from analyzed sample} / 1{,}000{,}000))$$

3.12. Immunoglobulin Gene Assembly

Most B-cell lymphomas express a B cell receptor (BCR) (16). The BCR is a membrane bound antibody composed of two identical heavy chain and two identical light chain immunoglobulin (Ig) polypeptides, non-covalently bound to the signaling components CD79A and CD79B. Immunoglobulins consist of variable (V) regions and constant (C) regions. The heavy chain V-regions are composed of three (V, D, and J) gene segments whereas the light chains are consisting of two gene segments (V and J). During B cell development, these gene segments are assembled by somatic DNA rearrangement to encode a functional Ig. This Ig gene segment assembly results in a large diversity of non-germ-line encoded immunoglobulin molecules. To map and assemble Ig transcript from RNA-seq, Ig germ-line segments including V, D, J and constant regions of heavy and light chain are downloaded from the Immunogentics database (IMGT). The sequences in fasta format are used to create a BLAST-compatible database with formatdb. To reduce computational complexity, only reads that were not aligned in the first alignment step (Subheading 3.7) are used for Ig gene assembly since Ig genes are not contained in the RefSeq database. Reads with a median Phred score of more than 20 are extracted and aligned separately to the reference database for V, D,

J and constant regions segments. The output of the BLAST alignment to the Ig V, D, J and C segments is parsed using PERL, and the mapping of each read to Ig segments is determined. The number of reads mapping to individual Ig segments is then determined. Reads for the most prevalent segments of V, D, J and constant regions are extracted, merged and assembled separately for heavy-chain and light-chain using the SSAKE assembly tool.

4. Notes

1. Kit should be used at room temperature and protected from light for the whole procedure.
2. Use prepared gel–dye mix within 1 day.
3. Aliquots can be stored at −70°C. Thaw aliquots on ice.
4. It is recommended to heat denature all RNA samples and RNA ladder before use.
5. The RIN number of total RNA used for library preparation should be greater or equal to 8.
6. Vortex the thawed RNA Purification Beads tube vigorously to completely resuspend the oligo-dT beads.
7. All beads should be attached to the side of the wells. The liquid should appear clear.
8. Remove the entire ethanol from the bottom of the wells. Residual ethanol can hamper downstream enzymatic reactions and may contain contaminants.
9. The Elute, Prime, Fragment Mix contains reaction buffer and random hexamers for priming of reverse transcription.
10. Leave the plate on the magnetic stand while performing the following 80% EtOH washes.
11. The gel–dye mix can be stored for 4 weeks at 4°C.
12. When pipetting the gel–dye mix, do not to draw up particles that may sit at the bottom of the gel–dye mix vial. Insert the tip of the pipette to the bottom of the chip well when dispensing. This prevents a large air bubble forming under the gel–dye mix. Placing the pipette at the edge of the well may lead to poor results.
13. Empty wells may cause improper running of the chip. Add 5 μL of DNA marker plus 1 μL of deionized water to each unused sample well.
14. Optimal quantification of cDNA libraries is achieved using an input DNA amount of 1 pM.

15. If the concentration for any sample is greater than 50 nM, the sample should be diluted and rerun in the qPCR.
16. The index of reference sequence database needs to be built just once unless new version of BWA is used.
17. Conversion of the SAM file format into BAM format can be performed in SAMtools with the command: samtools view -bS yourSAMfile > yourBAMfile. Sorting and Indexing: samtools yourBAMfile yourBAMfile.sorted and samtools index yourBAMfile.sorted.These processes typically take a few hours.

Acknowledgments

This work was supported by the by the Dr. Mildred Scheel Stiftung für Krebsforschung (Deutsche Krebshilfe). We are grateful to Yuliya Kriga, Jyoti Shetty, Yongmei Zhao, John Powell, and George Wright who were instrumental in establishing the protocols described here.

References

1. Morin RD, Mendez-Lago M, Mungall AJ et al (2011) Frequent mutation of histone-modifying genes in non-Hodgkin lymphoma. Nature 476:298–303
2. Pasqualucci L, Trifonov V, Fabbri G et al (2011) Analysis of the coding genome of diffuse large B-cell lymphoma. Nat Genet 43:830–837
3. Ngo VN, Young RM, Schmitz R et al (2011) Oncogenically active MYD88 mutations in human lymphoma. Nature 470:115–119
4. Puente XS, Pinyol M, Quesada V et al (2011) Whole-genome sequencing identifies recurrent mutations in chronic lymphocytic leukaemia. Nature 475:101–105
5. Kridel R, Meissner B, Rogic S et al (2012) Whole transcriptome sequencing reveals recurrent NOTCH1 mutations in mantle cell lymphoma. Blood 119:1963–1971
6. Schmitz R, Young RM, Ceribelli M et al (2012) Burkitt lymphoma pathogenesis and therapeutic targets from structural and functional genomics. Nature 490:116–120
7. Wang Z, Gerstein M, Snyder M (2009) RNA-Seq: a revolutionary tool for transcriptomics. Nat Rev Genet 10:57–63
8. Mortazavi A, Williams BA, McCue K et al (2008) Mapping and quantifying mammalian transcriptomes by RNA-Seq. Nat Methods 5:621–628
9. Frischmeyer PA, Dietz HC (1999) Nonsense-mediated mRNA decay in health and disease. Hum Mol Genet 8:1893–1900
10. Garber M, Grabherr MG, Guttman M et al (2011) Computational methods for transcriptome annotation and quantification using RNA-seq. Nat Methods 8:469–477
11. Li H, Durbin R (2009) Fast and accurate short read alignment with Burrows-Wheeler transform. Bioinformatics 25:1754–1760
12. Langmead B, Trapnell C, Pop M et al (2009) Ultrafast and memory-efficient alignment of short DNA sequences to the human genome. Genome Biol 10:R25
13. Larsson TP, Murray CG, Hill T et al (2005) Comparison of the current RefSeq, Ensembl and EST databases for counting genes and gene discovery. FEBS Lett 579:690–698
14. Robinson JT, Thorvaldsdottir H, Winckler W et al (2011) Integrative genomics viewer. Nat Biotechnol 29:24–26
15. Kumar P, Henikoff S, Ng PC (2009) Predicting the effects of coding non-synonymous variants on protein function using the SIFT algorithm. Nat Protoc 4:1073–1081
16. Rui L, Schmitz R, Ceribelli M et al (2011) Malignant pirates of the immune system. Nat Immunol 12:933–940

Index

Ralf Küppers (ed.), *Lymphoma: Methods and Protocols*, Methods in Molecular Biology vol. 971,
DOI 10.1007/978-1-62703-269-8, © Springer Science+Business Media, LLC 2013

MIX
Papier aus verantwortungsvollen Quellen
Paper from responsible sources
FSC® C105338

If you have any concerns about our products,
you can contact us on
ProductSafety@springernature.com

In case Publisher is established outside the EU,
the EU authorized representative is:
Springer Nature Customer Service Center GmbH
Europaplatz 3, 69115 Heidelberg, Germany

Printed by Libri Plureos GmbH
in Hamburg, Germany